Bayesian Statistics
and Marketing

WILEY SERIES IN PROBABILITY AND STATISTICS

Established by *Walter A. Shewhart and Samuel S. Wilks*

The *Wiley Series in Probability and Statistics* is well established and authoritative. It covers many topics of current research interest in both pure and applied statistics and probability theory. Written by leading statisticians and institutions, the titles span both state-of-the-art developments in the field and classical methods.

Reflecting the wide range of current research in statistics, the series encompasses applied, methodological and theoretical statistics, ranging from applications and new techniques made possible by advances in computerized practice to rigorous treatment of theoretical approaches.

This series provides essential and invaluable reading for all statisticians, whether in academia, industry, government, or research.

Bayesian Statistics and Marketing

Second Edition

Peter E. Rossi
Anderson School of Management
UCLA
Los Angeles
USA

Greg M. Allenby
Ohio State University
Fisher College of Business
Columbus
USA

Sanjog Misra
University of Chicago
Booth School of Business
Chicago
USA

Registered Offices
John Wiley & Sons, Inc., 111 River Street, Hoboken, NJ 07030, USA
John Wiley & Sons Ltd, The Atrium, Southern Gate, Chichester, West Sussex, PO19 8SQ, UK

For details of our global editorial offices, customer services, and more information about Wiley products visit us at www.wiley.com.

Wiley also publishes its books in a variety of electronic formats and by print-on-demand. Some content that appears in standard print versions of this book may not be available in other formats.

Library of Congress Cataloging-in-Publication Data:

Names: Rossi, Peter E. (Peter Eric), 1955- author. | Allenby, Greg M. (Greg
    Martin), 1956- author. | Misra, Sanjog, author.
Title: Bayesian statistics and marketing / Peter E. Rossi, Greg M. Allenby,
    Sanjog Misra.
Description: Second edition. | Hoboken, NJ : Wiley, [2024] | Series: Wiley
    series in probability and statistics | Includes index.
Identifiers: LCCN 2024003452 (print) | LCCN 2024003453 (ebook) | ISBN
    9781394219117 (hardback) | ISBN 9781394219131 (adobe pdf) | ISBN
    9781394219124 (epub)
Subjects: LCSH: Marketing research–Mathematical models. |
    Marketing–Mathematical models. | Bayesian statistical decision theory.
Classification: LCC HF5415.2 .R675 2024 (print) | LCC HF5415.2 (ebook) |
    DDC 658.8/3015118–dc23/eng/20240221
LC record available at https://lccn.loc.gov/2024003452
LC ebook record available at https://lccn.loc.gov/2024003453

Cover Design: Wiley
Cover Image: Courtesy of Peter Rossi, Greg Allenby and Sanjog Misra

Set in 10/12pt Galliard Std by Straive, Chennai, India
Printed and bound by CPI Group (UK) Ltd, Croydon, CR0 4YY

C9781394219117_030724

To Aimee, Tricia, and Debra

Contents

1

Introduction

Abstract

While the conceptual appeal of Bayesian methods has long been recognized, the recent popularity stems from computational and modeling breakthroughs that have made Bayesian methods attractive for many marketing problems. This book provides a self-contained and comprehensive treatment of Bayesian methods and the marketing problems for which these methods are especially appropriate. It presents a treatment of Bayesian methods that emphasizes the unique aspects of their application to marketing problems. The book emphasizes the unique aspects of the modeling problem in marketing and the modifications of method and models that researchers in marketing have devised. It also provides the requisite methodological knowledge and an appreciation of how these methods can be used to allow the reader to devise and analyze new models. The book takes a stand on customer differences by modeling differences via a probability distribution.

The past 30 years have seen a dramatic increase in the use of Bayesian methods in marketing. Bayesian analyses have been conducted over a wide range of marketing problems from new product introduction to pricing, and with a wide variety of different data sources. While the conceptual appeal of Bayesian methods has long been recognized, the recent popularity stems from computational and modeling breakthroughs that have made Bayesian methods attractive for many marketing problems. This book aims to provide a self-contained and comprehensive treatment of Bayesian methods and the marketing problems for which these methods are especially appropriate. There are unique aspects of important problems in marketing that make particular models and specific Bayesian methods attractive. We, therefore, do not attempt to provide a generic treatment of Bayesian methods. We refer the interested reader to classic treatments by Robert and Casella [2004], Gelman et al. [2004], and Berger [1985] for more general-purpose discussion of Bayesian methods. Instead, we provide a treatment of Bayesian methods that emphasizes the unique aspects of their application to marketing problems.

Until the mid-1980s, Bayesian methods appeared impractical since the class of models for which the posterior inference could be computed was no larger than the class of models for which exact sampling results were available. Moreover, the Bayes approach does require assessment of a prior which some feel to be an extra cost. Simulation methods, in particular Markov Chain Monte Carlo (MCMC) methods, have freed us from computational constraints for a very wide class of models. MCMC methods are ideally suited for models built from a sequence of conditional distributions, often called hierarchical models. Bayesian hierarchical models offer tremendous flexibility and modularity and are particularly useful for marketing problems.

There is an important interaction between the availability of inference methods and the development of statistical models. Nowhere has this been more evident than in the application of hierarchical models to marketing problems. Hierarchical models are those built up through a sequence of conditional distributions. These models match rather closely the various levels at which marketing decisions are made – from individual consumers to the marketplace. Bayesian researchers in marketing have expanded on the standard set of hierarchical models to provide models useful for marketing problems. Throughout this book, we will emphasize the unique aspects of the modeling problem in marketing and the modifications of method and models that researchers in marketing have devised. We hope to provide the requisite methodological knowledge and an appreciation of how these methods can be used to allow the reader to devise and analyze new models. This departs, to some extent, from the standard model of a treatise in statistics in which one writes down a set of models and catalogues the set of methods appropriate for analysis of these models.

1.1 A BASIC PARADIGM FOR MARKETING PROBLEMS

Ultimately, marketing data results from customers taking actions in a particular context and facing a particular environment. The marketing manager can influence some aspects of this environment. Our goal is to provide models of these decision processes and then make optimal decisions conditional on these models. Fundamental to this prospective is that customers are different in their needs and wants for marketplace offerings, thus expanding the set of actions that can be taken. At the extreme, actions can be directed at specific individuals. Even if one-on-one interaction is not possible, the models and system of inference must be flexible enough to admit nonuniform actions.

Once the researcher acknowledges the existence of differences between customers, the modeling task expands to include a model of these differences. Throughout this book, we will take a stand on customer differences by modeling differences via a probability distribution. Those familiar with standard econometric methods will recognize this as related to a random coefficients approach. The primary difference is that we do not regard the customer level parameters as nuisance parameters but, instead, regard these parameters as the goal of inference. Inferences about customer differences are required for any marketing action, from strategic decisions associated with formulating offerings to tactical decisions of customizing prices. Individuals who are most likely to respond to these variables are those that find highest value in the offering's attributes and those that are most price sensitive, neither of whom are well described by parameters such as the mean of the random coefficients distribution.

Statistical modeling of marketing problems consists of three components:

(i) Within-unit behavior

(ii) Across-unit behavior

(iii) Action

"Unit" refers to the particular level of aggregation dictated by the problem and data availability. In many instances, the unit is the consumer. However, it is possible to consider both less and more aggregate levels of analyses. For example, one might consider a particular consumption occasion or survey instances as the "unit" and consider changes in preferences across occasions or over time as part of the model (an example of this is in Yang et al. [2002]). In marketing practice, decisions are often made at a much higher level of aggregation such as the "key account" or sales territory. In all cases, we consider the "unit" as the lowest level of aggregation considered explicitly in the model.

The first component of problem is the conditional likelihood for the "unit-level behavior." We condition on unit-specific parameters that are regarded as the sole source of between-unit differences. The second component is a distribution of these unit-specific parameters over the population of units. Finally, the decision problem is the ultimate goal of modeling exercise. We typically postulate a profit function and ask – what is the optimal action conditional on the model and the information in the data? Given this view of marketing problems, it is natural to consider the Bayesian approach to inference, which provides a unified treatment of all three components.

1.2 A SIMPLE EXAMPLE

As an example of the components outlined in Section 1.1, consider the case of consumers observed making choices between different products. Products are characterized by some vector of choice attribute variables that might include product characteristics, prices, and advertising. Consumers could be observed to make choices either in the marketplace or in a survey/experimental setting. We want to predict how consumers will react to a change in the marketing mix variables or in the product characteristics. Our ultimate goal is to design products or vary the marketing mix so as to optimize profitability.

We start with the "within-unit" model of choice conditional on the observed attributes for each of the choice alternatives. A standard model for this situation is the Multinomial Logit model.

$$\Pr\left[i \middle| x_1, \ldots, x_{-p}, \beta\right] = \frac{\exp\left(x_i'\beta\right)}{\sum_{j=1}^{p} \exp\left(x_j'\beta\right)} \qquad (1.2.1)$$

If we observe more than one observation per consumer, it is natural to consider a model that accommodates differences between consumers. That is, we have some information about each consumer's preferences and we can start to tease out these differences. However, we must recognize that in many situations, we have only a small amount of information about each consumer. To allow for the possibility that each consumer has different preferences for attributes, we index the β vectors by c for consumer c.

Given the small amount of information for each consumer, it is impractical to estimate separate and independent logits for each of the C consumers. For this reason, it is useful to think about a distribution of coefficient vectors across the populations of consumers. One simple model would be to assume that the βs are distributed normally over consumers.

$$\beta_c \sim N\left(\mu, V_\beta\right) \tag{1.2.2}$$

One common use of logit models is to compute the implication of changes in marketing actions for aggregate market shares. If we want to evaluate the effect on market share for a change in x for alternative i, then we need to integrate over the distribution in (1.2.1). For a market with a large number of consumers, we might view the expected probability as market share and compute the derivative of market share with respect to an element of x.[1]

$$\frac{\partial MS(i)}{\partial x_{i,j}} = \frac{\partial}{\partial x_{i,j}} \int \Pr\left[i\,\middle|\,x_1,\ \dots\ ,x_p,\beta\right]\,\varphi\left(\beta\,\middle|\,\mu, V_\beta\right)d\beta \tag{1.2.3}$$

Here $\varphi()$ is the multivariate normal density.

The derivatives given in (1.2.3) are necessary to evaluate uniform marketing actions such as changing price in a situation in which all consumers face the same price. However, many marketing actions are aimed at a subset of customers or, in some cases, individual customers. In this situation, it is desirable to have a way of estimating not only the common parameters that drive the distribution of βs across consumers but the individual βs as well.

Thus, our objective is to provide a way of inferring about $\{\beta_1,\ \dots\ ,\beta_C\}$ as well as μ, V_β. We also want to use our estimates to derive optimal marketing policies. This will mean to maximize expected profits over the range of possible marketing actions.

$$\max_a E\left[\pi\left(a\,\middle|\,\Omega\right)\right] \tag{1.2.4}$$

Ω represents the information available about the distribution of the outcomes resulting from marketing actions. Clearly, information about both the distribution of choice given the model parameters as well as information about the parameters will be relevant to selecting the optimal action. Our goal, then, is to adopt a system of inference and decision-making that will make it possible to solve (1.2.4). In addition, we will require that there will be practical ways of implementing this system of inference. By practical, we mean computable for problems of the size which practitioners in marketing encounter.

Through this book, we will consider models similar to the simple case considered here and develop these inference and computational tools. We hope to convince the reader that the Bayesian alternative is the right choice.

---

[1] Some might object to this formulation of the problem as the aggregate market shares are deterministic functions of x. It is a simple matter to add an additional source of randomness to the shares. We are purposely simplifying matters for expositional purposes.

1.3 BENEFITS AND COSTS OF THE BAYESIAN APPROACH

In the beginning of Chapter 2, we outline the basics of the Bayesian approach to inference and decision-making. There are really no other approaches that can provide a unified treatment of inference and decision as well properly account for parameter and model uncertainty. However compelling the logic is behind the Bayesian approach, it has not been universally adopted. The reason for this is that there are nontrivial costs of adopting the Bayesian perspective. We will argue that some of these "costs" have been dramatically reduced and further that some "costs" are not really costs but are actually benefits.

The traditional view is that Bayesian inference provides the benefits of exact sample results, integration of decision-making, "estimation," "testing," and model selection, and a full accounting of uncertainty. Somewhat more controversial is the view that the Bayesian approach delivers the answer to the right question in the sense that Bayesian inference provides answers conditional on the observed data and not based on the distribution of estimators or test statistics over imaginary samples not observed. Balanced against these benefits are three costs: 1. Formulation of a prior; 2. Requirement of a Likelihood function; and 3. Computation of various integrals required in Bayesian paradigm. Development of various simulation-based methods in recent years has drastically lowered the computational costs of the Bayesian approach. In fact, for many of the models considered in this book, non-Bayesian computations would be substantially more difficult or, in some cases, virtually impossible. Lowering of the computational barrier has resulted in a huge increase in the amount of Bayesian applied work.

In spite of increased computational feasibility or, indeed, even computational superiority of the Bayesian approach, some are still reluctant to use Bayesian methods because of the requirement of a prior distribution. From a purely practical point of view, the prior is yet another requirement that the investigator must meet and this imposes a cost to the use of Bayesian approaches. Others are reluctant to utilize prior information based on concerns of scientific "objectivity." Our answer to those with concerns about "objectivity" is twofold. First, to our minds, scientific standards require that replication is possible. Bayesian inference with explicit priors meets this standard. Secondly, marketing is an applied field which means that the investigator is facing a practical problem often in situations with little information and should not neglect sources of information outside of the current data set.

For problems with substantial data information, priors in a fairly broad range will result in much the same a posteriori inferences. However, in any problem in which the data information to "parameters" ratio is low, priors will matter. In models with unit-level parameters, there is often relatively little data information so that it is vital that the system of inference incorporates even small amounts of prior information. Moreover, many problems in marketing explicitly involve multiple information sets so that the distinction between the sample information and prior information is blurred.

High-dimensional parameter spaces arise due to either the large numbers of "units" or the desire to incorporate flexibility in the form of the model specification. Successful solution of problems with high-dimensional parameter spaces requires additional structure. Our view is that prior information is one exceptionally useful way to impose structure on high-dimensional problems. The real barrier is not the philosophical concern over the use of prior information but the assessment of priors in high-dimensional spaces. We need devices for inducing priors on high-dimensional spaces that incorporate the desired structure with a minimum of effort in assessment. Hierarchical models are one particularly useful method for assessing and constructing priors over parameter spaces of the sort which routinely arise in marketing problems.

Finally, some have argued that any system of likelihood-based inference is problematic due to concerns regarding mis-specification of the likelihood. Tightly parameterized likelihoods can be misspecified, although the Bayesian is not required to believe that there is a "true" model underlying the data. In practice, a Bayesian can experiment with a variety of parametric models as way of guarding against mis-specification. Modern Bayesian computations and modeling methods make the use of a wide variety of models much easier than in the past. Alternatively, more flexible "non" or "semi" parametric models can be used. All nonparametric models are just high-dimensional models to the Bayesian and this simply underscores the need for prior information and Bayesian methods in general. However, there is a school of thought prominent in econometrics that proposes estimators that are consistent for the set of models outside one parametric class (method of moments procedures are the most common of this type). However, in marketing problems, parameter estimates without a probability model are of little use. In order to solve the decision problem, we require the distribution of outcome measures conditional on our actions. This distribution requires not only point estimates of parameters but a specification of their distribution. If we regard the relevant distribution as part of the parameter space, then this statement is equivalent to the need for estimates of all rather than a subset of model parameters.

In a world with full and perfect information, revealed preference should be the ultimate test of the value of a particular approach to inference. The increased adoption of Bayesian methods in marketing shows that the benefits do outweigh the costs for many problems of interest. However, we do feel that the value of Bayesian methods for marketing problem is underappreciated due to lack of information. We also feel that many of the benefits are as yet unrealized since the models and methods are still to be developed. We hope that this book provides a platform for future work on Bayesian methods in marketing.

1.4 AN OVERVIEW OF METHODOLOGICAL MATERIAL AND CASE STUDIES

Chapters 2 and 3 provide a self-contained introduction to the basic principles of Bayesian inference and computation. A background in basic probability and statistics on the level of Casella and Berger [2002] is required to understand this material. We assume a familiarity with matrix notation and basic matrix operations, including the Cholesky root. Those who need a refresher or a concise summary of the relevant material might examine appendices A and B of Koop [2003]. We will develop some of the key ideas regarding joint, conditional, and marginal densities in the beginning of Chapter 2 as we have found that this is an area not emphasized sufficiently in standard mathematical statistics or econometrics courses.

We recognize that a good deal of the material in Chapters 2 and 3 is available in many other scattered sources but we have not found a reference that puts it together in a way that is useful for those interested in marketing problems. We also will include some of the insights that we have obtained from the application of these methods.

Chapters 4 and 5 develop models for within-unit and across-unit analysis. We pay extensive attention to models for discrete data as much disaggregate marketing data involve aspects of discreteness. We also develop the basic hierarchical approach to modeling differences across units and illustrate this approach with a variety of different hierarchical structures and priors.

The problem of model selection and decision theory is developed in Chapter 6. We consider the use of the decision-based metric in valuing information sources and show the importance of loss functions in marketing applications.

Chapter 7 treats the important problem of simultaneity. In models with simultaneity, the distinction between dependent and independent variables is lost as the models are often specified as a system of equations which jointly or simultaneously determine the distribution of a vector of random variables conditional on some set of explanatory or exogenous variables. In marketing applications, the marketing mix variables and sales are joint determined given a set of exogenous demand or cost shifters.

Chapter 8 develops a Bayesian perspective on the Machine Learning literature. Basic concepts of shrinkage and model selection are extremely important in making the highly parameterized models used in the ML literature practical and avoid over-fitting. The main distinction between the ML literature and a classical Bayesian approach is that most ML models are fit with approximate methods that are not fully Bayesian. In addition, the ML literature takes an unabashedly predictive approach rather than an inference approach. These are fundamentally different objectives as is discussed in the chapter.

Chapter 9 takes up the important question of how to conduct inference with text data and develops a number of popular models for analysis of text data. This chapter provides an introduction to Bayesian analysis of text data using the Latent Dirichlet Allocation (LDA) model that summarizes respondent text data by way of topic probabilities, which are unique for each respondent, and word probabilities for each topic that are the same across respondents. Topic probabilities summarize the themes present in textual responses and the word probabilities are used to interpret each topic. We discuss variations of the LDA model that can improve the interpretation of topics and show how the vector of topic probabilities can be used to form integrated models of textual response, choice, and scaled response data. A conjoint dataset is used to illustrate the model. We find that the text data helps clarify the origin of demand.

These core chapters are followed by five case studies from our research agenda. These case studies illustrate the usefulness of the Bayesian approach by tackling important problems that involve extensions or elaborations of the material covered in the first eight chapters. Each of the case studies have been rewritten from their original journal form to use a common notation and emphasize the key points of differentiation for each article. Data and code are available for each of the case studies.

1.5 APPROXIMATE BAYES METHODS AND THIS BOOK

Bayesian methods have seen a number of developments over the last decade or two. Among these developments is the development and use of approximate Bayesian techniques. This sub-field has been fueled by the widespread availability of large datasets as well as the emergence of Machine learning which afford complex modeling of such data. In particular, the need for approximate Bayes methods arises from the computational challenges associated with exact Bayesian inference. Traditional Bayesian methods (such as MCMC) are known for providing a principled way to represent uncertainty in statistical modeling, but they often require intense computational resources. With the advent of big data and complex models, exact computations at this scale have become intractable in many cases, which has led to the search for more efficient and scalable approaches. In essence, the goal of approximate Bayes methods is to develop techniques that can provide tractable solutions without significantly compromising the quality of the inference.

Several approximate techniques have emerged to address the challenges in exact Bayesian inference. For example, approximation ideas such as Bootstrap approaches which use sampling and averaging to approximate posterior quantities have become widely used. In Chapter 8, we will discuss many of these methods in some detail. A notable exception that we do not discuss in this text is variational inference (VI). The core idea of VI is to turn the problem of Bayesian inference into an optimization problem. This is done by introducing a family of distributions (known as the variational family) and finding the distribution within this family that is closest to the true posterior. VI aims to find a tractable distribution that is closest to the true posterior, providing faster convergence but sometimes at the cost of accuracy. The breadth of ideas in this area and limited applications in marketing have kept us from including this topic in our discussions. In part, we also do not cover VI topics here since there are excellent reviews already available (Blei et al. [2017]).

The field of approximate Bayes methods continues to be an active area of research, and the future directions are promising. There is a growing interest in developing techniques that can balance computational efficiency with accuracy, especially in the context of deep learning and complex hierarchical models. The integration of approximate Bayesian methods with other machine learning paradigms and the development of software packages that make these methods accessible to non-experts are also important trends. Research is also focusing on the theoretical understanding of these methods, providing guarantees on their performance, and exploring their applicability in various scientific and industrial domains. The ongoing collaboration between statisticians, computer scientists, and domain experts ensures that approximate Bayes methods will continue to evolve and play a vital role in statistical modeling and data analysis. Future editions of this text may indeed cover these topics in a lot more depth.

1.6 COMPUTING AND THIS BOOK

It is our belief that no book on practical statistical methods can be credible unless the authors have computed all the methods and models contained therein. For this reason, we have imposed the discipline on ourselves that nothing will be included we haven't computed. It is impossible to assess the practical value of a method without applying it in a realistic setting. Far too often, treatises on statistical methodology gloss over the implementation. This is particularly important with modern Bayesian methods applied to marketing problems. The complexity of the models and the dimensionality of the data can render some methods impractical. MCMC methods can be theoretically valid but of little practical value. Computations lasting more than a day can be required for adequate inference due to high autocorrelation and slow computation of an iteration of the chain.

If a method takes more than 3 or 4 hours of computing time on standard equipment, we deem it impractical in the sense that most investigators are unwilling to wait much longer than this for results. However, what is practical depends not only on the speed of computing equipment but on the quality of the implementation. The good news is that in 2024, even the most pedestrian computing equipment is capable of truly impressive computations, unthinkable at the beginning of the MCMC revolution in the late 1980s and early 1990s. Achieving the theoretical capabilities of the latest CPU chip may require much specialized programming, use of optimized BLAS libraries and the use of a low-level language such as C or FORTRAN. Most investigators are not willing to make this investment unless their primary focus is on the development of methodology. Thus, we view a method as "practical" if it can be computed in a relatively high-level computing environment which mimics as closely as possible the mathematical formulas for the methods and models. For even wider dissemination of our methods, some sort of pre-packaged set of methods and models is also required.

For these reasons, we decided to program the models and methods of this book in the R language. In addition, we provide a web site for the book which provides further data and code for models discussed in the case studies. R is free, widely accepted in the statistical community, and offers much of the basic functionality needed and support for optimized matrix operations. Originally, our supporting code was written primarily

R in R with only a few functions translated into C. Since version 3.0 of *bayesm*, we have converted all computations into C++ and use the Armadillo matrix class. R is only used as a wrapper for these functions and to provide rudimentary checks on the validity of arguments. We have not implemented the use of parallelization or GPUs to enhance the speed of execution for our code. It is entirely possible that very large improvements in speed could be achieved with such improvements. Unfortunately, there are some issues at present that make it difficult to implement these approaches in a way that is transparent to hardware and software architecture and will supports the basic UNIX, Windows, and MacOS set of machines.

CPU speed is not the only resource that is important in computing. Memory is another resource which can be a bottleneck. Our view is that memory is so cheap that we do not want to modify our code to deal with memory constraints. All of our programs are designed to work entirely in memory. All of our applications use less than 10 GBs of memory.

Our experiences coding and profiling the applications in this book have changed our views on statistical computing. We were raised to respect minor changes in the speed of computations via various tricks and optimization of basic linear algebra operations. When we started to profile our code, we realized that, to a first approximation, linear algebra is free. The mainstay of Bayesian computations is the Cholesky root. These are virtually free on modern equipment (for example, one can compute the Cholesky root of 1000×1000 matrices at the rate of at least 200 per minute on standard-issue laptop computers). We found conversions from vectors to matrices and other "minor" operations to be more computationally demanding. Minimizing the number of matrix decompositions or taking advantage of the special structure of the matrices involved often has only minor impact. Optimization frequently involves little more than avoiding loops over the observations in the data set.

Computing also has an important impact on those who wish to learn from this book. We recognized, from the start, that our audience may be quite diverse. It is easy to impose a relatively minimal requirement regarding the level of knowledge of Bayesian statistics. It is harder to craft a set of programs which can be useful to readers with differing computing expertise and time to invest in computing. We decided that a two-pronged attack was necessary: 1. for those who want to use models pretty much "off-the-shelf," we have developed an R package to implement most of the models developed in the book; and 2. for those who want to learn via programming and who wish to extend the methods and models, we provide detailed code and examples for each of the chapters

R of the book and for each of the case studies. Our R package, *bayesm*, is available on the Comprehensive R Archive Network (CRAN, google "R language" for the URL). *bayesm* implements all of the models and methods discussed in the first seven chapters (see Appendix A for an introduction to R and *bayesm*). The book's website provides documented code, data sets, and additional information for those who which to adapt our models and methods.

We provide this code and examples with some trepidation. In some sense, those who really want to learn this material intimately will want to write their own code from scratch, using only some of our basic functions. We hope that providing the "answers" to the problem will not discourage study. Rather, we hope many of our readers will take our code as a base to improve on. We expect to see much innovation and improvement on what we think is a solid base.

ACKNOWLEDGMENTS

We owe a tremendous debt to our teachers, Dennis Lindley and Arnold Zellner, who impressed upon us the value and power of the Bayesian approach. Their enthusiasm is infectious. Our students (including Andrew Ainslie, Neeraj Arora, Peter Boatwright, Yancy Edwards, Tim Gilbride, Lichung Jen, Ling-Jing Kao, Jaehwan Kim, Alan Montgomery, Sandeep Rao, and Sha Yang) have been great collaborators as well as patient listeners as we have struggled to articulate this program of research. Junhong Chu read the manuscript very carefully and rooted out numerous errors and expositional problems. George Hochwarter provided invaluable assistance in converting the first edition manuscript into LaTex. Allenby would like to thank Vijay Bhargava for his encouragement in both personal and professional life. Rossi thanks the Anderson School of Management at UCLA for support. Misra thanks the Booth School of Business and the Kilts Center for Marketing at the University of Chicago for support.

2
Bayesian Essentials

Abstract

This chapter provides a self-contained introduction to Bayesian Inference. For those who need a refresher in distribution theory, Section 2.1 provides an introduction to marginal, joint, and conditional distributions and their associated densities. We then develop the basics of Bayesian inference, discuss the role of subjective probability and priors and provide some of the most compelling arguments for adopting the Bayesian point of view. Regression models (both univariate and multivariate) are considered along with their associated natural conjugate priors. Asymptotic approximations and Importance Sampling are introduced as methods for non-conjugate models. Finally, a simulation primer for the basic distributions/models in Bayesian Inference is provided. Those who want a basic introduction to Bayesian inference without many details should concentrate on Sections 2.2–2.6 and Section 2.10.1.

2.1 ESSENTIAL CONCEPTS FROM DISTRIBUTION THEORY

Bayesian inference relies heavily on probability theory and, in particular, distributional theory. This section provides a review of basic distributional theory with examples designed to be relevant to Bayesian applications.

A basic starting point for probability theory is a discrete random variable, X. X can take on a discrete number of values, each with some probability. The classic example would be a Bernoulli random variable. $X = 1$ with probability p and 0 with probability $p - 1$. X denotes some event such as whether a company will sell a product tomorrow. p represents the probability of a sale. For now, let us set aside the question of whether this probability can represent a long run frequency or whether it represents a subjective probability (note: it is hard to understand the long-run frequency argument for this example since it requires us to imagine an infinite number of "other-worlds" for the event of a sale tomorrow). We can easily extend this example to the number of units sold tomorrow. Then X is still discrete but can take on the values $0, 1, 2, \ldots, m$ with probabilities, $p_0, p_1, \ldots, p_m$. X now has a nontrivial probability distribution. With knowledge of this distribution, we can answer any question such as the probability that there will be at

Bayesian Statistics and Marketing, Second Edition. Peter E. Rossi, Greg M. Allenby, and Sanjog Misra
© 2024 John Wiley & Sons Ltd. Published 2024 by John Wiley & Sons Ltd.

least one sale tomorrow, the probability that there will be between 1 and 10 sales, etc. In general, we can compute the probability that sales will be in any set simply by summing over the probabilities of the elements in the set.

$$\Pr (X \in A) = \sum_{x \in A} p_x \tag{2.1.1}$$

We can also compute the *expectation* of the number of units sold tomorrow as the average over the probability distribution.

$$E[X] = \sum_{i=0}^{m} i p_i \tag{2.1.2}$$

If we are looking at aggregate sales of a popular consumer product, we might approximate sales as a *continuous* random variable that can take on any nonnegative real number. For this situation, we must summarize the probability distribution of X by a probability density. A density function is a *rate* function that tells us the probability per volume or unit of X. X has a density function, $p_X(x)$; p_X is a positive-valued function that integrates to one. The probability that X takes on any set of values we must integrate $p_X()$ over this set.

$$\Pr (X \in A) = \int_A p_X (x|\theta) \, dx \tag{2.1.3}$$

This is very much the analogue of the discrete sum in $(2.1.1)$. The sense in which p is a rate function is that the probability that $X \in (x_0, x_0 + dx)$ is approximately $p_X (x_0) \, dx$. Thus, the probability density function, $p_X( )$, plays the same role as the discrete probabilities (sometimes called probability mass function) in the discrete case. We can easily find the expectation of any function of X by computing the appropriate integral.

$$E\left[f(X)\right] = \int f(x) p(x|\theta) \, dx \tag{2.1.4}$$

In many situations, we will want to consider the *joint* distribution of two or more random variables, both of which are continuous. For example, we might consider the joint distribution of sales tomorrow in two different markets. Let X denote the sales in market A and Y denote the sales in market B. For this situation, there is a bivariate density function, $p_{X,Y}(x, y)$. This density gives the probability rate per unit of area in the plane. That is, the probability that both $X \in (x_0, x_0 + dx)$ and $Y \in (y_0, y_0 + dy)$ is approximately, $p_{X,Y}(x_0, y_0) \, dxdy$. With the joint density, we compute the probability of any set of (X, Y) values. For example, we can compute the probability that both X and Y are positive. This is the area of under the density for the positive orthant.

$$\Pr (X > 0 \ and \ Y > 0) = \int_0^\infty \int_0^\infty p_{X,Y}(x, y) \, dxdy \tag{2.1.5}$$

For example, the multinomial probit model, considered in Chapter 4, has choice probabilities defined the integrals of a multivariate normal density over various cones. If $p_{X,Y}()$ is a bivariate normal density, then $(2.1.5)$ is one such equation.

Given the joint density, we can also compute the *marginal* densities of each of the variables X and Y. That is to say, if we know everything thing about the joint distribution, we certainly know everything about the marginal distribution. The way to think of this is

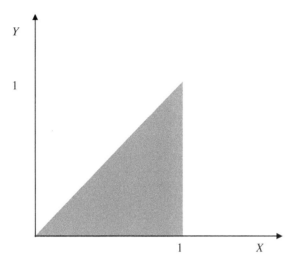

Figure 2.1   Support for the example of a bivariate distribution

via simulation. Suppose we were able to simulate from the joint distribution. If we look at the simulated distribution of either X or Y alone, we have simulated the marginal distribution.

To find the marginal density of X, we must average the *joint* density over all possible values of Y.

$$p_X(x) = \int p_{X,Y}(x,y)\, dy \qquad (2.1.6)$$

A simple example will help make this idea clear. Suppose X, Y are uniformly distributed over the triangle, $\{X, Y : 0 < X < 1 \text{ and } Y < X\}$, depicted in Figure 2.1. A uniform distribution means that the density is constant over the shaded triangle. The area of this triangle is $\frac{1}{2}$ so this means that the density must be 2 in order to insure that the joint density integrates to 1.

$$\int_0^1 \int_y^1 p_{X,Y}(x,y)\, dxdy = \int_0^1 \int_y^1 2\, dxdy = \int_0^1 \left(2x\big|_y^1\right) dy$$

$$= \int_0^1 (2 - 2y)\, dy = 2y - y^2\big|_0^1 = 1$$

This means that the joint density is a surface over the triangle with height 2.

We can use (2.1.6) to find the marginal distribution of X by integrating out Y.

$$p_X(x) = \int p_{X,Y}(x,y)\, dy = \int_0^x 2\, dy = 2y\big|_0^x = 2x$$

Thus, the marginal distribution of X is not uniform! The density increases as x increases toward 1. The marginal density of Y can easily be found to be of the "reverse" shape, $p_Y(y) = 2 - 2y$. This makes intuitive sense as the joint density is defined over the "widest" area with X near one and with Y near 0.

We can also define the concept of a conditional distribution and conditional density. If X, Y have a joint distribution, we can ask what is the conditional distribution of Y given

$X = x$? If X, Y are continuous random variables, then the conditional distribution of Y given $X = x$ is also a continuous random variable. The conditional density of $Y \,|X$ can be derived from the marginal and joint densities (the Borel paradox not withstanding).

$$p_{Y|X}\left(y|x\right) = \frac{p_{X,Y}\left(x, y\right)}{p_X\left(x\right)} \tag{2.1.7}$$

The argument of the conditional density on the left hand side of (2.1.7) is written $y|x$ to emphasize that there is a different density for every value of the conditioning argument x. We note that the conditional density is proportional to the joint! The marginal only serves to get the right normalization.

Let's return to our simple example. The conditional distribution of $Y|X = x$ is simply a slice of the joint density along a vertical line at the point x. This is clearly uniform but only extends from 0 to x. We can use (2.1.7) to get the right normalization.

$$p_{Y|X}\left(y|x\right) = \frac{2}{2x}; \quad y \in (0, x)$$

Thus, if $x = 1$, then the density is uniform over $(0,1)$ with height 1. The dependence between X and Y is only evidenced by the fact the range of Y is restricted by the value of x.

In many statistics courses, we are taught that correlation is a measure of the dependence between two random variables. This stems from the bivariate normal distribution that uses correlation to drive the shape of the joint density. Let's start with two independent standard normal random variables, Z and W. This means that their joint density factors (this is because of the product rule for independent events).

$$p_{Z,W}\left(z, w\right) = p_Z\left(z\right) p_W\left(w\right) \tag{2.1.8}$$

Each of the standard normal densities is given by:

$$p_Z\left(z\right) = \frac{1}{\sqrt{2\pi}} \exp\left(-\frac{1}{2}z^2\right) \tag{2.1.9}$$

If we create X and Y by an appropriate linear combination of Z and W, we can create correlated or dependent random variables.

$$X = Z$$
$$Y = \rho Z + \sqrt{\left(1 - \rho^2\right)} W$$

X and Y have a correlated bivariate normal density with correlation coefficient ρ.

$$p_{X,Y}\left(x, y\right) = \frac{1}{2\pi\sqrt{(1-\rho^2)}} \exp\left\{-\frac{1}{2(1-\rho^2)}\left[x^2 - 2\rho xy + y^2\right]\right\} \tag{2.1.10}$$

It is possible to show that $cov\left(X, Y\right) = E[XY] = \iint xy p_{X,Y}\left(x, y\right) dxdy$ is ρ. Both X, Y have marginal distributions that are standard normal and conditional distributions which are also normal but with a mean that depends on the conditioning argument.

$$X \sim N\left(0,1\right); \quad Y \sim N\left(0,1\right); \quad Y|X = x \sim N\left(\rho x, \left(1 - \rho^2\right)\right)$$

We will return to this example when we consider methods of simulation from the bivariate and multivariate normal distributions. We will also consider this situation when introducing the Gibbs Sampler in Chapter 3.

2.2 THE GOAL OF INFERENCE AND BAYES THEOREM

The goal of statistical inference is to use information to make inferences about unknown quantities. One important source of information is data but there is an undeniable role for non data-based information. Information can also come from theories of behavior (such as the information that, properly defined, demand curves slope downward). Information can also come from "subjective" views that there is a structure underlying the unknowns. For example, in situations with large numbers of different sets of parameters, an assumption that the parameters sets "cluster" or that they are drawn from some common distribution is often used in modeling. Less controversial might be the statement that we expect key quantities to be finite or even in some range (such as a price elasticity is not expected to be less than −50). Information can also be derived from prior analyses of other data, including data that is only loosely related to the dataset under investigation.

An unknown quantity is a generic term referring to any value not known to the investigator. Certainly, parameters can be considered unknown since these are purely abstractions that index a class of models. In situations in which decisions are made, the unknown quantities can include the, as yet unrealized, outcomes of marketing actions. Even in a passive environment, predictions of "future" outcomes are properly regarded as unknowns. There should be no distinction between a parameter and an unknown such as an unrealized outcome in the sense that the system of inference should treat each symmetrically.

Our goal, then, is to make inferences regarding unknown quantities *given* the information available. We have concluded that the information available can be partitioned into information obtained from the data as well as other information obtained independently or *prior* to the data. Bayesian inference utilizes probability statements as the basis for inference. What this means is that our goal is to make probability statements about unknown quantities *conditional* on the sample and prior information.

In order to utilize the elegant apparatus of conditional probability, we must encode the prior information as a probability distribution. This requires the view that probability can represent subjective beliefs and is not some sort of long run frequency. There is much discussion in the statistics and probability theory literature as to whether or not this is a reasonable thing to do. We take a somewhat more practical view – there are many kinds of non-data-based information to be incorporated into our analysis. A subjective interpretation of probability is a practical necessity rather than a philosophical curiosity.

It should be noted that there are several paths which lead to the conclusion that Bayesian inference is a sensible system of inference. Some start with the view that decision makers are expected utility maximizers. In this world, decision makers must be "coherent" or act in accordance with Bayes theorem in order to avoid exposing themselves to sure losses. Others start with the view that the fundamental primitive is not utility but subjective probability. Still others adhere to the view that the likelihood principle (Section 2.3) more or less forces you to adopt the Bayesian form of inference. We are more of the subjectivist stripe but we hope to convince the reader, by example, that there is tremendous practical value to the Bayesian approach.

2.2.1 Bayes Theorem

Denote the set of unknowns as θ. Our prior beliefs are expressed as a probability distribution, $p(\theta)$. $p(\bullet)$ is a generic notation for the appropriate density. In most cases, this represents a density with respect to standard Lebesgue measure but it can also represent a probability mass function for discrete parameter spaces or a mixed continuous-discrete measure. The information provided by the data is introduced via the probability distribution for the data, $p(D|\theta)$, where "D" denotes the observable data. In some classical approaches, modeling is the art of choosing appropriate probability models for the data. In the Bayesian paradigm, the model for prior information is also important. Much of the work in Bayesian statistics is focused on developing a rich class of models to express prior information and devices to induce priors on high dimensional spaces. In our view, the prior is very important and often receives insufficient attention.

To deliver on the goal of inference, we must combine the prior and likelihood to produce the distribution of the observables conditional on the data and the prior. Bayes Theorem is nothing more than an application of standard conditional probability to this problem.

$$p(\theta|D) = \frac{p(D,\theta)}{p(D)} = \frac{p(D|\theta)\,p(\theta)}{p(D)} \tag{2.2.1}$$

$p(\theta|D)$ is called the *posterior* distribution and reflects the combined data and prior information. (2.2.1) is often expressed using the likelihood function. Given D, any function that is proportional to $p(D|\theta)$ is call the "likelihood," $\ell(\theta)$. The shape of the posterior is determined entirely by the likelihood and prior in the numerator of (2.2.1) and this is often emphasized by rewriting the equation.

$$p(\theta|D) \propto \ell(\theta)\,p(\theta) \tag{2.2.2}$$

If $\ell(\theta) = p(D|\theta)$, then the constant of proportionality is the marginal distribution of the data, $p(D) = \int p(D,\theta)\,d\theta = \int p(D|\theta)\,p(\theta)\,d\theta$. Of course, we are assuming here that this normalizing constant exists. If $p(\theta)$ represents a proper distribution (i.e., it integrates to one), then this integral will likely exist. With improper priors, it will be necessary to show that the integral exists, which will involve the tail behavior of the likelihood functions.

2.3 CONDITIONING AND THE LIKELIHOOD PRINCIPLE

The likelihood principle states that the likelihood function, $\ell(\theta)$, contains all relevant information from the data. Two samples (not necessarily even from the same "experiment" or method of sampling/observation) have equivalent information regarding θ if their likelihoods are proportional (see Berger and Wolpert [1984] for extensive discussion and derivation of the LP from conditioning and sufficiency principles). The likelihood principle, by itself, is not sufficient to build a method of inference but should be regarded as a minimum requirement of any viable form of inference. This is a controversial point of view for anyone familiar with the modern econometric literature. Much of this literature is devoted to methods that do not obey the likelihood principle. For example, the phenomenal success of estimators based on the Generalized Method of Moments procedure is driven by the ease of implementing these estimators even though, in most instances, GMM estimators violate the likelihood principle.

Adherence to the likelihood principle means that inferences are *conditional* on the observed data as the likelihood function is parameterized by the data. This is worth contrasting to any sampling-based approach to inference. In the sampling literature, inference is conducted by examining the sampling distribution of some estimator of θ $\hat{\theta} = f(D)$. Some sort of sampling experiment[1] results in a distribution of D and, therefore, the estimator is viewed as a random variable. The sampling distribution of the estimator summarizes the properties of the estimator *prior* to observing the data. As such, it is irrelevant to making inferences given the data we actually observe. For any finite sample, this distinction is extremely important. One must conclude that, given our goal for inference, sampling distributions are simply not useful.

While sampling theory does not seem to deliver on the inference problem, it is possible to argue that it is relevant to the choice of estimating procedures. Bayesian inference procedures are simply one among many possible methods of deriving estimators for a given problem. Sampling properties are relevant to choice of procedures before the data is observed. As we will see in Section 2.6, there is an important sense in which one need never look farther than Bayes estimators even if the sole criterion is the sampling properties of the estimator.

2.4 PREDICTION AND BAYES

One of the appeals of the Bayesian approach is that all unknowns are treated the same. Prediction is defined as making probability statements about the distribution of as yet unobserved data, denoted by D_f. The only real distinction between "parameters" and unobserved data is that D_f is potentially observable.

$$p(D_f|D) = \int p(D_f, \theta|D)\; d\theta = \int p(D_f|\theta, D)\; p(\theta|D)\; d\theta \qquad (2.4.1)$$

(2.4.1) defines the "predictive" distribution of D_f given the observed data. In many cases, we assume that D and D_f are independent, conditional on θ. In this case, the predictive distribution simplifies.

$$p(D_f|D) = \int p(D_f|\theta)\; p(\theta|D)\; d\theta \qquad (2.4.2)$$

In (2.4.2), we average the likelihood for the unobserved data over the posterior of θ. This averaging properly accounts for uncertainty in θ when forming predictive statements about D_f.

2.5 SUMMARIZING THE POSTERIOR

For any problem of practical interest, the posterior distribution is a high dimensional object. Therefore, summaries of the posterior play an important role in Bayesian

---

[1] In the standard treatment, the sampling experiment consists of draws from the probability model for the data used in the likelihood. However, many other experiments are possible including samples experiments which involve additional assumptions regarding the data generation process.

statistics. Most schooled in classical statistical approaches are accustomed to reporting parameter estimates and standard errors. The Bayesian analogue of this practice is to report moments of the marginal distributions of parameters such as the posterior mean and posterior standard deviations. It is far more useful and informative to produce the marginal distributions of parameters or relevant functions of parameters as the output of the analysis. Simulation methods are ideally suited for this. If we can simulate from the posterior distribution of the parameters and other unknowns, then we can simply construct the marginal of any function of interest. Typically, we describe these marginals graphically. As these distributions are often very nonnormal, the mean and standard deviations are not particularly useful. One major purpose of this book is to introduce a set of useful simulation tools to achieve this goal of simulating from the posterior distribution.

Prior to the advent of powerful simulation methods, attention focused on the evaluation of specific integrals of the posterior distribution as a way of summarizing this high dimensional object. The general problem can be written as finding the posterior expectation of a function of θ. (We note that marginal posteriors, moments, quantiles, and probability of intervals are further examples of expectations of functions as in (2.5.1) with suitably defined h). For any interesting problem, only the un-normalized posterior, $\ell(\theta) p(\theta)$ is available so that two integrals must be performed to obtain the posterior expectation of $h(\theta)$

$$E_{\theta|D}\left[h(\theta)\right] = \int h(\theta)\, p(\theta|D)\, d\theta = \frac{\int h(\theta)\, \ell(\theta)\, p(\theta)\, d\theta}{\int \ell(\theta)\, p(\theta)\, d\theta} \tag{2.5.1}$$

For many years, only problems for which the integrals in (2.5.1) could be performed analytically were analyzed by Bayesians. Obviously, this restricts the set of priors and likelihoods to a very small set that produces posteriors of known distributional form and for which these integrals can be evaluated analytically. One approach would be to take various asymptotic approximations to these integrals. We will discuss the Laplace approximation method in Section 2.10. Unless these asymptotic approximations can be shown to be accurate, we should be very cautious about using them. In contrast, much of the econometrics and statistics literature uses asymptotic approximations to the sampling distributions of estimators and test statistics without investigating accuracy. In marketing problems, the combination of small amounts of sample information per parameter and the discrete nature of the data makes it very risky to use asymptotic approximations. Fortunately, we do not have to rely on asymptotic approximations in modern Bayesian inference.

2.6 DECISION THEORY, RISK, AND THE SAMPLING PROPERTIES OF BAYES ESTIMATORS

We started our discussion by posing the problem of obtaining a system of inference appropriate for marketing problems. We could just as well have started on the most general level – finding an appropriate framework for making decisions of any kind. Parameter estimation is only one of many such decisions that occur under uncertainty.

The general problem considered in decision theory is to search among possible actions for the action that minimizes expected loss. The loss function, $L(a, \theta)$, associates a loss with a state of nature (θ) and an action a. In Chapter 6, loss functions are derived for marketing actions from the profit function of the firm. We choose a decision that performs well, on average, where the averaging is taken across the posterior distribution of states of nature.

$$\min_{a} \left\{ \bar{L}\,(a) = E_{\theta|\,D}\,[L\,(a,\theta)] = \int L\,(a,\theta)\,p\,(\theta|\,D)\,d\theta \right\} \qquad (2.6.1)$$

In Chapter 6, we will explore the implications of decision theory for optimal marketing decisions and valuing of information sets. At this point it is important to note that (2.6.1) involves the entire posterior distribution and not just the posterior mean. With nonlinear loss functions, uncertainty or spread is just as important as location.

A special case of (2.6.1) is the estimation problem. If the action is the estimator and the state of nature is the unknowns to be estimated, then Bayesian decision theory produces a Bayes estimator. Typically, a symmetric function such as squared error or absolute error is used for loss. This defines the estimation problem as

$$\min_{\hat{\theta}} \left\{ L\,(\hat{\theta}) = E_{\theta|D}\left[L\,(\hat{\theta},\theta) \right] \right\}. \qquad (2.6.2)$$

For squared error loss, the optimal choice of estimator is the posterior mean.

$$\hat{\theta}_{Bayes} = E\,[\theta|\,D] = f\,(D|\tau) \qquad (2.6.3)$$

Here τ is the prior hyper-parameter vector (if any). If the prior is a parametric family of distributions, then the prior hyper-parameters are the parameters that describe this family. For example, if the prior is a normal distribution, then the prior mean and prior variance comprise the prior hyper-parameters.

What are the sampling properties of the Bayes estimator and how do these compare to those of other competing general purpose estimation procedures such as Maximum Likelihood? Recall the sampling properties are derived from the fact that the estimator is a function of the data and therefore is a random variable whose distribution is inherited from the sampling distribution of the data. We can use the same loss function to define the "risk" associated with an estimator, $\hat{\theta}$, as

$$r_{\hat{\theta}}(\theta) = E_{D|\theta}\left[L\,(\hat{\theta},\theta) \right] = \int L\,(\hat{\theta}(D),\theta)\,p\,(D|\theta)\,dD \qquad (2.6.4)$$

Note that the risk function for an estimator is a function of θ. That is, we have a different "risk" at every point in the parameter space.

An estimator is said to be *admissible* if there exists no other estimator with a risk function that is less than or equal to the risk of the estimator in question. That is, we cannot find another estimator that does better (or at least as well, as measured by risk, for every point in the parameter space.[2] Define expected risk, $E\,[r\,(\theta)] = E_{\theta}\left[E_{D|\theta}\left[L\,(\hat{\theta},\theta) \right] \right]$. The outer expectation on the right hand side is taken with respect to the prior distribution of θ. With a proper prior that has support over the entire parameter space, we can apply Fubini's theorem and interchange the order of integration and show that Bayes estimators

---

[2] Obviously if we have a continuous parameter space, we have to be a little more careful but we leave those niceties for those more mathematically inclined.

have the property of minimizing expected risk and, therefore, are admissible.

$$E\left[r\left(\theta\right)\right] = \quad E_{\theta}\left[E_{D|\theta}\left[L\left(\hat{\theta},\theta\right)\right]\right] = \iint L\left(\hat{\theta}(D),\theta\right)\, p\left(D|\theta\right) p\left(\theta\right) dDd\theta$$

$$= \quad E_{D}\left[E_{\theta|D}\left[L\left(\hat{\theta},\theta\right)\right]\right] = \iint L\left(\hat{\theta}(D),\theta\right)\, p\left(\theta|D\right) p\left(D\right) dDd\theta$$

(2.6.5)

The complete class theorem (see Berger [1985], Chapter 8) says even more – all admissible estimators are Bayes estimators. This provides a certain level of comfort and moral superiority but little practical guidance. There can be estimators that outperform Bayes estimators in certain regions of, but not all, of the parameter space. Bayes estimators perform very well if you are in the region of the parameter space you expect to be in as defined by your prior. These results on admissibility also don't provide any guidance as to how to choose among infinite number of Bayes estimators that are equivalent from the point of view of admissibility.

Another useful question to ask is what is the relationship between standard classical estimators such as the MLE and Bayes estimators? At least the MLE obeys the likelihood principle. In general, the MLE is not admissible so there can be no exact sample relationship. However, Bayes estimators are consistent, asymptotically normal and efficient as long as mild regularity conditions[3] hold and the prior is nondogmatic in the sense of giving support to the entire parameter space. The asymptotic "duality" between Bayes estimators and the MLE stems from the asymptotic behavior of the posterior distribution. As n increases, the posterior concentrates more and more mass in the vicinity of the "true" value of θ. The likelihood term dominates the prior and the prior becomes more and more uniform in appearance in the region in which the likelihood is concentrating. Thus, the prior has no asymptotic influence and the posterior starts to look more and more normal.

$$p\left(\theta|D\right) \sim N\left(\hat{\theta}_{MLE}, \left[-H_{\theta=\hat{\theta}_{MLE}}\right]^{-1}\right)$$

(2.6.6)

H_{θ} is the Hessian of the log-likelihood function. The very fact that, for asymptotics, the prior doesn't matter (other than its support) should be reason enough to abandon this method of analysis in favor of more powerful techniques.

2.7 IDENTIFICATION AND BAYESIAN INFERENCE

The set of models is only limited by the imagination of the investigator and the computational demands of the model and inference method. In marketing problems, we can easily write down a model that is very complex and may make extraordinary demands of data. A problem of identification is defined as the situation in which there is a set of different parameter values that give rise to the same distribution for the data. This set of parameter

---

[3] It should be noted that as the MLE is based on a maximum of a function while the Bayes estimator is based on an average, the conditions for asymptotic normality are different for the MLE than for the Bayes estimator. But both from a practical (i.e., computational) and theoretical perspective, averages behave more regularly than maxima.

values are said to observationally equivalent in the sense that the distribution of the data is the same for any member of this set.

Lack of identification is a property of the model and holds over all possible values of the data rather than just the data observed. Lack of identification implies that there will be regions over which the likelihood function is constant for any given data set. Typically, these can be flats or ridges in the likelihood.

From a purely technical point of view, identification is not a problem for a Bayesian analysis. First of all, the posterior may not have a "flat" or region of constancy as the prior can modify the shape of the likelihood. Even if there are regions for which the posterior is constant, the Bayesian will simply report, correctly, that the posterior is uniform over these regions.

Lack of identification is often regarded as a serious problem that must be dealt with by imposing some sort of restriction on the parameter space. Methods of inference, such as maximum likelihood, that rely on maximization will encounter severe problems with unidentified models. A maximizer may climb up and shut down anywhere in flat of the likelihood created by lack of identification. Bayesian computational methods that use simulation as the basis for exploring the posterior are not as susceptible to these computational problems.

Rather than imposing some sort of constraint on the parameter space, the Bayesian can deal with lack of identification through an informative prior. With the proper informative prior, the posterior may not have any region of constancy. However, it should be remembered that lack of identification means that there are certain functions of the parameters for which posterior is entirely driven by the prior. We can define a transformation function, $\tau = f(\theta)$, and a partition, $\tau' = (\tau'_1, \tau'_2)$, where $\dim(\tau'_1) = r$, such that $p(\tau_1|D) = p(\tau_1)$. This means that, for certain transformed coordinates, the posterior is the same as the prior and only prior information matters in a posteriori inference. r is the "dimension" of the redundancies in the θ parameterization. The "solution" to the identification problem can then be to report the marginal posterior for the parameters that are "identified," $p(\tau_2|D)$. We will see this idea is useful for analysis of certain non-identified models. It is important, however, to examine the marginal prior, $p_{\tau_2}(\bullet)$, which is induced by the prior on θ.

2.8 CONJUGACY, SUFFICIENCY, AND EXPONENTIAL FAMILIES

Prior to modern simulation methods, a premium was placed on models that would allow analytical expressions for various posterior summaries. Typically, this means that we choose models for which we can compute posterior moments analytically. What we want is for the posterior to be of a distributional form for which the posterior moments are available in analytical expressions. This requirement imposes constraints on both the choice of likelihood and prior distributions. One approach is to require that the prior distribution be conjugate to the likelihood. A prior is said to be *conjugate* to the likelihood if the posterior derived from this prior and likelihood is in the same class of distributions as the prior. For example, normal distributions have simple expressions for moments. We can get normal posteriors by combing normal priors and likelihoods based on normal sampling models as we will see in Section 2.8. However, there can be conjugate priors for which no analytic expressions exist for posterior moments.

The key to conjugacy is the form of the likelihood since we can always pick priors with convenient analytic features. Likelihoods in the exponential family of distributions have conjugate priors (Bernardo and Smith [1994], section 5.2). The exponential family is a family of distributions with a minimal sufficient statistic of fixed dimension equal to the dimension of the parameter space. The duality of sufficient statistics and the parameter vector is what drives this result. Moreover, the exponential form means that combining an exponential family likelihood and prior will result in an exponential family posterior that is tighter than either the likelihood or the prior. To see this, recall the form of the regular parameterization of the exponential family. If we have a random sample $y' = (y_1, \ldots, y_n)$ from the regular exponential family, the likelihood is given by

$$p(y|\theta) \propto g(\theta)^n \exp\left\{ \sum_{j=1}^k c_j \phi_j(\theta) \bar{b}_j(y) \right\} \tag{2.8.1}$$

$\bar{b}_j(y) = \sum_{i=1}^n b_j(y_i)$. $\{\bar{b}_1, \ldots, \bar{b}_k\}$ are the set of minimal sufficient statistics for θ. A prior in the same form would be given by:

$$p(\theta|\tau) \propto g(\theta)^{\tau_0} \exp\left\{ \sum_{j=1}^k c_j \phi_j(\theta) \tau_j \right\} \tag{2.8.2}$$

$\{\tau_0, \tau_1, \ldots, \tau_k\}$ are the prior hyperparameters. Clearly, the posterior is also in the exponential form with parameters $\tau_0^* = n + \tau_0, \tau_1^* = \tau_1 + \bar{b}_1, \ldots, \tau_k^* = \tau_k + \bar{b}_k$.

$$p(\theta|y) \propto g(\theta)^{\tau_0+n} \exp\left\{ \sum_{j=1}^k c_j \phi_j(\theta) (\bar{b}_j + \tau_j) \right\} \tag{2.8.3}$$

Since the conjugate prior in (2.8.2) is of the same form as the likelihood, we can interpret the prior as the posterior from some other sample of data with τ_0 observations.

A simple example will illustrate the functioning of natural conjugate priors. Consider the Bernoulli probability model for the data, $y_i \sim iidB(\theta)$. θ is the probability of one of two possible outcomes for each $y_i = (0,1)$. The joint density of the data is given by

$$p(y|\theta) = \theta^{\sum_i y_i}(1-\theta)^{n-\sum_i y_i} \tag{2.8.4}$$

The choice of prior for this simple Bernoulli problem has been the subject of numerous articles. Some advocate the choice of a "reference" prior or a prior that meets some sort of criteria for default scientific applications. Others worry about the appropriate choice of a "non-informative" prior. Given that this problem is one dimensional, one might want to choose a fairly flexible family of priors. The conjugate prior for this family is the Beta prior.

$$p(\theta) \propto \theta^{\alpha-1}(1-\theta)^{\beta-1} \sim Beta(\alpha, \beta) \tag{2.8.5}$$

This prior is reasonably flexible with regard to location, $E[\theta] = \alpha/(\alpha+\beta)$, spread, and shape (either uni-modal or "u-shaped"). We can interpret this prior as the posterior from another sample of $\alpha + \beta - 2$ observations and $\alpha - 1$ values of "1." The posterior is also of the Beta form

$$p\left(\theta|y\right) \propto \theta^{\alpha + \Sigma_i y_i - 1}(1 - \theta)^{\beta + n - \Sigma_i y_i - 1} \sim Beta\left(\alpha', \beta'\right) \tag{2.8.6}$$

with $\alpha' = \alpha + \Sigma_i y_i$; $\beta' = \beta + n - \Sigma_i y_i$. Thus, we can find the posterior moments from the Beta distribution.

Those readers who are familiar with numerical integration methods might regard this example as trivial and not very interesting since one could simply compute whatever posterior integrals are required by univariate numerical integration. This would allow for the use of any reasonable prior. However, the ideas of natural conjugate priors are most powerful when applied to vectors of regression parameters and covariance matrices, which we develop in the next section.

2.9 REGRESSION AND MULTIVARIATE ANALYSIS EXAMPLES

2.9.1 Multiple Regression

The regression problem has received a lot of attention in the Bayesian statistics literature and provides a very good and useful example of nontrivial natural conjugate priors. The standard linear regression model is a model for the conditional distribution of y given a vector of predictor variables in x.

$$y_i = x_i'\beta + \varepsilon_i \qquad \varepsilon_i \sim iidN\left(0, \sigma^2\right) \tag{2.9.1}$$

or

$$y \sim N\left(X\beta, \sigma^2 I_n\right) \tag{2.9.2}$$

$N\left(\mu, \Sigma\right)$ is the standard notation for a multivariate distribution with mean and variance-covariance matrix Σ. We have only modeled the conditional distribution of y given x rather than the joint distribution. In nonexperimental situations, it can be argued that we need to choose a model for the joint distribution of both X and y. In order to complete the model, we would need a model for the marginal distribution of x.

$$p\left(x, y\right) = p\left(x|\psi\right)p\left(y|x, \beta, \sigma^2\right) \tag{2.9.3}$$

If ψ is a priori independent[4] of (β, σ^2), then the posterior factors into two terms, the posterior of the x marginal parameters and the posterior for the regression parameters and we can simply focus on the rightmost term of $(2.9.4)$.

$$p\left(\psi, \beta, \sigma^2|y, X\right) \propto \left[p\left(\psi\right)p\left(X|\psi\right)\right]\left[p\left(\beta, \sigma^2\right)p\left(y|X, \beta, \sigma^2\right)\right] \tag{2.9.4}$$

---

[4] This rules out deterministic relationships between ψ and (β, σ^2) as well as stochastic dependence. In situations in which the x variables are set strategically by the marketer, this assumption can be violated. We explore this further in Chapter 7.

What sort of prior should be used for the regression model parameters? There are many possible choices for family of prior distributions and, given the choice of family, there is also the problem of assessing the prior parameters. Given that the regression likelihood is a member of the exponential family, a reasonable starting place would be to consider natural conjugate priors. The form of the natural conjugate prior for the regression model can be seen by examining the likelihood function.

To start, let us review the likelihood function for the normal linear regression model. We start from the distribution of the error terms which are multivariate normal. If $x \sim N(\mu, \Sigma)$, then the density of x is given by

$$p(x|\mu, \Sigma) = (2\pi)^{-k/2}|\Sigma|^{-1/2} \exp\left(-\tfrac{1}{2}(x - \mu)'\Sigma^{-1}(x - \mu)\right) \tag{2.9.5}$$

$\varepsilon \sim N(0, \sigma^2 I_n)$. The density of ε is then

$$p\left(\varepsilon|\sigma^2\right) = (2\pi)^{-n/2}\left(\sigma^2\right)^{-n/2} \exp\left(-\tfrac{1}{2\sigma^2}\varepsilon'\varepsilon\right) \tag{2.9.6}$$

Given that the Jacobian from ε to y is 1 and we are conditioning on X, we can write the density of y given X, β, and σ^2 as

$$p\left(y|X, \beta, \sigma^2\right) \propto \left(\sigma^2\right)^{-n/2} \exp\left(-\tfrac{1}{2\sigma^2}(y - X\beta)'(y - X\beta)\right) \tag{2.9.7}$$

The natural conjugate prior is a joint density for β and σ^2 that is proportional to the likelihood. Given that the likelihood has a quadratic form in β in the exponent, a reasonable guess is that the conjugate prior for β is normal. Rewrite the exponent so that the quadratic form is of the usual normal form. This can be done either by expanding out the existing expression and completing the square in β, or by recalling the usual trick of projecting y on the space spanned by the columns of X, $y = \hat{y} + e = X\hat{\beta} + (y - X\hat{\beta})$ and $\hat{y}'e = 0$.

$$(y - X\beta)'(y - X\beta) = (y - X\hat{\beta})'(y - X\hat{\beta}) + (\beta - \hat{\beta})'X'X(\beta - \hat{\beta})$$
$$= vs^2 + (\beta - \hat{\beta})'X'X(\beta - \hat{\beta})$$

$vs^2 = (y - X\hat{\beta})'(y - X\hat{\beta})$, $v = n - k$, and $\hat{\beta} = (X'X)^{-1}X'y$. If we substitute this expression in the exponent, we can factor the likelihood into two components keeping the terms needed for a proper normal density for β and then the balance for σ

$$p\left(y|X, \beta, \sigma^2\right) \propto \left(\sigma^2\right)^{-v/2} \exp\left\{-\frac{vs^2}{2\sigma^2}\right\}$$
$$\times \left(\sigma^2\right)^{-(n-v)/2} \exp\left\{-\frac{1}{2\sigma^2}(\beta - \hat{\beta})'(X'X)(\beta - \hat{\beta})\right\} \tag{2.9.8}$$

The first term in (2.9.8) suggests a form for the density[5] of σ^2 and the second for the density of β. We note that the normal density for β involves σ^2. This means that the conjugate

---

[5] Note that we can speak interchangeably about σ or $\theta = \sigma^2$. However, we must remember the Jacobian of the transformation from σ to θ when we re-write the density expressions converting from σ to θ (the Jacobian is 2σ).

prior will be specified as

$$p\left(\beta,\sigma^2\right) = p\left(\sigma^2\right) p\left(\beta|\sigma^2\right) \tag{2.9.9}$$

The first term in (2.9.8) suggests that the marginal prior on σ^2 has a density of the form $p\left(\theta\right) \propto \theta^{-\lambda} \exp\left(-\frac{\delta}{\theta}\right)$. This turns out to be in the form of the inverse-gamma distribution. The standard form of the inverse-gamma is

$$\theta \sim \text{IG}\left(\alpha,\beta\right)$$

$$p\left(\theta\right) = \frac{\beta^\alpha}{\Gamma\left(\alpha\right)} \theta^{-(\alpha+1)} \exp\left(-\frac{\beta}{\theta}\right) \tag{2.9.10}$$

The natural conjugate priors for σ^2 is of the form:

$$p\left(\sigma^2\right) \propto \left(\sigma^2\right)^{-(v_0/2+1)} \exp\left\{-\frac{v_0 s_0^2}{2\sigma^2}\right\} \tag{2.9.11}$$

The standard Inv-Gamma form with $\alpha = v_0/2$ and $\beta = v_0 s_0^2/2$. We note that Inv-Gamma density requires a slightly different power for σ^2 than found in the likelihood. There is an extra, $\left(\sigma^2\right)^{-1}$ in (2.9.11) that is not suggested by the form of the likelihood in (2.9.8). This term can be rationalized by viewing the conjugate prior as arising from the posterior of a sample of size v_0 with sufficient statistics, s_0^2, $\bar{\beta}$, formed with the noninformative prior, $p\left(\beta,\sigma^2\right) \propto \frac{1}{\sigma^2}$.

The conjugate normal prior on β is given by:

$$p\left(\beta|\sigma^2\right) \propto \left(\sigma^2\right)^{-k/2} \exp\left\{-\frac{1}{2\sigma^2}\left(\beta - \bar{\beta}\right)' A \left(\beta - \bar{\beta}\right)\right\} \tag{2.9.12}$$

Equations (2.9.11) and (2.9.12) can be expressed in terms of common distributions using the relationship between the Inv-Gamma and the Inverse of a Chi-squared random variable.

$$\sigma^2 \sim \frac{v_0 s_0^2}{\chi_{v_0}^2} \quad \text{and} \quad \beta|\sigma^2 \sim N\left(\bar{\beta},\sigma^2 A^{-1}\right) \tag{2.9.13}$$

The notation v_0, s_0^2 is suggestive of the interpretation of the natural conjugate prior as based on another sample with v_0 degrees of freedom and sum of squared errors, $v_0 s_0^2$.

Given the natural conjugate priors for the regression problem (2.9.13), the posterior is in the same form.

$$p\left(\beta,\sigma|y,X\right) \propto p\left(y|X,\beta,\sigma^2\right) p\left(\beta|\sigma^2\right) p\left(\sigma^2\right)$$

$$\propto \left(\sigma^2\right)^{-n/2} \exp\left\{-\frac{1}{2\sigma^2}\left(y - X\beta\right)'\left(y - X\beta\right)\right\}$$

$$\times \left(\sigma^2\right)^{-k/2} \exp\left\{-\frac{1}{2\sigma^2}\left(\beta - \bar{\beta}\right)' A \left(\beta - \bar{\beta}\right)\right\} \tag{2.9.14}$$

$$\times \left(\sigma^2\right)^{-(v_0/2+1)} \exp\left\{-\frac{v_0 s_0^2}{2\sigma^2}\right\}$$

The quadratic forms in the first two exponents on the right hand side can be combined by "completing the square." To remove some of the mystery from this operation that appears to require that you know the answer prior to obtaining the answer, we rewrite the problem as the standard problem of decomposing a vector into its projection on a subspace spanned by some column vectors and an orthogonal "residual." Since A is positive definite, we can find the upper triangular Cholesky root, U. $A = U'U$.

$$(y - X\beta)'(y - X\beta) + (\beta - \bar{\beta})'U'U(\beta - \bar{\beta}) = (v - W\beta)'(v - W\beta)$$

where $v = \begin{bmatrix} y \\ U\bar{\beta} \end{bmatrix}$ $W = \begin{bmatrix} X \\ U \end{bmatrix}$. We now project v onto the space spanned by the W columns using standard least squares.

$$(v - W\beta)'(v - W\beta) = ns^2 + (\beta - \tilde{\beta})'W'W(\beta - \tilde{\beta})$$

where

$$\tilde{\beta} = (W'W)^{-1}W'v = (X'X + A)^{-1}(X'X\hat{\beta} + A\bar{\beta}) \tag{2.9.15}$$

and

$$ns^2 = (v - W\tilde{\beta})'(v - W\tilde{\beta}) = (y - X\tilde{\beta})'(y - X\tilde{\beta}) + (\tilde{\beta} - \bar{\beta})'A(\tilde{\beta} - \bar{\beta}). \tag{2.9.16}$$

Using these results, we can write (2.9.14).

$$p(\beta, \sigma^2 | y, X) \propto (\sigma^2)^{-k/2} \exp\left\{ -\frac{1}{2\sigma^2}(\beta - \tilde{\beta})'(X'X + A)(\beta - \tilde{\beta}) \right\}$$
$$\times (\sigma^2)^{-((n+v_0)/2+1)} \exp\left\{ -\frac{(v_0 s_0^2 + ns^2)}{2\sigma^2} \right\} \tag{2.9.17}$$

or

$$\beta | \sigma^2, y, X \sim N(\tilde{\beta}, \sigma^2(X'X + A)^{-1})$$
$$\sigma^2 | y, X \sim \frac{v_1 s_1^2}{\chi_{v_1}^2} \text{ with } v_1 = v_0 + n; \ s_1^2 = \frac{v_0 s_0^2 + ns^2}{v_0 + n} \tag{2.9.18}$$

The Bayes estimator corresponding to the posterior mean is $\tilde{\beta}$.

$$E[\beta|y] = E_{\sigma^2|y}\left[E_{\beta|\sigma^2,y}[\beta] \right] = E_{\sigma^2|y}[\tilde{\beta}] = \tilde{\beta}$$

$\tilde{\beta}$ is a weighted average of the prior mean and the least squares estimator, $\hat{\beta}$. The weights depend on the prior precision and the sample information (recall that the information matrix for all n observations in the regression model is proportional to $X'X$). There are two important practical aspects of this estimator:

1. The Bayes estimator is a "shrinkage" estimator in the sense that the least squares estimator is "shrunk" toward the prior mean. Similarly, the posterior distribution of σ^2 is "centered" over s_1^2 which is a weighted average of the prior parameter and a sample

quantity (however, the "sample" sum of squares includes a term (see (2.9.16)) which represents the degree to which the prior mean differs from the least square estimator).

2. As we acquire more sample information in the sense of total X variation, the Bayes estimator converges to the least squares estimator (this insures consistency which we said is true, in general, for Bayes estimators).

These results for the posterior are special to the linear regression model and the natural conjugate prior set-up but if the likelihood function is approximately normal with mean equal to the MLE then these results suggest that the Bayes estimator will be a shrinkage estimator, which is a weighted average of the MLE and prior mean.

While the posterior mean of β is available by direct inspection of the conditional distribution of β given σ^2, the marginal distribution of β must be computed. It turns out that the marginal distribution of β is in the multivariate t form. The multivariate t distribution is in the elliptical class of distributions[6] and the marginal distribution of β will have the same density contours as the conditional distribution but will have fatter, algebraic tails rather than the thin exponential tapering normal tails. Moments of the multivariate t distribution are available.[7] However, the results in (2.9.18) can be used to devise a simple simulation strategy for making iid draws from the posterior of β. Simply draw a value of σ^2 from its marginal posterior, insert this value into the expression for the covariance matrix of the conditional normal distribution of $\beta|\sigma^2, y$ and draw from this multivariate normal.[8]

Equation (2.9.18) illustrates one of the distinguishing characteristics of natural conjugate priors. The posterior is centered between the prior and the likelihood and is more concentrated than either. Thus, it appears that you always gain by combining prior and sample information. It should be emphasized that this is a special property of natural conjugate priors. The property comes, in essence, from the interpretation of the prior as the posterior from another sample of data. The posterior we calculate can be interpreted as the same as the posterior one would obtain by pooling both the prior "sample" and the actual sample and using a very diffuse prior. If the prior information is at odds with the sample information, the use of natural conjugate priors will gloss this over because of the implicit assumption that both prior and sample information are equally valid. Many investigators would like to be aware of the situation in which the prior information is very different from the sample information. It is often difficult to check on this if the parameter space is high dimensional. This is particularly a problem in hierarchical models in which natural conjugate results like the ones given here are used. At this point, we call attention to this as a potential weakness of natural conjugate priors. After we develop computational methods that can work with nonconjugate priors, we will return to this issue.

---

[6] Note that linear combinations of independent t distributions are not in the multivariate t form.

[7] $Var(\beta) = \frac{v_1}{v_1-2} s_1^2 (X'X + A)^{-1}$

[8] Note: you will simply multiply the appropriate draws from a $N(0, (X'XA)^{-1})$ by the square root of this draw, so that this will be computationally nearly the same as multivariate normal draws. In Section 2.12, we will discuss some of the details for simulation from the distributions required for natural conjugate Bayes models.

2.9.2 Assessing Priors for Regression Models

Equation (2.9.13) is one implementation of a particular view regarding prior information on the regression coefficients. The idea here is that our views about coefficients are dependent on the error variance or equivalently the scale of the regression error terms (σ). If σ is large, the prior on the regression coefficients spreads out to accommodate larger values or reflects greater uncertainty. This is very much driven by the view that prior information comes from data. In situations where the "sample" creating the prior information comes from a regression with highly variable error terms, this sample information will be less valuable and the prior should spread out. However, it is entirely possible to imagine prior information that directly addresses the size of the regression coefficients. After all, the regression coefficients are designed to measure the effects of changes in the x variables. We may have information about the size of effects of a one unit price change on sales, for example, that is independent of the percentage of variation in sales explained by price. For this reason, the natural conjugate prior may not be appropriate for some sorts of information.

The more serious practical problem with the prior in (2.9.13) is the large number of prior hyper-parameters that must be assessed, 2 for prior distribution of σ, and $k + (k(k + 1))/2$ for the conditional prior on β. Moreover, as the likelihood is of order n, we may find that for any standard regression dataset with a modest number of observations, the hard work of prior assessment will have a low return as the prior may make little difference. Obviously, the prior will only matter with a small amount of sample information that typically arises in situations with small datasets or lack of independent variation in the x variables. The full Bayesian machinery is rarely used directly on one dataset regression problems but is increasingly being used with sets of related regression equations. We will develop this further in Chapter 5 when we discuss hierarchical models.

One approach to the problem of assessment of the natural conjugate prior is to use prior settings that are very diffuse. This would involve a "large" value of s_0^2, "small" value of v_0, and a "small" value of A, the prior precision. If the prior is diffuse, the mean of β, $\bar{\beta}$ is not very critical and can be set to zero. What constitutes a "diffuse" or spread out prior is a relative statement – relative to the diffusion of the likelihood. We can be sure that the prior has "small" influence if we make the prior diffuse relative to the likelihood. If $v_0 =$ some small fraction of n like max(3, 0.01 n) and

$$A = v_0 S_X \quad \text{where} \quad S_X = diag\left(s_1^2, \ \dots \ , s_k^2\right) \quad \text{and} \quad s_j^2 = 1/(n-1)\sum_i \left(x_{ij} - \bar{x}_j\right)^2$$

$$(2.9.19)$$

then we have assessed a relatively diffuse prior. We have introduced the scaling of x into the prior as would seem reasonable. If we change the units of X, we might argue that our prior information should remain the same and this sort of prior is invariant to scale changes. The g-prior of Zellner [1986] takes this idea further and sets $A = gX'X$. This means that we are using not only the scale of the x variables but their observed correlation structure as well. There is some controversy as to whether this is "coherent" and/or desirable but for relatively diffuse priors, the difference between the prior given by (2.9.19) and the g-prior will be minimal.

Our scheme for assessing a relatively diffuse prior is not complete without method for assessing the value of s_0^2. s_0^2 determines the location of the prior in the sense that mean of σ^2 is

$$E\left[\sigma^2\right] = \frac{v_0 s_0^2}{v_0 - 2}, \text{ for } v_0 > 2; \quad \text{Var}\left(\sigma^2\right) = \frac{2v_0^2}{\left(v_0 - 2\right)^2 \left(v_0 - 4\right)} \left(s_0^2\right)^2, \text{ for } v_0 > 4$$

(2.9.20)

As v_0 increases, the prior becomes tighter and centered around s_0^2. Many investigators are rather cavalier about the value of s_0^2 and set it to some rather arbitrary number like 1. If the prior is barely proper then the argument is that there is sufficient prior mass over very large values of σ and this is all that is relevant. However, these prior settings often mean that little or no prior mass is put on small values of σ below 0.5 or so. Depending on the scale of y and the explanatory power of the X variables, this can be a very informative prior! A somewhat more reasonable, but still controversial view, is to take into account of the scale of y in assessing the value of s_0^2. For example, we could use the sample variance of y for s_0^2. At the most extreme, we could use the mean error sum of squares from the regression of y on X. Purists would argue that this is violating Bayes theorem in that the data is being "used" twice – once in the assessment of the prior and again in the likelihood. In the absence of true prior information on σ, it seems preferable to use some scale information on y rather than to put in an arbitrary value for and hope that the prior is diffuse enough so this doesn't affect the posterior inferences in an undesirable manner.

An alternative to assessment of an informative prior would be to use one of the many candidates for "noninformative" priors. There are many possible priors, each derived from different principles. For the case of continuous-valued parameters, it is important to understand that there can be no such thing as an uninformative prior. For example, if we have a uniform but improper[9] prior on a uni-dimensional parameter $\theta, p(\theta) \propto$ constant then $\tau = e^\theta$ has a nonuniform density $1/_\tau$. Various invariance principles have been proposed in which priors are formulated under the constraint that they are invariant to certain types of transformations. In many situations, prior information should not be invariant to transformations. Our view is that prior information is available and should be expressed through proper priors assessed in the parameterization which yields maximum interpretability. Moreover, the use of proper priors avoids mathematical pathologies and inadvertent rendering of extreme prior information. However, for sake of completeness, we present the standard noninformative prior for the regression model.

$$p(\beta, \sigma) = p(\beta) p(\sigma) \propto \frac{1}{\sigma}$$

(2.9.21)

Or

$$p(\beta, \sigma^2) \propto \frac{1}{\sigma^2}$$

(2.9.22)

In contrast to the natural conjugate prior, β is independent of σ. The uniform distribution of over all of β over all of R^k means that prior in (2.9.21) implies that we think

---

[9] Improper means nonintegrable, i.e., the integral $\int p(\theta) d\theta$ diverges.

β is large with very high prior probability. In our view, this exposes the absurdity of this prior. We surely have prior information that the regression coefficients are not arbitrarily large! The prior on σ^2 can be motivated by appeal to the notion of scale invariance – that is the prior should be unchanged when y is multiplied by any positive constant.

2.9.3 Bayesian Inference for Covariance Matrices

A key building block in many Bayesian models is some sort of covariance structure. The workhorse model in this area is a multivariate normal sample and the associated natural conjugate prior. Consider the likelihood for a random sample from an m-dimensional multivariate normal (for expositional purposes, we omit the mean vector; in Section 2.9.5, we will consider the general case where the mean vector is a set of regression equations).

$$p\left(y_1, \ldots, y_n | \Sigma\right) \propto \prod_{i=1}^{n} |\Sigma|^{-1/2} \exp\left\{-\tfrac{1}{2} y_i' \Sigma^{-1} y_i\right\} = |\Sigma|^{-n/2} \exp\left\{-\tfrac{1}{2} \sum_{i=1}^{N} y_i' \Sigma^{-1} y_i\right\}$$
(2.9.23)

We can rewrite the exponent of (2.9.23) in a more compact manner using the trace operator.

$$\sum tr\left(y_i' \Sigma^{-1} y_i\right) = \sum tr\left(y_i y_i' \Sigma^{-1}\right) = tr\left(S\Sigma^{-1}\right) \text{ with } S = \sum y_i y_i'.$$

Using the notation, $etr\,() \equiv \exp\left(tr\,()\right)$

$$p\left(y_1, \ldots, y_n | \Sigma\right) \propto |\Sigma|^{-n/2} \, etr\left\{-\tfrac{1}{2} S\Sigma^{-1}\right\}$$
(2.9.24)

Equation (2.9.24) suggests that a natural conjugate prior for Σ is of the form

$$p\left(\Sigma | v_0, V_0\right) \propto |\Sigma|^{-(v_0+m+1)/2} \, etr\left(-\tfrac{1}{2} V_0 \Sigma^{-1}\right)$$
(2.9.25)

with $v_0 > m$ required for an integrable density.

Equation (2.9.25) is the expression for an Inverted Wishart density. We use the notation

$$\Sigma \sim IW\left(v_0, V_0\right)$$
(2.9.26)

If $v_0 \geq m + 2$, then $E\left[\Sigma\right] = (v_0 - m - 1)^{-1} V_0$. The full density function for the IW distribution is given by

$$p\left(\Sigma | v_0, V_0\right) = \left(2^{v_0 m/2} \pi^{\{m(m-1)/4\}} \prod_{i=1}^{m} \Gamma\left(\frac{v_0 + 1 - i}{2}\right)\right)^{-1}$$
$$\times |V_0|^{v_0/2} |\Sigma|^{\{-(v_0+m+1)/2\}} etr\left(-\tfrac{1}{2} V_0 \Sigma^{-1}\right)$$
(2.9.27)

This implies that Σ^{-1} has a Wishart distribution, $\Sigma^{-1} \sim W\left(v_0, V_0^{-1}\right)$. $E\left[\Sigma^{-1}\right] = v_0 V_0^{-1}$.

We can interpret V_0 as determining the "location" of the prior and v_0 as determining the spread of the distribution. However, some caution should be exercised in interpreting V_0 as a location parameter, particularly for small values of v_0. The IW is a highly skewed distribution that can be thought of as the matrix-valued generalization of the inverted chi-squared prior for the variance in the single parameter case.[10] As with all highly skewed distributions, there is a close relationship between the spread and the location. As we increase V_0 for small v_0, then we also increase the spread of the distribution dramatically.

The Wishart and Inverted Wishart distributions have a number of additional drawbacks. The most important drawback is that the Wishart has only one tightness parameter. This means that we can't be very informative on some elements of the covariance matrix and less informative on others. We would have to use independent Inverted Wisharts or some more richly parameterized distribution to handle this situation. In some situations, we wish to condition on some elements of the covariance matrix and use a conditional prior. Unfortunately, the conditional distribution of a portion of the matrix given some other elements is not in the Inverted Wishart form.

As is usual with natural conjugate priors, we can interpret the prior as the posterior stemming from some other data set and with a diffuse prior. If we use the diffuse prior suggested by Zellner via the Jeffreys' invariance principle (Zellner [1971]), p. 225, $p(\Sigma) \propto |\Sigma|^{-(m+1)/2}$, then we can interpret v_0 as the effective sample size of the sample underlying the prior and V_0 as the sum of squares and cross products matrix from this sample. This interpretation helps to assess the prior hyper-parameters. Some assess V_0 by appeal to the sum of squares interpretation. Simply scale up some sort of prior notation of the variance-covariance matrix by v_0, $V_0 = v_0 \hat{\Sigma}$. A form of "cheating" would be to use the exact or a stylized version of covariance matrix of$\{y_1, \ldots, y_n\}$. In situations where a relatively diffuse prior is desired, many investigators use $v_0 = m + 3$; $V_0 = v_0 I$. This is an extremely spread out prior that should be used with some caution.

If we combine the likelihood in (2.9.24) with the conjugate prior (2.9.25), then we have a posterior in the IW form:

$$p(\Sigma|Y) \propto p(\Sigma)\, p(Y|\Sigma)$$

$$= |\Sigma|^{-(v_0+m+1)/2}\, etr\left(-\tfrac{1}{2} V_0 \Sigma^{-1}\right) |\Sigma|^{-n/2}\, etr\left(-\tfrac{1}{2} S \Sigma^{-1}\right) \qquad (2.9.28)$$

$$= |\Sigma|^{-(v_0+m+n+1)/2}\, etr\left(-\tfrac{1}{2}\left(V_0 + S\right)\Sigma^{-1}\right)$$

or

$$p(\Sigma|Y) \sim IW\left(v_0 + n,\, V_0 + S\right) \qquad (2.9.29)$$

with mean $E[\Sigma] = \frac{V_0+S}{v_0+n-m-1}$ which converges to the MLE as n approaches infinity.

2.9.4 Priors and the Wishart Distribution

Although it seems more natural and interpretable to place a prior directly on Σ, there is a tradition in some parts of the Bayesian literature of choosing the prior on Σ^{-1} instead.

---

[10] IW distributions have the property that the marginal distribution of a square block along the diagonal is also IW with the same degrees of freedom parameter and the appropriate submatix of V_0. Therefore, an IW distribution with one degree of freedom is proportional to the inverse of a chi-squared variate.

In the Σ^{-1} parameterization, the prior (2.9.26) is a Wishart distribution.

$$p\left(G = \Sigma^{-1}\right) \propto p\left(\Sigma | v_0, V_0\right) |_{\Sigma = G^{-1}} \times J_{\Sigma \to G}$$

or

$$
\begin{aligned}
p(G) &\propto |G|^{(v_0 + m + 1)/2} \, etr\left(-\tfrac{1}{2} V_0 G\right) \times |G|^{-(m+1)} \\
&= |G|^{(v_0 - m - 1)/2} \, etr\left(-\tfrac{1}{2} A_0^{-1} G\right)
\end{aligned}
\tag{2.9.30}
$$

where $A_0 = V_0^{-1}$. This is denoted $G \sim W\left(v_0, A_0\right)$. If $v_0 \geq m + 2$, then $E[G] = v_0 A_0$.

The real value of the Wishart parameterization comes from its use to construct simulators. This comes from a result in multivariate analysis (c.f. Muirhead [1982], theorem 3.2.1, p. 85). If we have a set of iid random vectors $\{e_1, \ldots, e_v\}$ and $e_i \sim N(0, I)$, then

$$E = \sum_{i=1}^{v} e_i e_i' \sim W(v, I) \tag{2.9.31}$$

if $\Sigma = U'U$, then

$$W = U'EU \sim W(v, \Sigma) \tag{2.9.32}$$

We note that we do not use (2.9.32) directly to simulate Wishart random matrices. In Section 2.12, we will provide an algorithm for simulating from Wisharts.

2.9.5 Multivariate Regression

One useful way to view the multivariate regression model is as a set of regression equations related through common X variables and correlated errors.

$$
\begin{aligned}
y_1 &= X\beta_1 + \varepsilon_1 \\
&\vdots \\
y_c &= X\beta_c + \varepsilon_c \\
&\vdots \\
y_m &= X\beta_m + \varepsilon_m
\end{aligned}
\tag{2.9.33}
$$

The subscript "c" denotes a vector of n observations on equation c; we use the subscript "c" to suggest the columns of a matrix, which will be developed later. The model in (2.9.33) is not complete without a specification of the error structure. The standard multivariate regression model specifies that the errors for observation c are correlated across equations. Coupled with a normal error assumption, we can view the model as a direct generalization of the multivariate normal inference problem discussed in Section 2.9.3. To see this, we will think of a row vector of one observation on each of the m regression equations. This row vector will have a multivariate normal distribution with means given by the appropriate regression equation.

$$y_r = B'x_r + \varepsilon_r$$
$$\varepsilon_r \sim iid\, N\,(0, \Sigma)$$

(2.9.34)

The "r" subscript refers to observation r and B is a $k \times m$ matrix whose columns are the regression coefficients in (2.9.33). y_r and ε_r are m vectors of the observations on each of the dependent variables and error term.

Equation (2.9.34) is a convenient way of expressing the model for the purpose of writing down the likelihood function for the model.

$$p\,(\varepsilon_1, \ldots, \varepsilon_n | \Sigma) \propto |\Sigma|^{-n/2} \exp\left\{-\tfrac{1}{2} tr S_\varepsilon \Sigma^{-1}\right\}$$
$$S_\varepsilon = \sum_{r=1}^{n} \varepsilon_r \varepsilon_r'$$

(2.9.35)

Since the Jacobian from ε to y is 1, we can simply substitute in to obtain the distribution of the observed data. Clearly, an IW prior will be the conjugate prior for Σ. To obtain the form for the natural conjugate prior for the regression coefficients, it will be helpful to write down the likelihood function using the matrix form of the multivariate regression model.

$$Y = XB + E$$
$$B = [\beta_1, \ldots, \beta_c, \ldots, \beta_m]$$

(2.9.36)

Both Y and E are $n \times m$ matrices of observations whose (i, j) elements are the ith observation on equation j. X is a $n \times k$ matrix of observations on k common independent variables. The columns of Y and E are $\{y_c\}$ and $\{\varepsilon_c\}$, respectively, given in (2.9.33). The rows of the E matrix are the $\{\varepsilon_r\}$ given in (2.9.35). Observing that $E'E = S_\varepsilon$, we can use (2.9.35) and (2.9.36) to write down the complete likelihood.

$$p\,(E|\Sigma) \propto |\Sigma|^{-n/2} etr\left\{-\tfrac{1}{2} E'E\Sigma^{-1}\right\}$$

$$p\,(Y|X, B, \Sigma) \propto |\Sigma|^{-n/2} etr\left\{-\tfrac{1}{2}(Y - XB)'\,(Y - XB)\Sigma^{-1}\right\}$$

Again, we can decompose the sum of squares using the least squares projection.

$$p\,(Y|X, B, \Sigma) \propto |\Sigma|^{-n/2} etr\left\{-\tfrac{1}{2}\left(S + \left(B - \hat{B}\right)'X'X\left(B - \hat{B}\right)\right)\Sigma^{-1}\right\}$$

with $S = \left(Y - X\hat{B}\right)'\left(Y - X\hat{B}\right)$ and $\hat{B} = (X'X)^{-1}X'Y$. To suggest the form of the natural conjugate prior, we can break up the two terms in the exponent.

$$p\,(Y|X, B, \Sigma) \propto |\Sigma|^{-(n-k)/2} etr\left\{-\tfrac{1}{2} S\Sigma^{-1}\right\}$$
$$\times |\Sigma|^{-k/2} etr\left\{-\tfrac{1}{2}\left(B - \hat{B}\right)'X'X\left(B - \hat{B}\right)\Sigma^{-1}\right\}$$

(2.9.37)

Equation (2.9.37) suggests that the natural conjugate prior is a IW on Σ and a prior on B which is conditional on Σ. The term involving B is a density expressed as a function of an arbitrary $k \times m$ matrix. We can convert this density expression from a function of B to a function of $\beta = vec(B)$ using standard results on vec operators[11] (see Magnus and Neudecker, p. 30).

$$tr\left(\left(B - \hat{B}\right)' X'X \left(B - \hat{B}\right) \Sigma^{-1}\right) = vec\left(B - \hat{B}\right)' vec\left(X'X\left(B - \hat{B}\right)\Sigma^{-1}\right)$$

and

$$vec\left(X'X\left(B - \hat{B}\right)\Sigma^{-1}\right) = \left(\Sigma^{-1} \otimes X'X\right) vec\left(B - \hat{B}\right)$$

Thus,

$$tr\left(\left(B - \hat{B}\right)' X'X \left(B - \hat{B}\right) \Sigma^{-1}\right) = vec\left(B - \hat{B}\right)' \left(\Sigma^{-1} \otimes X'X\right) vec\left(B - \hat{B}\right)$$

$$= \left(\beta - \hat{\beta}\right)' \left(\Sigma^{-1} \otimes X'X\right) \left(\beta - \hat{\beta}\right)$$

Thus, the second term on the rhs of (2.9.37) is a normal kernel. This means that the natural conjugate prior for β is a normal prior conditional on this specific covariance matrix which depends on Σ.

The natural conjugate priors for the multivariate regression model are of the form.

$$p(\Sigma, B) = p(\Sigma)\,p(B|\Sigma)$$
$$\Sigma \sim IW(v_0, V_0) \tag{2.9.38}$$
$$\beta|\Sigma \sim N\left(\bar{\beta}, \Sigma \otimes A^{-1}\right)$$

Just as in the univariate regression model, the prior on the regression coefficients is dependent on the scale parameters and the same discussion applies. If we destroy natural conjugacy by using independent priors on β and Σ, we will not have analytic expressions for the posterior moments. The posterior can be obtained by combining terms from the natural conjugate prior to produce a posterior, which is a product of an IW and a "matrix" normal kernel.

$$p(\Sigma, B|Y, X) \propto |\Sigma|^{-(v_0+m+1)/2}\ \text{etr}\left(-\tfrac{1}{2}V_0\Sigma^{-1}\right)$$
$$\times |\Sigma|^{-k/2}\ \text{etr}\left(-\tfrac{1}{2}\left(B - \bar{B}\right)' A \left(B - \bar{B}\right)\Sigma^{-1}\right) \tag{2.9.39}$$
$$\times |\Sigma|^{-n/2}\ \text{etr}\left(-\tfrac{1}{2}(Y - XB)'(Y - XB)\Sigma^{-1}\right)$$

We can combine the two terms involving B, using the same device we used for the univariate regression model.

$$\left(B - \bar{B}\right)' A \left(B - \bar{B}\right) + (Y - XB)'(Y - XB)$$
$$= (Z - WB)'(Z - WB) \tag{2.9.40}$$
$$= \left(Z - W\tilde{B}\right)'\left(Z - W\tilde{B}\right) + \left(B - \tilde{B}\right)' W'W\left(B - \tilde{B}\right)$$

with $W = \begin{bmatrix} X \\ U \end{bmatrix};\ \ Z = \begin{bmatrix} Y \\ U\bar{B} \end{bmatrix};\ \ A = U'U$

---

[11] $tr(A'B) = (vec(A))'vec(B)$ and $vec(ABC) = (C' \otimes A)\,vec(B)$.

The posterior density can now be written

$$
\begin{aligned}
p\left(\Sigma, B | Y, X\right) &\propto |\Sigma|^{-(v_0 + n + m + 1)/2}\, etr\left(-\tfrac{1}{2}\left(V_0 + \left(Z - W\tilde{B}\right)'(\bullet)\Sigma^{-1}\right)\right) \\
&\times |\Sigma|^{-k/2}\, etr\left(-\tfrac{1}{2}\left(B - \tilde{B}\right)' W' W \left(B - \tilde{B}\right)\Sigma^{-1}\right)
\end{aligned}
\tag{2.9.41}
$$

with

$$
\begin{aligned}
\tilde{B} &= \left(X'X + A\right)^{-1}\left(X'X\hat{B} + A\bar{B}\right) \\
\left(Z - W\tilde{B}\right)'\left(Z - W\tilde{B}\right) &= \left(Y - X\tilde{B}\right)'\left(Y - X\tilde{B}\right) + \left(\tilde{B} - \bar{B}\right)' A \left(\tilde{B} - \bar{B}\right)
\end{aligned}
\tag{2.9.42}
$$

Thus, the posterior is in the form the conjugate prior: IW x conditional normal.

$$
\begin{aligned}
\Sigma | Y, X &\sim IW\left(v_0 + n, V_0 + S\right) \\
\beta | Y, X, \Sigma &\sim N\left(\tilde{\beta}, \Sigma \otimes \left(X'X + A\right)^{-1}\right) \\
\tilde{\beta} &= vec\left(\tilde{B}\right);\quad \tilde{B} = \left(X'X + A\right)^{-1}\left(X'X\hat{B} + A\bar{B}\right); \\
S &= \left(Y - X\tilde{B}\right)'\left(Y - X\tilde{B}\right) + \left(\tilde{B} - \bar{B}\right)' A \left(\tilde{B} - \bar{B}\right)
\end{aligned}
\tag{2.9.43}
$$

2.9.6 The Limitations of Conjugate Priors

Up to this point, we have considered some standard problems for which there exist natural conjugate priors. Although the natural conjugate priors have some features that might not always be very desirable, convenience is a powerful argument. However, the set of problems for which there exist useable expressions for conjugate priors is very small as was pointed out in Section 2.8. In this section, we will illustrate that a seemingly minor change in the multivariate regression model destroys conjugacy. We will also examine the logistic regression model (one of the most widely used models in marketing) and see that there are no nice conjugate priors there. One should not conclude that conjugate priors are useless. Conjugate priors such as the normal and Wishart are simple useful representations of prior information that can be used even in nonconjugate contexts. Finally, many models are *conditionally conjugate* in the sense that conditional on some subset of parameters, we can use these conjugate results. This is exploited heavily in the hierarchical models literature.

The multivariate regression model in (2.9.33) with different regressors in each equations is called the Seemingly Unrelated Regression model (SUR) by Zellner. This minor change to the model destroys conjugacy. We can no longer utilize the matrix form in (2.9.36) to write the model likelihood as in (2.9.37).[12] The best we can do is

---

[12] This should not be too surprising since we know from the standard econometrics treatment that the MLE for the SUR is not the same as equation by equation least squares.

stack up the regression equations into one large regression.

$$y = X\beta + \varepsilon$$

$$
y = \begin{bmatrix} y_1 \\ y_2 \\ \vdots \\ y_m \end{bmatrix}; \quad
X = \begin{bmatrix} X_1 & 0 & 0 & 0 \\ 0 & X_2 & 0 & 0 \\ 0 & 0 & \ddots & 0 \\ 0 & 0 & 0 & X_m \end{bmatrix}; \quad
\varepsilon = \begin{bmatrix} \varepsilon_1 \\ \varepsilon_2 \\ \vdots \\ \varepsilon_m \end{bmatrix}
\tag{2.9.44}
$$

with

$$Var(\varepsilon) = \Sigma \otimes I_n \tag{2.9.45}$$

Conditional on Σ, we can introduce a normal prior, standardize the observations to remove correlation and produce a posterior. However, we cannot find a convenient prior to integrate out Σ from this conditional posterior. We can also condition on β, use an IW prior on Σ and derive an IW posterior for Σ. The reason we can do this is that, given β, we "observe" the ε directly and we are back to the problem of Bayesian inference for a correlation matrix with zero mean. In Chapter 3, we will illustrate a very simple simulation method based on the so-called Gibbs Sampler for exploring the marginal posteriors of the SUR model with a normal prior on β and an IW prior on Σ.

A workhorse model in the marketing literature is the Multinomial Logit model. The dependent variable is a multinomial outcome whose probabilities are linked to independent variables that are alternative specific. $y_i = \{0, 1, \ldots, J\}$ with probability p_{ij}.

$$p_{i,j} = \frac{\exp\left(x'_{ij}\beta\right)}{\sum_{k=0}^{J} \exp\left(x'_{ik}\beta\right)} \tag{2.9.46}$$

x_{ij} represent alternative specific attributes. Here the subscript "i" refers to observation i. Thus, the likelihood for the data (assuming independence of the observations) can be written as

$$p(y|\beta) = \prod_{i=1}^{n} p_{i,y_i} = \prod_{i=1}^{n} \frac{\exp\left(x'_{iy_i}\beta\right)}{\sum_{k=0}^{J} \exp\left(x'_{ik}\beta\right)} \tag{2.9.47}$$

Given that this model is in the exponential family, there should be a natural conjugate prior.[13] However, all this means is that the posterior will be in the same form as the likelihood. This does not assure us that we can integrate that posterior against any interesting functions. Nor does the existence of a natural conjugate prior insure that it is interpretable and, therefore, easily assessable. Instead, we might argue that a normal prior on β would be reasonable. The posterior, of course, is not in the form of a normal or, for that matter, any other standard distribution. While we might have solved the problem of prior assessment, we are left with the integration problem. Since it is known that the log-likelihood is globally concave, we might expect asymptotic methods to work reasonably well on this problem. We explore these in the next section.

---

[13] See Robert and Casella [2004], p. 146, for an example of the conjugate prior for $J = 2$.

2.10 INTEGRATION AND ASYMPTOTIC METHODS

Outside the realm of natural conjugate problems, we will have to resort to numerical methods to compute the necessary integrals for Bayesian inference. The integration problem is of the form

$$I = \int h(\theta)\, p(\theta)\, p(D|\theta)\, d\theta \qquad (2.10.1)$$

This includes the computation of normalizing constants, moments, marginals, credibility intervals and expected utility.

There are three basic sorts of methods for approximating integrals of the sort given in (2.10.1): (1). Approximate the integrand by some other integrand that can be integrated numerically; (2). Approximate the infinite sum represented by the integral through some finite sum approximation such as a quadrature method; or (3). View the integral as an expectation with respect to some density and use simulation methods to approximate the expectation by a sample average over simulations.

The first approach is usually implemented via resort to an asymptotic approximation of the likelihood. We can expand the log-likelihood in a second order Taylor series about the MLE and use this as a basis of a normal kernel.

$$I \doteq \int h(\theta) p(\theta) \exp\left\{ L(\theta)\,|_{\theta=\hat\theta} + \frac{1}{2}(\theta - \hat\theta)' H (\theta - \hat\theta) \right\} d\theta \qquad (2.10.2)$$

$H = \left[\frac{\partial^2 L}{\partial\theta\partial\theta'}\right]$. This is the classic Laplace approximation. If we use the "asymptotic" natural conjugate normal prior, we can compute approximate normalizing constants and moments.

$$I \doteq e^{L(\hat\theta)} \int h(\theta)(2\pi)^{-1/2}|A|^{1/2} \exp\left\{ -\frac{1}{2}(\theta - \hat\theta)' A(x) \right\}$$

$$\times \exp\left\{ -\frac{1}{2}(\theta - \hat\theta)' H^*(X) \right\} d\theta \qquad (2.10.3)$$

$H^* = -H$ (negative of the Hessian of the log-likelihood). Completing the square in the exponent,[14] we obtain

$$I \doteq e^{L(\hat\theta)-\frac{1}{2}SS}|A|^{1/2}|A + H^*|^{-1/2}$$

$$\times \int h(\theta)(2\pi)^{-k/2}|A + H^*|^{1/2} \exp\left\{ -\frac{1}{2}(\theta - \bar\theta)'(A + H^*)(\theta - \bar\theta) \right\} d\theta \qquad (2.10.4)$$

where $SS = (\bar\theta - \hat\theta)' A(\bullet) + (\bar\theta - \hat\theta)' H^*(\bullet)$ and $\bar\theta = (A + H^*)^{-1}(A\bar\theta + H^*\hat\theta)$. (2.10.4) implies that the normalizing constant and first two posterior moments ($h(\theta) = \theta$;

---

[14] Recall the standard completing the square result.

$$(x - \mu_1)' A_1 (x - \mu_1) + (x - \mu_2)' A_2 (x - \mu_2) =$$
$$(x - \tilde\mu)' (A_1 + A_2)(x - \tilde\mu) + (\mu_1 - \tilde\mu)' A_1 (\mu_1 - \tilde\mu)$$
$$+ (\mu_2 - \tilde\mu)' A_2 (\mu_2 - \tilde\mu)$$
$$where \ \tilde\mu = (A_1 + A_2)^{-1} (A_1\mu_1 + A_2\mu_2).$$

$h(\theta) = \theta^2$) are given by

$$\int p(\theta) p(\theta|D) \, d\theta \doteq e^{L(\hat{\theta}) - \frac{1}{2}SS} |A|^{1/2} |A + H^*|^{-1/2}$$

$$E[\theta|D] \doteq \tilde{\theta} \tag{2.10.5}$$

$$Var(\theta|D) \doteq \left(A + H^*\right)^{-1}$$

While quadrature methods can be very accurate for integrands that closely resemble a normal kernel multiplied by a polynomial, many posterior distributions are not of this form. Moreover, quadrature methods suffer from the curse of dimensionality (the number of computations required to achieve a given level of accuracy increases to the power of the dimension of the integral) and are not useful for more than a few dimensions. For this reason, we will have to develop specialized simulation methods that will rely on approximating (2.10.1) with a sample average of simulated values. One such method is importance sampling.

2.11 IMPORTANCE SAMPLING

While we indicated that it is not feasible to devise efficient algorithms for producing iid samples from the posterior, it is instructive to consider how such samples could be used if available. The problem is to compute

$$E_{\theta|D}\left[h(\theta)\right] = \int h(\theta) \, p(\theta|D) \, d\theta.$$

If $\{\theta_1, \ldots, \theta_R\}$[15] are random sample from the posterior density, then we can approximate this integral by a sample average.

$$\bar{h}_R = \frac{1}{R}\sum_r h\left(\theta_r\right) \tag{2.11.1}$$

If $Var\left(h(\theta)\right)$ is finite, we can rely on the standard theory of sample averages to obtain an estimate of the accuracy of $\bar{h}_R$.

$$STD\left(\bar{h}_R\right) = \frac{\sigma}{\sqrt{R}}; \quad \sigma^2 = E_{\theta|D}\left[\left(\bar{h}_R - E_{\theta|D}\left[h(\theta)\right]\right)^2\right]$$

$$STDERR\left(\bar{h}_R\right) = \frac{s}{\sqrt{R}}; \quad s^2 = \frac{1}{R}\sum_r \left(h_r - \frac{\sum_r h_r}{R}\right)^2 \tag{2.11.2}$$

STD denotes standard deviation and STDERR denotes the estimated standard deviations or standard error. The formula in (2.11.2) has tremendous appeal since the accuracy of

---

[15] We use R to denote the size of the simulation sample in order to distinguish this from the sample size (n).

the integral estimate is *independent* of k, the dimension of θ. This is true in a strict technical sense. However, in many applied contexts, the variation of the h function increases as the dimension of θ increases. This means that in some situations σ in (2.11.2) is a function of k. The promise of Monte Carlo integration to reduce or eliminate the curse of dimensionality is somewhat deceptive.

However, since we don't have any general purpose methods for generating samples from the posterior distribution, the classic method of Monte Carlo integration is not of much use in practice. The method of importance sampling uses an importance sampling density to make the Monte Carlo integration problem practical. Assume that we only have access to the un-normalized posterior density (the product of the prior and the likelihood). We then must estimate

$$E_{\theta|D}\left[h(\theta)\right] = \frac{\int h(\theta)\, p(\theta)\, p(D|\theta)\, d\theta}{\int p(\theta)\, p(D|\theta)\, d\theta}$$

Suppose we have an importance distribution with density g. We can rewrite the integrals in the numerator and denominator as expectations with respect to this distribution and then use a ratio of sample averages to approximate the integral.

$$E_{\theta|D}\left[h(\theta)\right] = \frac{\int h(\theta)\left[{}^{p(\theta)p(D|\theta)}/{}_{g(\theta)}\right] g(\theta)\, d\theta}{\int \left[{}^{p(\theta)p(D|\theta)}/{}_{g(\theta)}\right] g(\theta)\, d\theta} \tag{2.11.3}$$

This quantity can be approximated by sampling $\{\theta_1, \ldots, \theta_R\}$ from the importance distribution.

$$\bar{h}_{IS,R} = \frac{\frac{1}{R}\sum_r h(\theta_r)\, w_r}{\frac{1}{R}\sum_r w_r} = \frac{\sum_r h(\theta_r)\, w_r}{\sum_r w_r} \tag{2.11.4}$$

$w_r = p(\theta_r)\, p(D|\theta_r)\, /g(\theta_r)$. (2.11.4) is a ratio of weighted averages. Note that we do not have to use the normalized density g in computing the importance sampling estimate as the normalizing constants will cancel out from the numerator and denominator. The kernel of g is often called the *importance function*.

Assuming that $E_{\theta|d}\left[h(\theta)\right]$ is finite, then the law of large numbers implies that $\bar{h}_{IS,R}$ converges to the value of (2.11.3). This is true for any importance density that has the same support as the posterior. However, the choice of importance sampling density is critical for proper functioning of importance sampling. In particular, the tail behavior of the posterior relative to the importance density is extremely important. If the ratio $p(\theta)\, p(D|\theta)\, /g(\theta)$ is unbounded, then the variance of $\bar{h}_{IS,R}$ can be infinite (see Geweke [1989], theorem 2, or p. 82 of Robert and Casella [2004]). Geweke [1989] gives sufficient conditions for a finite variance. These conditions are basically satisfied if the tails of the importance density are thicker than the tails of posterior distribution.[16] The standard

---

[16] One of the conditions is that the ratio of the posterior to the importance density is finite. Robert and Casella point out that this means that g could be used in an rejection method. This is a useful observation only in very small dimensional parameter spaces. However, the real value of importance sampling methods are for moderate to reasonably high dimensional parameter spaces. In any more than a few dimensions, rejection methods based on the importance density would grind to a halt, rejecting a huge fraction of draws.

error of the importance sampling estimate (sometimes called the "numerical standard error") can be computed as follows:

$$STDERR\left(\bar{h}_{IS,R}\right) = \sqrt{\frac{\sum_r \left(h\left(\theta_r\right) - \frac{\sum_r h(\theta_r)}{R}\right)^2 w_r^2}{\left(\sum_r w_r\right)^2}} \qquad (2.11.5)$$

One useful suggestion for an importance density is to use the asymptotic approximation to the posterior developed in (2.10.4). The thin normal tails can be fattened by scale mixing to form a multivariate student t distribution with low degrees of freedom. This strategy was suggested by Zellner and Rossi [1984]. Specifically, we develop an importance function as $MSt\left(v, \hat{\theta}_{MLE}, s\left(-H|_{\theta=\hat{\theta}_{MLE}}\right)^{-1}\right)$. One could also use the posterior mode and Hessian evaluated at the posterior mode.

Draws from this distribution can be obtained via (2.12.6). "Trial" runs can be conducted with different degrees of freedom values. The standard error in (2.11.5) can be used to help "tune" the degrees of freedom and the scaling of the covariance matrix (s). The objective is to fatten the tails sufficiently to minimize variation in the weights. It should be noted that very low degree of freedom student t distributions ($v < 5$) become very "peaked" and do not make very good importance functions as the "shoulders" of the distribution are too narrow. We recommend moderate values of the degrees of freedom parameter and more emphasis on choice of the scaling parameter to broaden the "shoulders" of the distribution.

Geweke [1989] provides an important extension to this idea by developing a "split-t" importance density. The "split-t" importance function can handle posterior distributions that are highly skewed. The standard MSt importance function will be inefficient for skewed integrands as considerable fattening will be wasted in one "tail" or principal axis of variation.

A fat tailed importance function based on the posterior mode and Hessian can be very useful in solving integration problems in moderate to large dimension problems (2 to 20). For example, the extremely regular shape of the multinomial logit likelihood means that importance sampling will work very well for this problem. Importance Sampling has the side advantage of only requiring a maximizer and its associated output. Thus, importance sampling is almost a free good to anyone who is using the MLE.

But, there is also a sense that this exposes the limitations of importance sampling. If the posterior looks a lot like its asymptotic distribution, then importance sampling with the importance density proposed here will work well. However, it is in these situations that finite sample inference is apt to be least valuable.[17] If you want to tackle problems with very nonnormal posteriors, then greater care must be taken in the choice of importance function. The variance formulas can be deceiving. For example, suppose we situate the importance density far from the posterior mass. Then the weights will not vary much and we may convince ourselves using the standard error formulas that we have estimated the integral very precisely. Of course, we have completely missed the mass we

---

[17] One classical econometrician is rumored to have said – "importance sampling just adds fuzz to our standard asymptotics."

want to integrate against. In high dimensional problems, this situation can be very hard to detect. In very high dimensions, it would be useful to impose greater structure on the parameter space and break the integration problem down into more manageable parts. This is precisely what the methods and models discussed in Chapter 3 are designed to do. Finally, there are many models for which direct evaluation of the likelihood function is computationally intractable. Importance sampling will be of no use in these situations. However, importance sampling can be used as part of other methods to tackle many high dimensional problems, even those with intractable likelihoods.

2.11.1 GHK Method for Evaluation of Certain Integrals of MVN

In many situations, the evaluation of integrals of a multivariate normal distribution over a rectangular region may be desired. The rectangular region A is defined by $A = \{x : a < x < b\}$ where a and b are vectors of endpoints which might include infinity. Let $P = \Pr(x \in A)$ with $x \sim N(0, \Sigma)$. The GHK method (Keane [1994]; Hajivassiliou et al. [1996]) uses an importance sampling method to approximate this integral. The idea of this method is to construct the importance function and draw from it using univariate truncated normals.

We can define P either in terms of the correlated normal vector, x, or in terms of a vector of z of uncorrelated unit normals, $z \sim N(0, I)$.

$$x = Lz; \quad \Sigma = LL'$$

$$P = \Pr(x \in A) = \Pr(a < x < b) = \Pr(a < Lz < b)$$

$$= \Pr\left(L^{-1}a < z < L^{-1}b\right) = \Pr(z \in B)$$

$$B = \left\{z : L^{-1}a < z < L^{-1}b\right\}$$

L is the lower triangular Cholesky root of Σ. We express the density of z as the product of a series of conditional densities that will allow us to exploit the lower triangular array which connects z and x.

$$p(z|B) = p(z_1|B)\, p(z_2|z_1, B) \cdots p(z_m|z_1, \dots, z_{m-1}, B) \qquad (2.11.6)$$

The region defined by A can be expressed in terms of elements of z as follows

$$a_j < x_j < b_j$$
$$a_j < l_{j,1}z_1 + l_{j,2}z_2 + \cdots + l_{j,j-1}z_{j-1} + l_{j,j}z_j < b_j$$

$l_{i,j}$ are the elements of the lower Cholesky root, L. Given $z_{<j} = (z_1, z_2, \dots, z_{j-1})$, this inequality implies that z is a univariate normal truncated to a particular interval given by

$$\frac{a_j - \mu_j(z_{<j})}{l_{jj}} < z_j < \frac{b_j - \mu_j(z_{<j})}{l_{jj}} \qquad (2.11.7)$$

$\mu_j(z_{<j}) = l_{j,1}z_1 + l_{j,2}z_2 + \cdots + l_{j,j-1}z_{j-1}$. Equation (2.11.7) provides an algorithm for drawing from $p(z|B)$. We simply draw $z_1|B$ and then $z_2|z_1, B$ and so on to fill out the

z vector. The equation also provides us with a way to evaluate the conditional density of $z|B$.

$$p\left(z_j|z_{<j}, B\right) = \frac{\phi\left(z_j\right)}{D_j\left(z_{<j}\right)} \text{ with } D_j\left(z_{<j}\right) = \Phi\left(\frac{b_j-\mu_j}{l_{jj}}\right) - \Phi\left(\frac{a_j-\mu_j}{l_{jj}}\right)$$

Using (2.11.6), $p(z|B) = \dfrac{\prod \phi\left(z_j\right)}{\prod D_j\left(z_{<j}\right)} = \dfrac{f(z)}{D(z)}$

We now write the integral defining P in terms of the density of z and use $p(z|B)$ as an importance function.

$$P = \Pr\left(z \in B\right) = \int_B f\left(z\right) dz$$

$$= \int_B \frac{f\left(z\right)}{p\left(z|B\right)} p\left(z|B\right) dz \tag{2.11.8}$$

$$= \int_B D\left(z\right) p\left(z|B\right) dz$$

Thus, the GHK algorithm can be constructed as follows:

GHK ALGORITHM
Draw z from $p(z|B)$ using (2.11.7) and truncated univariate normal draws
Evaluate $D(z)$
Repeat R times and form the estimate $\hat{P} = \frac{1}{R}\sum_r D\left(z_r\right)$

2.12 SIMULATION PRIMER FOR BAYESIAN PROBLEMS

If we could construct an iid sample directly from the posterior, the problem of summarizing the posterior could be solved to any desired degree of simulation accuracy. Unfortunately, the problem of generating random variables from an arbitrary (and possibly very high dimensional) distribution has no general purpose and computationally tractable solution. We will have to exploit the special structure of Bayesian models in order to develop useful methods. The basis for all of these methods are methods of simulating random variates from a set of frequently used distributions.

2.12.1 Uniform, Normal, and Gamma Generation

All methods of continuous random variate generation start with a uniform pseudo random number generator. Univariate pseudo random number generators generate deterministic (conditional on a seed) sequences of numbers which pass various "tests" for distributional accuracy and appearance of randomness. In particular, the sequence should have a very long "period" before it repeats itself, exhibit minimal time dependence (sometimes measured by autocorrelation), and do a good job of "filling" a k-dimensional hyper-cube constructed from sub-sequences of length k. R uses the

Mersenne Twister (Matsumoto and Nishimura [1998]) as the default method. Some might argue that the KISS generator by Marsaglia [1999] is faster.

Any standard computing environment will also supply methods for generating normal as well as gamma distributed random variates. By default, R uses the Inverse CDF method to generate normal random variates. This certainly produces draws with excellent properties but it may be slightly inefficient compared to method such as the Ziggurat method of Marsaglia and Tsang [2000]. However, we have not found the computations required for normal draws to be a computational bottleneck in Bayesian computations.

R also provides methods to draw Gamma and chi-squared random variates using the Inverse CDF method. If your computing environment does not provide high quality Gamma random variates, the method of Marsaglia and Tsang [2000] can be programmed in a low level language such as C.

We have seen in Section 2.9 that draws from the Inverted Gamma prior and posterior for σ^2 are needed. Recall that $IG(\alpha = v/2, \beta = \frac{vs^2}{2}) \sim vs^2 / \chi_v^2$

$$IG\left(\alpha = v/2, \quad \beta = \frac{vs^2}{2}\right) = \frac{vs^2}{\text{Gamma}\left(\frac{v}{2}, \frac{1}{2}\right)} \tag{2.12.1}$$

The uniform, normal and Gamma methods available in R and many other computing environments can be used to construct simulators for many of the distributions needed for Bayesian inference. We will now provide a simulation primer for these distributions.

2.12.2 Truncated Distributions

The Inverse CDF method can be used to draw from truncated distributions, provided that a computationally efficient method is available to evaluate the inverse CDF or quantile function. To review the inverse CDF method, let $F_X^{-1}(x)$. be the inverse of the cdf of random variable X.[18] Then $X = F_X^{-1}(U)$, $U \sim unif(0,1)$ has distribution with cdf F_X. An important example is the truncated normal distribution. Consider the normal distribution truncated to the interval (a,b).

$$Y = X \times I_{(a,b)}(X), \ I_{(a,b)}(X) = 1 \ if \ X \in (a, b), \ 0 \ otherwise$$

The cdf of Y can be obtained from the cdf of X.

$$G_Y(y) = \frac{F(y) - F(a)}{F(b) - F(a)}$$

Let $p = G_Y(y)$ and solve for G^{-1}.

$$y = F^{-1}\left(p\left(F(b) - F(a)\right) + F(a)\right) \tag{2.12.2}$$

Recall that $F(x) = \Phi\left(\frac{x - \mu_X}{\sigma_X}\right)$ and $F^{-1}(p) = \mu_X + \sigma_X \Phi^{-1}(p)$. We have a simple algorithm for simulating from the truncated normal.

---

[18] To avoid any technical difficulties, we must assume that there are no jumps in F or that the distribution of X is absolutely continuous wrt to Lebesgue measure.

TRUNCATED NORMAL ALGORITHM

Draw $U \sim Unif\,(0,1)$

$$y = \mu + \sigma \Phi^{-1} \left(U \left(\Phi \left(\frac{b-\mu}{\sigma} \right) - \Phi \left(\frac{a-\mu}{\sigma} \right) \right) + \Phi \left(\frac{a-\mu}{\sigma} \right) \right)$$

$$Y \sim trun_{(a,b)} N\,(\mu,\sigma)$$

There is a legitimate argument that this algorithm is computationally inefficient due to the evaluation of the normal CDF and inverse CDF functions. A combination of rejection sampling and other methods can be used to develop more efficient algorithms (McCulloch and Rossi [1994]). However, for a language such as R or MATLAB where vectorization is essential for efficiency, the algorithm based on the cdf method is more **R** efficient. In our R package *bayesm*, we provide the routine rtrun to simulate a vector of truncated normals.

2.12.3 Multivariate Normal and Student t Distributions

Given unit or standard normal draws, we can create the general Normal variate by scale and location transform. Multivariate normal draws can be calculated via the following algorithm:

$$z' = (z_1, \dots , z_k);\;\; z_i \sim iidN\,(0,1)$$
$$x = U'z + \mu \sim N\,(\mu,\Sigma);\;\; \Sigma = U'U \tag{2.12.3}$$

U is the upper triangular Cholesky root of Σ (the so-called "LU" decomposition). Computation of roots of positive definite real matrices has been studied closely and there are reliable and computationally efficient algorithms for doing so (c.f. the implementation in LAPACK used by R).

In some cases, we must simulate from various the conditional distribution of a subvector of a multivariate normal given the remainder of the normal random vector. The fact that the conditional mean and conditional covariance matrix can be computed directly from the elements of the inverse of Σ is useful.

$$x = \begin{pmatrix} x_1 \\ x_2 \end{pmatrix} \sim N \left(\begin{pmatrix} \mu_1 \\ \mu_2 \end{pmatrix}, \Sigma = \begin{bmatrix} \Sigma_{11} & \Sigma_{12} \\ \Sigma_{21} & \Sigma_{22} \end{bmatrix} \right)$$

If

$$\Sigma^{-1} = V = \begin{bmatrix} V_{11} & V_{12} \\ V_{21} & V_{22} \end{bmatrix}$$

Then

$$x_1 | x_2 \sim N \left(\mu_1 - V_{11}^{-1} V_{12} \,(x_2 - \mu_2)\,, V_{11}^{-1} \right) \tag{2.12.4}$$

The multivariate t distribution is an elliptically symmetric distribution closely related to the multivariate normal distribution. The major difference is that the multivariate t has algebraic tails (not exponential like the normal) and it can be very peaked for very low degrees of freedom. A k dimensional multivariate t distribution with degrees of freedom parameter v, location parameter μ and scale parameter Σ has density

$$p(x|v, \mu, \Sigma) \propto |\Sigma|^{-1/2} \left[v + (x - \mu)'\Sigma^{-1} (x - \mu) \right]^{-(k+v)/2} \tag{2.12.5}$$

$E[x] = \mu; \; Var(x) = \frac{v}{v-2}\Sigma$. We can simulate from this distribution by fattening the tails of a multivariate normal.

$$X = \frac{Y}{\left(Z/_v\right)^{1/2}} + \mu; \; Y \sim N(0, \Sigma); \; Z \sim \chi_v^2 \tag{2.12.6}$$

$$X \sim MSt(v, \mu, \Sigma)$$

R Our R package, bayesm, includes rmvst to simulate from the multivariate Student t distribution.

2.12.4 The Wishart and Inverted Wishart Distributions

In Section 2.9, we observed that prior and posteriors for the covariance matrix of the multivariate normal distribution are of the Inverted Wishart form. The correspondence between the Wishart and Inverted Wishart distribution can be exploited to develop a simulator. Let $G \sim W(v, V^{-1})$ and factor V^{-1} into the product of upper Cholesky roots, $V^{-1} = U'U$. Then $G = U'BU$ where $B \sim W(v, I_m)$. To simulate from the standard Wishart, we construct the following lower triangular array of random variates (note all nonzero elements are independent. $Z_{i,j} \sim N(0,1)$.

$$T = \begin{bmatrix} \sqrt{\chi_v^2} & 0 & \cdots & 0 \\ Z_{2,1} & \sqrt{\chi_{v-1}^2} & & \vdots \\ \vdots & & \ddots & 0 \\ Z_{m,1} & \cdots & Z_{m,m-1} & \sqrt{\chi_{v-m+1}^2} \end{bmatrix} \tag{2.12.7}$$

$Z_{i,j} \sim N(0,1)$ and $v > m$.

$$TT' \sim W(v, I_m)$$

G can be constructed from the pieces.

$$G = U'TT'U = C'C$$
$$C = T'U \tag{2.12.8}$$

The Inverted Wishart can be computed from the draw of the corresponding Wishart using (2.12.8) by taking the inverse of C.

$$\Sigma = G^{-1} = C^{-1}(C^{-1})' \tag{2.12.9}$$

Note that (2.12.9) is the UL decomposition of Σ.

WISHART/INVERTED WISHART DRAW ALGORITHM

To draw $\Sigma \sim IW(v, V)$ or $\Sigma^{-1} \sim W(v, V^{-1})$:

1. Factor $V^{-1} = U'U$

2. Draw random variates and compute T as in (2.12.7)

3. Compute $C^{-1} = (T'U)^{-1}, \Sigma = C^{-1}(C^{-1})', \Sigma^{-1} = C'C$

R *bayesm* includes the function, `rwishart`, to simulate from both the Inverted Wishart and Wishart distribution.

2.12.5 Multinomial Distributions

The most general discrete distribution is the multinomial distribution. If θ can take on a d values, $S = \{\theta^1, \theta^2, \ldots, \theta^d\}$, each with probability p_i, then $\theta \sim MN(p, S)$.

MULTINOMIAL DRAW ALGORITHM

To draw from the multinomial, we only need to draw the index into S with the appropriate probability.
Draw $U \sim Unif(0,1)$
Find k such that $\sum_{i=0}^{k-1} p_i < U \leq \sum_{i=0}^{k} p_i$ with $p_0 = 0$.
$\theta = S(k)$.

2.12.6 Dirichlet Distribution

The natural conjugate prior for the multinomial distribution is called the Dirichlet distribution with density

$$p(\theta|\alpha_1, \ldots, \alpha_k) = \frac{\Gamma(\alpha_1 + \ldots \alpha_k)}{\Gamma(\alpha_1) \cdots \Gamma(\alpha_k)} \theta_1^{\alpha_1 - 1} \cdots \theta_k^{\alpha_k - 1} \quad \alpha_i > 0 \qquad (2.12.10)$$

θ is a k dimensional vector that must be in the unit simplex.

DIRICHLET DRAW ALGORITHM

Draw $x_i \sim ind$ Gamma(α_i, α_i)

$$\theta_i = \frac{x_i}{\sum_j x_j}$$

R Our R package, *bayesm*, includes `rdirichlet` to simulate from the Dirichlet distribution.

2.13 SIMULATION FROM POSTERIOR OF MULTIVARIATE REGRESSION MODEL

The Multivariate Regression model discussed in Section 2.9.5 is a very useful model not only because of direct application to situations with sets of related linear regressions but also in various hierarchical model settings in which the model is used as part of the prior structure. To analyze these hierarchical models, it will be necessary to use simulation-based methods. For this reason, it will be useful to have an efficient algorithm for sampling from the posterior of this model given in (2.9.43). Recall the basic set-up.

Model:

$$Y = XB + U \quad U = \begin{bmatrix} u_i' \end{bmatrix} \quad u_i \sim N(0, \Sigma) \tag{2.13.1}$$

Here Y is $n \times m$, X is $n \times k$, B is $k \times m$ where each row of B contains the regression coefficients for one of the m equations.

Prior:

$$\Sigma \sim IW(\nu, V)$$
$$\beta = vec(B) | \Sigma \sim N\left(vec(\bar{B}), \Sigma \otimes A^{-1}\right) \tag{2.13.2}$$

and
To draw from the posterior, we first draw Σ and then draw B given Σ.

Draw of Σ

$$\Sigma | Y, X \sim IW(\nu + n, V + S)$$
$$S = E'E$$
$$E = Y - X\tilde{B} + \left(\tilde{B} - \bar{B}\right)' A \left(\tilde{B} - \bar{B}\right) \tag{2.13.3}$$
$$\tilde{B} = \left(X'X + A\right)^{-1}\left(X'Y + A\bar{B}\right)$$

S and $\tilde{B}$ can be computed using the QR decomposition. The QR decomposition of a matrix X is $X = QR$ where Q is an $n \times k$ matrix whose columns are orthogonal and R is a $k \times k$ upper triangular matrix that is the Cholesky root of $X'X$ up to sign differences. We can compute the relevant quantities in (2.13.3) by forming following augmented matrices.

$$W = \begin{bmatrix} X \\ R_A \end{bmatrix} \quad Z = \begin{bmatrix} Y \\ R_A\bar{B} \end{bmatrix} \tag{2.13.4}$$

If we take the QR decomposition of W, $W = Q_W R_W$, then

$$\tilde{B} = R_W^{-1} Q_W Z \quad \text{and} \quad S = \left(Z - W\tilde{B}\right)'\left(Z - W\tilde{B}\right) = Z'\left(I - Q_W Q_W'\right)Z \tag{2.13.5}$$

However, timing experiments in R suggest that it is faster to compute these quantities by using the Cholesky root approach (between 30% and 100% faster than using the LAPACK QR method in R). We take the Cholesky root of $W'W$, invert this, and use this to compute $\tilde{B}$ and E.[19] The inverse of the root can be computed by efficiently in R and the root can also be used to draw from the posterior as we see below.

$$W'W = R'_{W'W}R_{W'W} \text{ and } IR_{W'W} = R^{-1}_{W'W}$$
$$\tilde{B} = IR_{W'W}\left(IR_{W'W}\right)'W'Z$$
$$E = Z - W\tilde{B} \tag{2.13.6}$$
$$S = E'E$$

$R_{W'W}$ is the upper triangular Cholesky root of $W'W$. We note that $IR_{W'W}$ defines the UL decomposition of $\left(W'W\right)^{-1}$ whereas $R_{W'W}$ forms the LU decomposition of $W'W$

To draw Σ, we draw from appropriate Wishart and then invert this matrix to obtain the Σ draw.

$$\Sigma^{-1} \sim W\left(v + n,(V + S)^{-1}\right)$$

The Bartlett strategy outlined in Section 2.12.4 produces a draw of the Cholesky root of Σ^{-1}.

$$\Sigma^{-1} = C'C$$
$$\Sigma = \left(C^{-1}\right)\left(C^{-1}\right)' = CICI' \tag{2.13.7}$$

We note that we have the LU decomposition of Σ^{-1} and the UL decomposition of Σ in (2.13.7).

Draw of $\beta|\Sigma$

The direct, but naïve, approach would be to draw β from a $N\left(vec\left(\bar{B}\right),\Sigma \otimes \left(X'X+A\right)^{-1}\right)$ distribution. We can exploit the special structure of the covariance matrix to develop an efficient strategy for making this draw.

$$Var\left(\beta\right) = \Omega = \Sigma \otimes \left(X'X + A\right)^{-1}$$
$$= CICI' \otimes \left(R'_{W'W}R_{W'W}\right)^{-1}$$
$$= CICI' \otimes IR_{W'W}IR'_{W'W} \tag{2.13.8}$$
$$= \left(CI \otimes IR_{W'W}\right)\left(CI \otimes IR_{W'W}\right)'$$

$IR_{W'W} = R^{-1}_{W'W}$. Note that (2.13.8) is the UL decomposition of the covariance matrix. Thus, we can use this root directly to produce normal variates with the right covariance.

$$v = \left(CI \otimes IR_W\right)z \quad z \sim N\left(0, I_{m\times k}\right)$$
$$Var\left(v\right) = \Omega$$

However, we can simplify this even further by using the identity $vec\left(ABC\right) = \left(C' \otimes A\right)vec\left(B\right)$.

$$B = \tilde{B} + IR_{W'W}ZCI' \quad vec\left(Z\right) = z \tag{2.13.9}$$

R rmultireg in our R package, bayesm, implements this strategy.

3

MCMC Methods

Abstract

This chapter provides an introduction to Markov Chain Monte Carlo (MCMC) methods and provides detailed discussion of the MCMC methods that have proved especially useful for marketing and micro-econometric problems. In addition, several key examples are developed which help set the stage for more complicated models covered in later chapters. These include the Gibbs samplers for binary probit, mixture of normals and hierarchical linear models. In addition, Metropolis methods are introduced and illustrated with the multinomial logit model. Those readers who desire an introduction to MCMC methods without much theoretical background should skip section 3.3 on Markov Chain theory and only skim the beginning of 3.9 on Metropolis methods (skip over the proof following the introduction of the "Continuous State Space Metropolis" in section 3.10).

Given a model (prior and likelihood) or set of models, the computational phase of Bayesian inference requires practical methods for summarizing/exploring the posterior distribution. In many cases, the posterior distribution is represented by an un-normalized density, $\pi * (\theta)$, and the problem is to construct simulation-based estimates of various aspects of this distribution. For any problem outside the conjugate family, the posterior density will be of a form for which analytical results on marginals or moments will be unavailable. More importantly, problems in which the dimension of θ is much more than 100 can easily arise in marketing applications. For example, suppose we want to compute response to marketing instruments for each of $100-200$ customers. We might formulate a group of $100-200$ regression models each with $5-10$ independent variables. If there are any linkages between these models through correlated unobservables,[1] then we must explore a parameter space of dimension $500-2000$. Problems of this dimension are considerably beyond the scope of importance-sampling methods. In order to tackle

---

[1] The regression errors could be correlated (as in the SUR model of Zellner) or the prior on the regression coefficients could have correlations or dependencies across equations. The latter is often formulated as a hierarchical model which is introduced below in Section 3.7 and in Chapter 5.

---

these problems, we will have to exploit the structure of the model and introduce new methods for simulation from arbitrary distributions.

3.1 MCMC METHODS

The idea behind Markov Chain Monte Carlo (MCMC) methods is to formulate a Markov Chain on the parameter space. If care is taken to insure that this chain has $\pi()$ as the equilibrium or "long-run" distribution of the chain, then the chain can be used to construct simulation-based estimates of the required integrals. Starting from some point in the parameter space, we simulate the chain forward. A sub-sequence of these draws can be used to construct simulation-based estimates of the posterior distribution of θ or any function of θ. For example, we can simply focus on one element of θ to simulate its marginal or we can average $h(\theta)$ over sequences of draws from the chain to estimate, $E_\pi[h(\theta)]$.

A Markov Chain specifies a method for generating a sequence of random variables $\{\theta_1, \theta_2, \dots, \theta_r, \dots\}$ starting from initial point θ_0 This sequence is created by specifying a way of transitioning or moving from θ_r to θ_{r+1}. Since we are dealing with random variables, this transition process is specified by choosing the conditional distribution, $\theta_{r+1}|\theta_r$, or $\theta_{r+1}|\theta_r \sim F(\theta_r)$. F denotes this conditional distribution that can be discrete, continuous, or a mixture of discrete and continuous distributions. By iterating the conditional distribution forward we construct a joint distribution on the sequence. The fact that this conditional distribution only depends on the last θ (or that θ_r completely summarizes all information up to this point) is the Markov property that greatly simplifies simulation and analysis of the chain.

Clearly the Markov Chain was well-designed for simulation. To simulate from the Markov Chain:

START from θ_0
DRAW $\theta_1 \sim F(\theta_0)$
Replace θ_0 with θ_1 and REPEAT, a total of R times

This will create a set of realizations of the chain given the starting point. Under some conditions on the conditional distribution F, the distribution of $\theta_r|\theta_0$ will converge to a fixed and unique distribution as r goes to infinity. This distribution is called the stationary, invariant,[2] or equilibrium distribution. If we can construct a Markov Chain with stationary distribution $\pi()$ and the conditions for convergence are met, then we can use the MC method to construct a simulation method for estimating the posterior expectation of any function.[3]

$$E_\pi[h(\theta)] \doteq \frac{1}{R}\sum h(\theta_r) \tag{3.1.1}$$

Use of (3.1.1) is asymptotic in the sense that we are relying on the fact averages of the Markov Chain converge (again under some conditions) to the expectation under the stationary distribution.

$$\lim_{R\to\infty} \frac{1}{R}\sum h(\theta_r) = E_\pi[h(\theta)] \tag{3.1.2}$$

If a Markov Chain satisfies (3.1.2), it is called *ergodic*.

---

[2] The meaning of the term "invariant" will be explained later in Section 3.3.
[3] Assuming, of course, that the posterior expectation of this function exists.

As a practical matter, we will be using large but finite R. Although the theory does not require it, most practitioners will discard some set of initial draws out of concern for the effects of the "initial condition" or value of θ_0. This is sometimes called the "burn-in" period. The idea is that it will take a while for the chain to "equilibrate" or shrug off the effects of the initial condition. We then use the draws after the burn-in period to create simulation estimates. If we "burn-in" for B draws, then the MCMC estimate is given by

$$E_\pi\left[h(\theta)\right] \doteq \frac{1}{R-B} \sum_{r=B+1}^{R} h(\theta_r) \tag{3.1.3}$$

Of course, this is an example of a Monte Carlo integration estimate. The difference now is that the draw sequence is not an IID draw sequence. Since we are allowed to condition on the last value in generating the next value in a Markov Chain, the draws can be dependent. To some, this dependence is unfamiliar and troublesome. We are relying on something like the law of large numbers to insure the sample average converges to the "population" average. Most standard versions of the law of large numbers assume independence. As long as the dependence is not pathologically strong, then the intuition behind the law of large numbers still goes through.[4] We are still getting more information on the stationary distribution with every draw. The practical problem is that these simulation-based estimates may have large sampling errors.

We are relying on asymptotic theory to justify our use of simulation methods with large samples. That is, we are using the fact that long-run averages of draws from the Markov Chain will converge to the appropriate integral over the posterior distribution. More generally, the posterior distribution constructed from draws from the MC will closely approximate the true posterior distribution for large enough samples. This does not mean we are using sampling theory as the basis for our inferences regarding the data. The data is of fixed size, giving rise to a posterior for that sample. It is only when we approximate the posterior that we appeal to asymptotics or long-run behavior. We should recognize that, up to computational limitations, the sample size used in any Monte Carlo method is under our control. In particular, we can increase the sample size as necessary and, for most problems, we can generate truly huge simulation samples at moderate computational cost. This means that the practical use of MCMC methods comes much closer to the long-run sampling experiments envisaged by the inventors of asymptotic theory than the usual application of these methods to small and fixed size samples.

The purpose of this introductory section was to give the reader an overview of the basic idea behind MCMC methods. We are a long way from practical application in the sense that we need to provide:

1. methods or algorithms for specifying chains with the right stationary distribution. This amounts to specifying the conditional distribution of $\theta_{r+1}|\theta_r$ using information about $\pi(\cdot)$.

2. Theoretical assurance that the methods in (1) will produce ergodic chains.

3. Practical guidance on convergence including some notion of how long we should run the chain and how long the "burn-in" period should be.

---

[4] Those trained in time series analysis will recall that there are versions of the Law of Large Numbers for dependent but stationary sequences.

There is good news and bad news regarding the answers to these questions. The good news is that there are families of algorithms that can deliver chains for arbitrary posterior distributions. Further, these algorithms enjoy very strong theoretical convergence properties under relatively mild and verifiable conditions. The "bad" news is that even a theoretically convergent chain may be very slow to converge and/or highly dependent. This means that some care and experimentation must be used in the selection and use of the MCMC algorithms. However, it is fair to say that MCMC methods have been applied to problems of dimension (exceeding 1000) and complexity (problems for which the likelihood is intractable) well beyond the original developers' wildest dreams.

3.2 A SIMPLE EXAMPLE: BIVARIATE NORMAL GIBBS SAMPLER

The ideas introduced in Section 3.1 can be better understood by considering one of the most useful and well-used MCMC methods, the Gibbs Sampler. We will treat the general case of the Gibbs sampler in Section 3.4 and we will see many non-trivial examples. However, it is best to start with a very simple problem.

Consider the problem of simulating from the bivariate normal distribution.

$$\begin{pmatrix} \theta_1 \\ \theta_2 \end{pmatrix} \sim N\left(\begin{pmatrix} 0 \\ 0 \end{pmatrix}, \begin{bmatrix} 1 & \rho \\ \rho & 1 \end{bmatrix} \right) \tag{3.2.1}$$

In Section 2.10.4, the standard method for drawing from the multivariate normal distribution was presented. Recall that we only have to compute the Cholesky root of the covariance matrix and use this to induce the proper level of correlation. Let Z be an $R \times 2$ matrix of iid $N(0,1)$ draws, then we simply compute

$$\Theta = ZU \quad \text{where } U = \begin{bmatrix} 1 & \rho \\ 0 & \sqrt{1-\rho^2} \end{bmatrix} \tag{3.2.2}$$

The $R \times 2$ matrix, Θ, is a matrix whose rows are iid draws from (3.2.1).

The Gibbs sampler specifies a Markov Chain whose stationary distribution is the bivariate normal. The transition mechanism in the Gibbs sampler is specified through iterative sampling from conditional distributions. At point $\theta_r = \begin{pmatrix} \theta_{r,1} \\ \theta_{r,2} \end{pmatrix}$ (here $\theta_{r,i}$ is the rth draw of component i), the next random variables $\begin{pmatrix} \theta_{r+1,1} \\ \theta_{r+1,2} \end{pmatrix}$ are constructed by drawing from the two conditional distributions associated with the bivariate normal.

$$\theta_2 | \theta_1 \sim N\left(\mu_2 + \rho\frac{\sigma_2}{\sigma_1}(\theta_1 - \mu_1), \sigma_2^2(1-\rho^2) \right)$$

and

$$\theta_1 | \theta_2 \sim N\left(\mu_1 + \rho\frac{\sigma_1}{\sigma_2}(\theta_2 - \mu_2), \sigma_1^2(1-\rho^2) \right)$$

BIVARIATE NORMAL GIBBS SAMPLER:

Start at point θ_0.

Draw θ_1 in two steps

$$\begin{aligned}
\theta_{1,2} &\sim N\left(\rho\theta_{0,1}, 1-\rho^2\right) \\
\theta_{1,1} &\sim N\left(\rho\theta_{1,2}, 1-\rho^2\right)
\end{aligned} \qquad (3.2.3)$$

Repeat as long as desired to draw $\theta_2|\theta_1, \ldots, \theta_{r+1}|\theta_r, \ldots$

We draw first from the conditional distribution of the second component given the previous draw of the first component. We then draw the first component given the most recent drawn value of the second component. This means that we move from one point to another in the two-dimensional parameter space in a sequence of two moves along the coordinate axes. In this particular sampler, we draw $\theta_{r+1,2}|\theta_{r,1}$ and then $\theta_{r+1,1}|\theta_{r+1,2}$. This means that we only "use" the initial value $\theta_{0,1}$ and do not use $\theta_{0,2}$ (check the code!). Of course, we could have swapped the order of these "intermediate" steps and still have a valid (but different) sampler.

This simple example shows that one of the problems with MCMC algorithms is that they don't lend themselves to vectorization that is required for the most efficient use of interpreted languages such as R. There is no avoiding the "loop" here since the arguments in the loop are recursively dependent and cannot be computed prior to vector operations. This means that there is at least one "loop" even in the most efficient R code to implement MCMC methods. Our experience, however, has shown that this is not a serious computational constraint for many problems of interest.

Figure 3.1 illustrates this algorithm for $\rho = 0.9$ and a starting value $\begin{pmatrix} \theta_{0,1} \\ \theta_{0,2} \end{pmatrix} = \begin{pmatrix} 2 \\ -2 \end{pmatrix}$. To move away from the initial condition, we first draw the second (vertical)

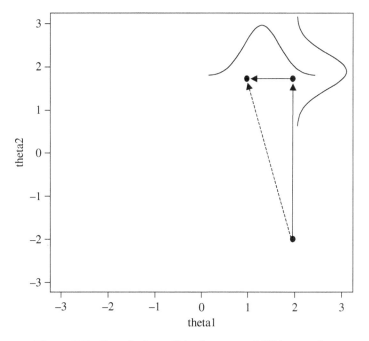

Figure 3.1   Functioning of bivariate normal Gibbs sampler

component value. Given that we start from the first component value of 2, we draw (according to (3.2.3)) from a $N(0.9 \times 2, 0.19)$ distribution, illustrated by the vertically oriented density curve in Figure 3.1. This distribution is centered at 1.8 but with a fairly large standard deviation of 0.44. A realized value is shown on the figure by the "•" symbol. This realized value is about 1.7, near the mean of the conditional distribution. We now draw the first component given the second value of 1.7 by drawing from a $N(0.9 \times 1.7, 0.19)$ distribution. These two draws combine to move the chain from the initial value of $(2, -2)$ to a new value of approximately $(1, 1.7)$. This move is given by the dotted arrow on the figure.

Figure 3.2 shows a realization of the first 20 draws from the same initial condition as in Figure 3.1. The figure shows both the "intermediate" moves internal to the sampler (i.e., the component by component updatings) as well as the final result. Each of these points are connected by lines to trace out the movements. The fact the intermediate moves are included is the reason why the trace consists of only vertical and horizontal line segments of random length. The chain moves quickly away from the initial condition to the region where the bivariate normal has substantial mass. The reason for this is that the initial value of $(2, -2)$ is highly unusual for this bivariate normal. The conditional distributions capture this immediately by moving the second component to a value more in line with the strong positive correlation.

The average size of the move along any given coordinate axis is constant since the standard deviation of normal conditional distributions, $\sqrt{1 - \rho^2}$, does not depend on the conditioning argument. Clearly as ρ gets even closer to one in magnitude, the size of the moves of the sampler will be very limited. However, for all but extremely large values of ρ, the chain will dissipate the effects of the initial conditions rapidly although it may move slowly once it gets to the areas of high mass. After only 40 draws, the chain

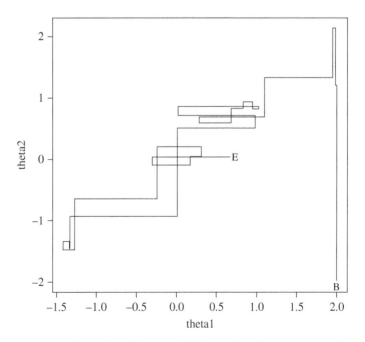

Figure 3.2 Twenty draws from bivariate Gibbs sampler showing intermediate moves

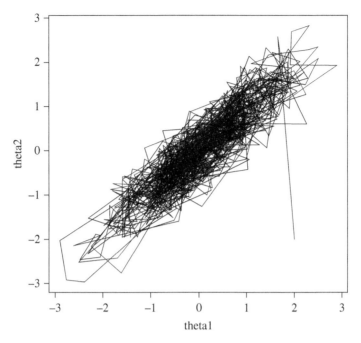

Figure 3.3 One thousand draws from bivariate Gibbs sampler

has started to navigate the regions where the bivariate normal distribution puts mass, but nowhere near adequately for the purpose of estimating probabilities of sets or moments. For example, we certainly would be foolish to estimate the probability of a set by simply computing the proportion of draws falling in this set. The theory of MCMC would tell us that as the number of draws tends to infinity this answer can be computed to any desired degree of accuracy. Figure 3.3 shows 1000 draws (without the intermediate moves on the lattice) of the sampler. Clearly, the chain seems to navigate freely and the regions of high mass are dark with ink.

Figure 3.2 also shows that there is substantial dependence for replicates close together in the sequence. If the chain is in the positive orthant along the "ridge" created by the positive correlation, then it may take several (or more) moves to navigate into the negative orthant. One way of measuring this serial or "time" dependence is to compute the sample autocorrelation function.

$$s_{\theta_1}(k) = \frac{\sum_{r=k}^{R}\left(\theta_{r,1}-\bar{\theta}_1\right)\left(\theta_{r-k,1}-\bar{\theta}_1\right)}{\sum_{r=1}^{R}\left(\theta_{r,1}-\bar{\theta}_1\right)^2} \tag{3.2.4}$$

The top panel of Figure 3.4 shows the autocorrelation function for the first component of θ. The lag or "order" of the autocorrelation is the number of "periods" or draws separating random variables in the chain (k in (3.2.4)). As we might expect, there is a fair amount of autocorrelation in the sequence of draws. However, by lag 10 or 12, there is no appreciable dependence.

If this sampler is ergodic, we can use sample averages of functions evaluated on the chain draws to estimate the expectations of those functions with respect to the stationary distribution of the chain. Figure 3.3 suggests that the stationary distribution for this

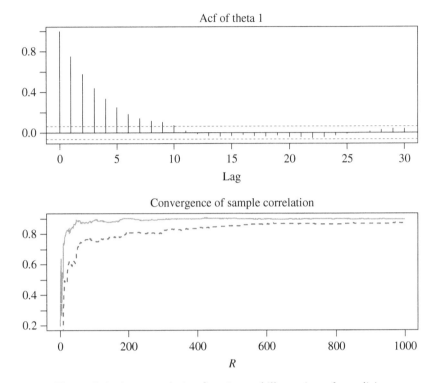

Figure 3.4   Autocorrelation function and illustration of ergodicity

chain is actually the bivariate normal distribution. We can use the sample correlation of the draws of θ_1 and the draws of θ_2 to estimate the "population" correlation or the expectation of the function $h(\theta) = \theta_1\theta_2$ with respect to the bivariate normal. In this example, we know that $E[h(\theta)] = E[\theta_1\theta_2] = \rho$. The bottom panel shows the sample correlation based on the Gibbs sampler draw sequence (dotted line) for samples of successively larger size starting from the beginning of the chain. For example, the value of the line corresponding to the horizontal axis value of 100 is the sample correlation based on replicates 0, 1, ..., 100. The figure illustrates the ergodicity of this chain. These sample averages rapidly converge to the true value of 0.9. The figure also illustrates the importance of a "burn-in" period. It takes a good 100 or so iterations to "work off" the effects of our extreme initial condition. Of course, with modern algorithms for normal random number generators, we can generate in excess of 50 million univariate normals per second on garden variety PCs. This means we can run out this sampler to a million or more draws in less time than it takes us to graph the results. However, the problems of dissipation of initial conditions and serial dependence demonstrated for this example can be found in higher dimensional situations even with tens or even hundreds of thousands of draws.

The serial dependence in the Gibbs sampler draws is the price of this method. The estimates of integrals using these dependent draws can be substantially less efficient (in the sense of sampling error) than estimates based on iid draw sequences. For the bivariate normal, we have an iid sampler and the solid line in the bottom panel of Figure 3.4 shows the convergence of estimates of ρ based on a sequence of iid draws. These estimates converge much more rapidly than those based on the Gibbs sampler draws. This, of course, is not really a fair comparison since the Gibbs Sampler is used on problems for which iid draw algorithms are not available or are computationally infeasible.

The Gibbs Sampler for the bivariate normal problem is a nice illustration of the general idea of MCMC as well as an introduction to one important method. However, this method exploits the very special structure of the normal distribution and the ease with which the conditional distributions can be drawn from. It remains to be seen how generalizable this approach is and what alternatives are available if a strict Gibbs Sampler cannot easily be constructed.

3.3 SOME MARKOV CHAIN THEORY

Before outlining some of the more useful algorithms for constructing Markov chains, we will discuss some of the basic theory. While Markov chains are easy to invent, analysis of the convergence and distributional properties of these chains can be involved. This literature has a complex notation and requires at least some familiarity with measure-theoretic probability for full access. Tierney [1994] and Robert and Casella ([2004], Chapters 7 and 10) have distilled much of the relevant theory from this literature. However, both Tierney and Robert and Casella still require measure theory and a considerable time investment to digest. One view is that theory is largely irrelevant for the practitioner and, therefore, one should only provide a menu of algorithms along with assurances that they will work. Our view is that a practitioner should understand the basic intuition as to why his methods work. This basic intuition can help diagnose algorithmic and programming errors. In addition, some notation and vocabulary may make it easier to follow the practical implications of the burgeoning MCMC literature, some of which is highly technical. For this reason, we provide a minimal set of concepts and illustrate these with discrete and continuous state space chains.

In most instances, the parameters in models (remember this includes all unobservables) will be continuous random variables and $\pi(\cdot)$ will be a standard density. Thus, most of the Markov chains we will consider generate random variables with a continuous component in their transition distribution and the chain navigates in some subset of $\mathfrak{R}^k$. This sort of Markov chain is called a continuous state space Markov chain. We will start, however, by considering discrete state space[5] chains. Much of the intuition we develop for discrete state space chains carries over to the continuous case with some technical difficulties, which we will note.

Interest in discrete state space chains can be motivated by considering a discrete approximation to the posterior distribution. We could lay a grid down along each of the coordinate axes and, therefore, construct a discrete approximation to the posterior distribution using the heights of the posterior density on this grid of points. Let g_i be a grid of values $(g_{i,1}, \ldots, g_{i,m})$ for the ith component of the parameter vector, θ where m is the number of grid points. If we lay a grid of points on each of the k coordinate axes in the parameter space, Θ, then we have constructed the product set, $G = g_1 \times g_2 \times \cdots g_k$. G has m^k elements in it. An element $\theta \in G$ takes on one of the values of the grid for each of the k axes, $\theta'_{i_1, i_2, \ldots, i_k} = \left(g_1\left(i_1\right), g_2\left(i_2\right), \ldots, g_k\left(i_k\right)\right)$. i_j is the index of the grid

---

[5] The term state space itself conjures up a discrete world in which the chain can only take on a finite or, at least countable, number of values or "states." The current "state" of the chain is nothing more than the current realized value that corresponds to one of the finite number of possibilities that define the state space.

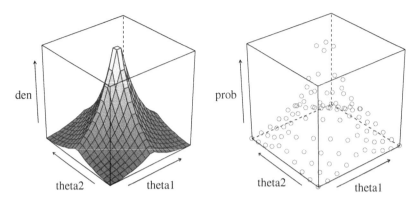

Figure 3.5 Double exponential density and discrete approximation

for θ component j and $g_d\left(i_d\right)$ is the i_d value of the g_d grid at this index. For example, consider the two dimensional case, $\theta_{2,3}$ is value corresponding to the 2nd element of the grid on the first coordinate and the 3rd element of the grid on the second component. $\theta_{2,3}' = \left(g_1\left(2\right), g_2\left(3\right)\right)$; as both indices range over the m possible values, all possible discrete values for θ are enumerated.

Figure 3.5 illustrates discretization for the two dimension case with a continuous bivariate $\pi\left(\cdot\right)$. In the left panel, the bivariate double exponential density is plotted, $\pi\left(\theta\right) \propto \exp\left(-\left|\theta_1\right|-\left|\theta_2\right|\right)$. This density has a mode at $(0, 0)$ and a scale of 1. If we lay down an equal spaced grid of 10 values between -2 and 2 on each coordinate axis, we have the basis for a 10×10 or 100 point discretization of $\pi()$. The right panel of Figure 3.5 shows the discrete approximation to the density based on this grid. A simple discrete approximation would be to normalize the 100 values of π on the grid.

We have seen that we can motivate an interest in discrete state space Markov chains by discretizing the parameter space. If grids of m points are used for each component of θ, then any Markov Chain would be defined on a state space with m^k elements. This grid is the basis for a discrete approximation to π and an IID sampler from this discrete approximation:

DISCRETE APPROXIMATION ALGORITHM

Lay down grids on each axis, $\left(g_1, g_2, \dots, g_k\right)$.
Develop a mapping[6] from the integers $j = 1, 2, \dots, m^k$ to each of the m^k points in $G = g_1 \times g_2 \times \cdots g_k$, $i_1\left(j\right), \dots, i_k\left(j\right)$.
Evaluate the un-normalized posterior,[7] π^*, at each of the m^k grid points and normalize this vector

$$p_j = \left(\pi^*\left(\theta_{i_1(j), \dots, i_k(j)}\right)\right) \Big/ \sum_{l=1}^{m^k} \pi^*\left(\theta_{i_1(l), \dots, i_k(l)}\right)$$

$\theta \sim$ Multinomial with probabilities given above.

---

[6] Usually this is done by establishing an order in which the subscripts denoting the components of j are allowed to vary and then considering a base m representation of the integer j that gives the grid elements for each of the k components as digits.

[7] Not only is it unnecessary to evaluate the normalized posterior but to the extent that any normalizing constants are expensive to compute, this could be very inefficient. Recall that the normalizing constants can be functions of the data, making this a real possibility.

This seems like a very appealing general idea to construct IID samplers. In particular, it would work extremely well for the two dimensional example in Figure 3.5 and a grid of 100 points on each axis. With $k = 10$ and a grid of 100 points, we would have to make 10 billion evaluations of the posterior density. For simple densities and small datasets, this is very feasible on modern 1 G FLOP computers. However, for complicated densities or larger dimension this quickly becomes a computational nightmare since the computations required rise to the power of k. Moreover, there is an assumption that we would know where $\pi(\cdot)$ concentrates its mass. Since the point of doing MCMC in the first place is to explore the posterior and see where its mass is, it may be difficult to lay down grids so that there is enough detail where the posterior is concentrated. Even if the posterior is uni-modal, we don't know much about its spread and shape except by resort to asymptotic approximations.

Even on discrete parameter spaces, we need methods of exploring that are capable of handling high dimensional problems. Markov Chains are one class of these methods. We now introduce general notation for a discrete space Markov Chain. Let $S = \{\theta^1, \theta^2, \dots, \theta^d\}$ be the state space, we define a Markov Chain as the sequence of random variables, $\{\theta_1, \theta_2, \dots, \theta_r, \dots\}$ given θ_0 generated by the following transition.

$$\Pr\left[\theta_{r+1} = \theta^j | \theta_r = \theta^i\right] = p_{i,j} \tag{3.3.1}$$

Given the current realization of the chain, the rows of the matrix, P, formed from p_{ij} specify the conditional distribution of the chain at the next iteration. The Markov property states that conditional distribution of θ_{r+1} depends only on θ_r and nothing from the "earlier" history of the chain Up to this point, we have conditioned on a specific initial value θ_0. This is certainly how we simulate chains. One way of thinking about the chain is how this initial value is transformed. However, in analyzing the behavior of Markov chains, it is more useful to consider the more general case in which we specify a distribution for the initial value, $\theta_0 \sim \pi_0$. By convention in the Markov Chain literature, distributions over states are denoted by the *row* vector of probabilities. The chain transforms this distribution into a new distribution in each iteration of the chain. Consider the distribution of θ_1 given that $\theta_0 \sim \pi_0$.

$$\Pr\left[\theta_1 = \theta^j\right] = \sum_{i=1}^{d} \Pr\left[\theta_0 = \theta^i\right] p_{ij} = \sum_{i=1}^{d} \pi_{0,i} p_{ij}$$

In matrix form, the above equation states that $\pi_1 = \pi_0 P$. After r iterations, we have $\theta_r \sim \pi_0 P^r$.

As the number of iterations increases, we might expect that the effects of the initial distribution π_0 will "wear off." In addition, we might expect some chains to "settle down" to some sort of equilibrium distribution. Here we are ruling out chains that have "absorbing" states or sets of states that they get trapped in and never get out of or, the converse, that they never visit some states (these are called "reducible" chains). If $p_{ij} > 0$ for all i and j, then all states will communicate with one and other and there can be no subset of states to get trapped in. If you get into state i, there is some positive probability that you will get out of it. However, we can immediately see the distinction between the theory and practice. If p_{ij} is small for all i and a specific j, then the chain might only visit state j very infrequently in a finite sequence of draws[8] even though, in theory, this state will be visited infinitely often!

If $p_{ij} > 0$ for all i, j, then the chain is called *irreducible* and there exists a stationary distribution, π, such that

$$\lim_{r \to \infty} \pi_0 P^r = \pi \qquad (3.3.2)$$

Equation (3.3.2) states that, if we start from any distribution, we will get to π eventually. If we start in π, then we must stay in π otherwise π would not be the stationary distribution.

$$\pi P = \pi \qquad (3.3.3)$$

Equation (3.3.3) is the reason that the stationary distribution is also called the *invariant* distribution. For discrete state space chains, irreducibility also implies ergodicity (chain averages of functions converge to their expectation under π).

If presented with a discrete Markov chain that claims to have π as the stationary distribution, it is straightforward to check that (3.3.3) holds. However, it will turn out that for more general state space chains, it will be useful to have an equivalent property called *time reversibility*. Time reversibility states that, if we reverse the sequence order of a Markov chain, the resulting chain will have the same transition behavior. First, we will reverse the order of the chain and check to see that it is still Markov. Then we will compute the transition probabilities for the reversed chain in terms of the standard forward chain. We want to compute the probability of being in state j at "time" r given the future history.

Using the standard definition of conditional probability, we can write the "backward" transition probability as follows:

$$
\Pr\left[\theta_r = \theta^j \,\middle|\, \theta_{r+1} = \theta^{i_1}, \theta_{r+2} = \theta^{i_2}, \ \ldots \ , \theta_{r+s} = \theta^{i_s}\right]
$$
$$
= \frac{\Pr\left[\theta_r = \theta^j, \theta_{r+1} = \theta^{i_1}, \theta_{r+2} = \theta^{i_2}, \ \ldots \ , \theta_{r+s} = \theta^{i_s}\right]}{\Pr\left[\theta_{r+1} = \theta^{i_1}, \theta_{r+2} = \theta^{i_2}, \ \ldots \ , \theta_{r+s} = \theta^{i_s}\right]}
$$

To see that the "reversed" or "backwards" chain is Markov, we can write this ratio using terms that involve the future $r + 2$ periods and beyond and other terms that involve only the rth and $(r+1)$th period.

$$
= \frac{\Pr\left[\theta_r = \theta^j\right] \Pr\left[\theta_{r+1} = \theta^{i_1} \,\middle|\, \theta_r = \theta^j\right] \Pr\left[\theta_{r+2} = \theta^{i_2}, \ \ldots \ , \theta_{r+s} = \theta^{i_s} \,\middle|\, \theta_r = \theta^j, \theta_{r+1} = \theta^{i_1}\right]}{\Pr\left[\theta_{r+1} = \theta^{i_1}\right] \Pr\left[\theta_{r+2} = \theta^{i_2}, \ \ldots \ , \theta_{r+s} = \theta^{i_s} \,\middle|\, \theta_{r+1} = \theta^{i_1}\right]}
$$

The Markov property of the forward chain implies that

$$
\Pr\left[\theta_{r+2} = \theta^{i_2}, \ \ldots \ , \theta_{r+s} = \theta^{i_s} \,\middle|\, \theta_r = \theta^j, \theta_{r+1} = \theta^{i_1}\right]
$$
$$
= \Pr\left[\theta_{r+2} = \theta^{i_2}, \ \ldots \ , \theta_{r+s} = \theta^{i_s} \,\middle|\, \theta_{r+1} = \theta^{i_1}\right]
$$

---

[8] Resulting in a very poor estimate of the marginal probability of state j.

The conditional probabilities for periods $r + 2, \ldots, r + s$, cancel from the numerator and denominator. This means that the reversed chain is also Markov and, further, that

$$\Pr\left[\theta_r = \theta^j \mid \theta_{r+1} = \theta^{i_1}, \theta_{r+2} = \theta^{i_2}, \ldots, \theta_{r+s} = \theta^{i_s}\right] = \Pr\left[\theta_r = \theta^j \mid \theta_{r+1} = \theta^{i_1}\right]$$

$$= \frac{\Pr\left[\theta_r = \theta^j\right] \Pr\left[\theta_{r+1} = \theta^{i_1} \mid \theta_r = \theta^j\right]}{\Pr\left[\theta_{r+1} = \theta^{i_1}\right]}$$

If P^* represents the transition matrix of the reverse chain, then the above equation is the relationship

$$p_{ij}^* = \frac{\pi_j p_{ji}}{\pi_i} \tag{3.3.4}$$

Time reversibility requires that $p_{ij}^* = p_{ij}$. This means that time reversibility is equivalent to

$$p_{ij} = \frac{\pi_j p_{ji}}{\pi_i} \quad \text{or} \quad \pi_i p_{ij} = \pi_j p_{ji} \tag{3.3.5}$$

Roughly speaking, the property of time reversibility implies that chance of seeing a transition from state i to state j is the same as the chance of seeing a transition from state j to state i. Some say that (3.3.5) means that the chain described by P is "reversible with respect to π."

There is a complete equivalence between reversibility and the stationarity of π in the sense that if a chain is reversible with respect to some distribution ω then ω is also the stationary distribution of the chain. Reversibility wrt to ω means $\omega_i p_{ij} = \omega_j p_{ji}$. Summing both sides over i, we obtain

$$\sum_i \omega_i p_{ij} = \sum_i \omega_j p_{ji} = \omega_j \sum_i p_{ji} = \omega_j \times 1$$

or

$$\omega P = \omega$$

and ω is the stationary distribution of the chain.

The bivariate normal Gibbs sampler discussed in Section 3.2 is an example of a continuous state space chain. It will, therefore, be important to extend the ideas we have developed for discrete state space chains to the continuous case. Fortunately, the basic ideas of reversibility and invariant distributions extend without much difficulty. There are some technical difficulties in establishing convergence and ergodic results but central intuition that the chain must freely navigate is at the core of these results just as in the discrete case.

In the continuous state space case, the transitional conditional distribution of $\theta_{r+1} | \theta_r$ must have a continuous component. Rather than specifying this distribution via the probabilities of each of the singletons $\{\theta^i\}$ that comprise the state space, we must specify the conditional distribution by associating probabilities with sets in the state space. For example, consider a set $A \in \Theta$,[9] then the chain is specified by the probabilities of

---

[9] Technically, we can only assign probabilities to certain subsets, but we will gloss over this.

the set A given the value of the chain on the previous iteration. This is sometimes called the Kernel of the chain. $K(\theta, A)$ is the probability of set A given the chain is at value θ. Some Kernels (such as the one corresponding to the Gibbs Sampler, but not all Kernels) can be represented using a standard density.

$$K(\theta, A) = \int_A p(\theta, \vartheta)\, d\vartheta.$$

$p(\theta, \vartheta)$ is a density for fixed θ. To distinguish p from K, we will call $p(\bullet, \bullet)$ the transition function of the Kernel.

Analogous to the discrete case (see (3.3.3)), we can define the concept of an invariant distribution. A distribution with density $\pi(\cdot)$ is an invariant distribution if the probability of $A \in \Theta$ computed under $\pi(\cdot)$ is the same as one-step ahead probability of A given that $\theta \sim \pi$.

$$\int_A \pi(\theta)\, d\theta = \int_\Theta K(\theta, A)\, \pi(\theta)\, d\theta = \int_\Theta \left[\int_A p(\theta, \vartheta)\, d\vartheta \right] \pi(\theta)\, d\theta$$

The principle of time reversibility can also be defined for the continuous state space chain. The chain is said to be reversible with respect to a distribution with density $\omega(\theta)$ if the transition function satisfies

$$\omega(\theta)\, p(\theta, \vartheta) = \omega(\vartheta)\, p(\vartheta, \theta) \tag{3.3.6}$$

If (3.3.6) is satisfied, then the stationary distribution of the chain with transition function $p(\bullet, \bullet)$ has density $\omega(\theta)$.

$$\int_\Theta \omega(\theta)\, K(\theta, A)\, d\theta = \int_\Theta \omega(\theta) \int_A p(\theta, \vartheta)\, d\vartheta d\theta$$

$$= \int_\Theta \int_A \omega(\theta)\, p(\theta, \vartheta)\, d\vartheta d\theta$$

Reversing the order of integration and using the reversibility condition,

$$= \int_A \int_\Theta \omega(\vartheta)\, p(\vartheta, \theta)\, d\theta d\vartheta = \int_A \omega(\vartheta) \left[\int_\Theta p(\vartheta, \theta)\, d\theta \right] d\vartheta$$

$$= \int_A (\omega(\vartheta) \times 1)\, d\vartheta$$

ω, therefore, is the invariant distribution of the chain. Thus, reversibility and invariance are equivalent.

Finally, the concept of *irreducibility* can be extended to the continuous state space setting. Irreducibility requires that the chain navigate the state space freely so that it cannot get trapped in a subset of the entire state space. In the discrete case, all that is required is strict positivity of the transition probabilities, $P > 0$. In the continuous case, the definition of irreducibility is straightforward but verification that a given Kernel produces an irreducible chain is not always a simple matter. A chain with kernel, K, is irreducible with respect to $\pi()$ if every set A with positive π probability can be reached with positive probability after a finite number of steps. $\int_A \pi(\theta)\, d\theta > 0 \Rightarrow$ there exists $n \geq 1$ such that $K^n(\theta, A) > 0$.

3.4 GIBBS SAMPLER

The Gibbs sampler is a Markov chain obtained by cycling through a set of conditional distributions of π. If we break θ into p separate "groups" or "blocks" of parameters, then the Gibbs sampler is defined by iterative sampling from each of these p conditional distributions.

GIBBS SAMPLER

$$\theta' = (\theta_1, \theta_2, \ldots, \theta_p) \tag{3.4.1}$$

(p groups or blocks[10])
Set θ_0
Sample from

$$\theta_{1,1} \sim f_1\left(\theta_1 \big| \theta_{0,2}, \ldots, \theta_{0,p}\right)$$

$$\theta_{1,2} \sim f_2\left(\theta_2 \big| \theta_{1,1}, \theta_{0,3}, \ldots, \theta_{0,p}\right)$$

$$\vdots \tag{3.4.2}$$

$$\theta_{1,p} \sim f_p\left(\theta_p \big| \theta_{1,1}, \ldots, \theta_{1,p-1}\right)$$

to obtain the 1st iterate
REPEAT as necessary.

$f_1, \ldots, f_p$ are the appropriate conditional densities derived from π.

$$f_i = \pi(\theta) \big/ {\textstyle\int} \pi(\theta) d\theta_{-i}, \quad \theta'_{-i} = (\theta_1, \ldots, \theta_{i-1}, \theta_{i+1}, \ldots, \theta_p)$$

Implementation of the Gibbs sampler requires the ability to sample from the set of conditional posterior distributions. In many situations, it is possible to define the groups or blocks of parameters so that the conditional distributions are of known form and can be sampled from very efficiently using the algorithms defined in Chapter 2. As a "fall-back" or default alternative, one could always define a Gibbs sampler based on the k univariate conditionals implied by π. This would require a generic method for making draws from univariate distributions whose un-normalized density can be evaluated.

The Gibbs sampler defined by (3.4.2) is clearly a Markov Chain. It is also easy to verify that the invariant distribution of this chain is π. To see this, consider the bivariate case. $\theta' = (\theta_1, \theta_2)$. The rth iteration of the bivariate Gibbs Sampler draws successively from two conditional distributions.

$$\theta_{r+1,2} \sim \pi_{2|1}\left(\theta_{r,1}\right)$$

$$\theta_{r+1,1} \sim \pi_{1|2}\left(\theta_{r+1,2}\right) \tag{3.4.3}$$

To check that π is the invariant distribution, we must verify that if $\theta_r \sim \pi\left(\bullet\right)$, then $\theta_{r+1} \sim \pi\left(\bullet\right)$. The notation $\pi_{i|j}\left(\theta\right)$ means the distribution of component i given that component j takes on the value θ.

---

[10] We will see examples where each block is only one dimensional and others in which each block corresponds to subsets of the complete set of unobservables.

Suppose $\theta_r \sim \pi\,()$ This means that $\theta_{r,1} \sim \pi_1\,(\theta_1) = \int \pi\,(\theta_1, \theta_2)\,d\theta_2$. $\theta_{r+1,2}$ is a draw from the conditional distribution $\pi_{2|1}$. Therefore, the distribution of $\theta_{r+1,2}$ from one iteration of the sampler is a draw from the marginal distribution of the second component, $\theta_{r+1,2} \sim \pi_2 = \int \pi_{2|1}\,(\theta_2\,|\theta_1)\,\pi_1\,(\theta_1)\,d\theta_1$. The same argument can be used to show that $\theta_{r+1,1} \sim \pi_1$ using the fact that $\theta_{r+1,2} \sim \pi_2$. Thus, the (r+1)th iteration of the chain reproduces a draw from the invariant distribution.

Convergence of the Gibbs sampler is assured under very mild conditions.[11] If the Gibbs sampler is irreducible,[12] then Theorem 1 of Tierney [1994] assures convergence of the n step ahead distribution to the invariant distribution for almost all starting points. Most examples of reducible Gibbs Samplers involve sort of constraint on the state space, Θ.[13] If the state space is the Cartesian product of intervals on each coordinate axis, then a Gibbs sampler whose conditional densities are strictly positive everywhere and whose marginal densities exist will be irreducible.[14] Furthermore, by Tierney Theorem 2, Corollary 1, this sampler will converge to the stationary distribution from all initial points and will be ergodic. This justifies the practical use of the Gibbs sampler – to start from an arbitrary initial condition and use sample averages to approximate integrals of the posterior. Marginal densities will not exist for improper posteriors. If proper priors are used and if the likelihood is bounded, we can avoid the problem of improper posteriors and Gibbs Samplers that attempt to approximate quantities that don't exist! We regard this as yet another (and not even the most important) reason to avoid improper, "reference" or "diffuse" priors.

An appreciation for the power of the Gibbs sampler as well as better feel for implementation can be obtained by considering some important and nontrivial examples. We will consider first the generalization of the Multivariate Regression Model (MRM) introduced in Chapter 2.

3.5 GIBBS SAMPLER FOR THE SUR REGRESSION MODEL

In the SUR model, a system of m regression equations are related through correlated error terms.

$$
\begin{aligned}
y_i &= X_i \beta_i + \varepsilon_i \\
(\varepsilon_{k,1}, \varepsilon_{k,2}, \; \dots \; , \varepsilon_{k,m})' &\sim N\,(0, \Sigma) \\
i &= 1, \; \dots \; , m \;\; k = 1, \; \dots \; , n
\end{aligned}
\tag{3.5.1}
$$

---

[11] These conditions are so mild that some, c.f. Liu [2001], assume convergence and focus attention on the rate of convergence, extent of autocorrelation and various ideas for improving performance.

[12] Also, it is required that the sampler be aperiodic. Periodic Chains require deterministic constraints on movement of the chain, something not found in MCMC algorithms.

[13] The classic example involves two disjoint disks located on the $45°$ line. The sampler would get "trapped" in one disk depending on the initial condition. A closely related is the example in Hobert et al. [1997] in which the state space consists of two boxes that are oriented along the $45°$ line and are tangent at one vertex. In this example, the state space is a subset of a product set that defines a larger box enclosing the two smaller boxes. Finally, an example in Geweke [2004] has a state space consisting of a solid polygon with an acute angle for one vertex and again, oriented on the $45°$ line. In this example, this vertex is an absorbing state. Again, the product set rules this example out while keeping what is needed for virtually all applications.

[14] This is equivalent to the positivity condition (c.f. Robert and Casella [2004], p. 345).

It will be convenient to stack up the m regressions in (3.5.1) into one large regression.

$$y = X\beta + \varepsilon \quad \varepsilon \sim N\left(0, \Sigma \otimes I_n\right) \tag{3.5.2}$$

with

$$y' = (y'_1, \ldots, y'_m) \quad X = \begin{bmatrix} X_1 & 0 & \cdots & 0 \\ 0 & X_2 & \ddots & \vdots \\ \vdots & \ddots & \ddots & 0 \\ 0 & \cdots & 0 & X_m \end{bmatrix} \quad \beta' = (\beta'_1, \ldots, \beta'_m) \quad \varepsilon' = (\varepsilon'_1, \ldots, \varepsilon'_m)$$

As discussed in Section 2.8, there is no convenient natural conjugate joint prior on $\{\beta_i\}, \Sigma$. Recall that the natural conjugate prior for the MRM has the prior on β depending on Σ This prior embodies the notion that information on β can never be *scale-independent*. In all three situations (SUR, MRM, and univariate regression), we may have prior information that is non data-based and, hence, would not always be scale dependent. A simple prior specification would be to make β and Σ a priori independent.

$$p(\beta, \Sigma) = p(\beta) p(\Sigma)$$
$$\beta \sim N(\bar{\beta}, A^{-1}) \tag{3.5.3}$$
$$\Sigma \sim IW(v_0, V_0)$$

These priors are not conjugate but they are *conditionally* conjugate. Given Σ, the SUR likelihood in (3.5.2) can be written in the standard normal regression form and is conjugate with a normal prior on the stacked vector of regression coefficients. Given β, the SUR likelihood is in a form that has an IW conjugate prior. The Gibbs sampler simply alternates between draws from these two sets of conjugate distributions.

Given Σ, we can transform (3.5.2) into a system with uncorrelated errors using root of the cross-equation covariance matrix. $\Sigma = LL'$ and $L^{-1}\Sigma(L^{-1})' = I_m$. This means that, if we premultiply both sides of (3.5.2) by $L^{-1} \otimes I_n$, the transformed system has uncorrelated errors

$$\tilde{y} = \tilde{X}\beta + \tilde{\varepsilon} \quad \text{Var}(\tilde{\varepsilon}) = E\left[\left(L^{-1} \otimes I_n\right) \varepsilon\varepsilon' \left(\left(L^{-1}\right)' \otimes I_n\right)\right] = I_m \otimes I_n,$$
$$\tilde{y} = \left(L^{-1} \otimes I_n\right) y \quad \tilde{X} = \left(L^{-1} \otimes I_n\right) X$$

A normal prior for β, $\beta \sim N(\bar{\beta}, A^{-1})$, is conjugate with the conditional likelihood for the transformed system. This means that we can apply the results from Section 2.8 and the posterior of β given Σ is normal.

$$\beta | \Sigma, y, X \sim N\left(\tilde{\beta}, \left(\tilde{X}'\tilde{X} + A\right)^{-1}\right) \quad \tilde{\beta} = \left(\tilde{X}'\tilde{X} + A\right)^{-1} \left(\tilde{X}'\tilde{y} + A\bar{\beta}\right) \tag{3.5.4}$$

As A gets small, the prior becomes flat and we recognize that the mean of this distribution is the Generalized Least Squares estimator.

The posterior of $\Sigma | \beta$ is in the IW form. To see this, first recognize that, given β, we "observe" or can compute the errors, ε. This means that, given β, the problem is the standard problem of inference regarding a covariance matrix using a multivariate normal sample. The IW prior is, therefore, conditional conjugate. If $\Sigma \sim IW(\nu_0, V_0)$, the posterior is in the form

$$\Sigma \mid \beta, y, X \sim IW(\nu_0 + n, S + V_{0,}) \quad S = E'E \;\; E = [\varepsilon_1, \ldots, \varepsilon_m] \tag{3.5.5}$$

Again, if we let the prior precision go to zero, the posterior on Σ is centered over the sum of squared residuals matrix.

GIBBS SAMPLER FOR SUR MODEL[15]
Pick starting values, β_0, Σ_0 (note: Σ_0 must be a positive definite matrix)
Draw $\beta_1 \mid \Sigma_0$ from (3.5.4)[16]
Draw $\Sigma_1 \mid \beta_1$ from (3.5.5)
Repeat

   This sampler can be related to the non-Bayesian approach to estimating this model. Zellner originally proposed a feasible GLS procedure in which an estimate of Σ is formed by using residuals from equation by equation least squares estimates, $\hat{\Sigma} = \frac{1}{n}\hat{E}'\hat{E}$, where $\hat{E} = [e_1, \ldots, e_m]$ and $e_i = y_i - X_i\hat{\beta}_{LS,i}$. If we start the Gibbs sampler at this point and if we have a very diffuse prior on β, then the first iteration on β will be a draw from a distribution centered on the Zellner feasible GLS estimator. The Gibbs sampler takes this a step further and uses simulation to capture the uncertainty in both β and Σ. The finite sample distribution of the feasible GLS estimator is a nightmare due to the nonlinearities introduced by matrix inversion and multiplication. For this reason, econometricians have had to resort to asymptotic approximations. The sampling error in $\hat{\Sigma}$ does not figure in the asymptotic distribution of the "plug-in" or two-stage feasible GLS estimator. This shows the weakness of asymptotics. However, we no longer have to utilize these approximations. The Gibbs sampler for the SUR model performs extremely well with relatively trivial computation costs. Finally, note that if $m = 1$, we have a sampler for the univariate regression model with a non-conjugate prior.

3.6 CONDITIONAL DISTRIBUTIONS AND DIRECTED GRAPHS

One of the most common applications of the Gibbs Sampler is to hierarchical models. Hierarchical models are models constructed from a sequence of conditional distributions. More generally, we can construct a model by "connecting" or piecing together a set of conditional distributions in some sort of network or "graph." In this section, we provide a brief introduction to the basics of directed graphs. We will explain how to write a model as a directed graph and how to "read off" the Gibbs sampler from a graph.
   The Bayesian paradigm starts with a prior and a likelihood. We can think of the prior as the first step and then we consider the distribution of the data given the model parameters. One way of remembering this "ordering" is to think about how we would simulate

---

[15] As in Chib and Greenberg [1995a].
[16] Some care should be taken in the computations to draw β. The transformation of y and X involves very sparse matrices and can be optimized dramatically by taking advantage of the structure of these matrices.

from the model (here model refers to the joint distribution of the unknowns and the data). First we would draw from the prior and then we would draw the data given the prior. This can be represented by a "directed acyclic" graph or DAG. A graph is a set of connected nodes. A directed graph has a notion of direction from node to node. An "acyclic" graph must have a direction from top to bottom with no "recirculation."

$$p(\theta) \qquad p(y|\theta)$$
$$\theta \quad \rightarrow \quad y \tag{3.6.1}$$

A hierarchical model is specified through a sequence of two or more conditional distributions that specify the prior. This case of two conditional distributions can be represented as a directed graph as follows.

$$p(\theta_2) \qquad\qquad p(\theta_1|\theta_2) \qquad p(y|\theta_1)$$
$$\text{1st Stage} \qquad \text{2nd Stage} \tag{3.6.2}$$
$$\theta_2 \quad\rightarrow\quad \theta_1 \quad\rightarrow\quad y$$

Typically, θ_2 is of much lower dimension than θ_1. The sequence of two prior distributions can be thought of as a device to induce a marginal prior over θ_1.

$$p(\theta_1) = \int p(\theta_1, \theta_2)\, d\theta_2 = \int p(\theta_2)\, p(\theta_1|\theta_2)\, d\theta_2 \tag{3.6.3}$$

The hierarchical model in (3.6.2) specifies that θ_2 and y are independent conditional on θ_1 or that all dependence comes through θ_1. We can easily verify this by writing down the joint distribution.

$$p(\theta_1, \theta_2, y) = p(\theta_2)\, p(\theta_1|\theta_2)\, p(y|\theta_1) = f(\theta_1, \theta_2)\, g(y, \theta_1)$$

There are two ways to see that this implies conditional independence. First, given θ_1, the joint distribution factors into two terms (represented by the functions f and g). Therefore, we have conditional independence. The other way to see this is to observe that there is no term involving all three variables, only a term (f) involving θ_1 and θ_2. This means that

$$p(\theta_2|\theta_1, y) \propto f(\theta_1, \theta_2) \quad\Rightarrow\quad p(\theta_2|\theta_1, y) = p(\theta_2|\theta_1)$$

The hierarchical structure in (3.6.2) immediately suggests a "two-stage" Gibbs Sampler to simulate from the distribution of (θ_1, θ_2) given y.

$$\theta_2|\theta_1$$
$$\theta_1|\theta_2, y \tag{3.6.4}$$

It is easily possible to write down more complicated directed graphs. However, some simple rules can help understand the structure of dependence implied by the graph. There are three sorts of local node arrangements.

The first is a linear set of three nodes

$$\theta_1 \rightarrow \theta_2 \rightarrow \theta_3 \qquad (3.6.5)$$

We have already seen an example of this in (3.6.2) (except that we do not "draw" y but condition on it). This structure has the basic conditional independence in it. A Gibbs sampler for (3.6.5) is given by

$$\begin{aligned} &\theta_1 \,|\, \theta_2 \\ &\theta_2 \,|\, \theta_1 \,, \theta_3 \\ &\theta_3 \,|\, \theta_2 \end{aligned} \qquad (3.6.6)$$

Here θ_1, θ_3 are independent conditional on θ_2

The next structure looks different but has the same feature of conditional independence.

$$\theta_1 \begin{array}{c} \nearrow\ \theta_2 \\ \\ \searrow\ \theta_3 \end{array} \qquad (3.6.7)$$

The joint distribution implied by the graph in (3.6.7) is

$$p\left(\theta_1,\theta_2,\theta_3\right) = p\left(\theta_1\right) p\left(\theta_2 \,|\, \theta_1\right) p\left(\theta_3 \,|\, \theta_1\right)$$

Again, we have $\theta_2, \theta_3 \perp \,|\, \theta_1$.

However, the structure formed from two nodes pointing into one node does not display conditional independence.

$$\begin{array}{c} \theta_1 \ \searrow \\ \qquad\quad \theta_3 \\ \theta_2 \ \nearrow \end{array} \qquad (3.6.8)$$

Here we have full dependence among all three random variables. The joint for (3.6.8) would be written $p\left(\theta_1,\theta_2,\theta_3\right) = p\left(\theta_1\right) p\left(\theta_2\right) p\left(\theta_3 \,|\, \theta_1, \theta_2\right)$. The Gibbs Sampler requires the full set of complete conditionals.

$$\begin{aligned} &\theta_1 \,|\, \theta_3, \theta_2 \\ &\theta_3 \,|\, \theta_1, \theta_2 \\ &\theta_2 \,|\, \theta_3, \theta_1 \end{aligned} \qquad (3.6.9)$$

It is obvious that the "middle" conditional in (3.6.9) belongs in the sampler. What is less obvious is that the "top" and the "bottom" conditionals depend on a node that is more than one node away. But inspection of the joint distribution that the graph represents indicates that there is no conditional independence at all as there is one term in the joint involving all three parameters.

All directed graphs are made up of some combination of three examples above. This suggests three rules for "reading" the dependence structure from a graph:

A node depends on:

(i) any node it points to

(ii) any node that points to it

(iii) any node that points to the node directly "downstream"

For example, consider the following graph.

$$\begin{array}{ccc} & \theta_2 & \searrow \\ & & \theta_4 \rightarrow \theta_5 \\ \theta_1 & \rightarrow \theta_3 & \nearrow \end{array} \qquad (3.6.10)$$

The Gibbs sampler for the graph in (3.6.10) is

$$\begin{array}{c} \theta_1 \mid \theta_3 \\ \theta_2 \mid \theta_4, \theta_3 \\ \theta_3 \mid \theta_1, \theta_4, \theta_2 \\ \theta_4 \mid \theta_2, \theta_3, \theta_5 \end{array} \qquad (3.6.11)$$

3.7 HIERARCHICAL LINEAR MODELS

In Section 3.5, we considered systems of regressions that are related through correlated errors. An alternative approach would be to relate regression equations through correlations in the regression coefficient vectors. This amounts to specifying a prior structure. Consider the set of regression equations.

$$y_i = X_i \beta_i + \varepsilon_i \quad \varepsilon_i \sim \text{iid } N\left(0, \sigma_i^2 I_{n_i}\right) \quad i = 1, \ldots, m \qquad (3.7.1)$$

We specify a different error variance for each equation but consider each regression to be independent of others. We tie together the equations by assuming that the $\{\beta_i\}$ have a common prior distribution.

$$\beta_i = \Delta' z_i + v_i \quad v_i \sim \text{iid } N\left(0, V_\beta\right) \qquad (3.7.2)$$

Equation (3.7.2) specifies a normal prior with mean $\Delta' z_i$ for each β. The d variables in the z vector represent characteristics of each of the m "cross-sectional" units or regression equations. A special case has $z_i = 1$ and $\Delta = \mu'$ that would have a common mean vector for all betas. This prior can be written as a MRM.

$$B = Z\Delta + V \quad B = \begin{bmatrix} \beta_1' \\ \vdots \\ \beta_m' \end{bmatrix} \quad Z = \begin{bmatrix} z_1' \\ \vdots \\ z_m' \end{bmatrix} \quad \Delta = \begin{bmatrix} \delta_1 & \cdots & \delta_k \end{bmatrix} \quad v_i' \sim N\left(0, V_\beta\right) \quad (3.7.3)$$

B is $m \times k$, Z is $m \times n_z$ where n_z is the number of z variables, Δ is $n_z \times k$. Each column of Δ has coefficients that describe how the mean of the k regression coefficients varies as a function of the variables in z. We also need a prior on the regression error variances. It is convenient to take a prior that specifies that each of the error variances is independent.

$$\sigma_i^2 \sim {}^{v_i s_{0,i}^2} / \chi_{v_i}^2 \tag{3.7.4}$$

The prior in (3.7.3) specifies a fixed Δ matrix that determines the mean of the β distribution and a fixed V_β matrix which specifies the variance. Assessment of these priors could be difficult. Early "empirical Bayes" approaches simply estimate these parameters and then, conditional on these estimates, perform an approximate Bayesian analysis of each regression. A Full Bayesian solution can be obtained by specifying a further "stage" of priors on Δ and V_β.

$$V_\beta \sim IW(v, V)$$
$$\text{vec}(\Delta)\,\big|\,V_\beta \sim N\left(\text{vec}(\bar{\Delta}), V_\beta \otimes A^{-1}\right) \tag{3.7.5}$$

The priors in (3.7.5) are the natural conjugate priors for the MRM.

The prior on the collection of βs is specified through a two-stage process. First, we specify a normal prior on β and then a second stage prior on the parameters of this distribution. We can write out this model as a sequence of conditional distributions.

$$y_i\,\big|\,X_i, \beta_i, \sigma_i^2$$
$$\beta_i\,\big|\,z_i, \Delta, V_\beta$$
$$\sigma_i^2\,\big|\,v_i, s_{0,i}^2 \tag{3.7.6}$$
$$V_\beta\,|\,v, V$$
$$\Delta\,\big|\,V_\beta, \bar{\Delta}, A$$

The model above can also be written as a directed graph. The rules of directed graphs given in Section 3.6 can be used to write down the Gibbs sampler for this model. In particular, a key observation is that all dependence between V_β, Δ and the data comes through the regression coefficients, $\{\beta_i\}$. The directed graph is shown below.

Equation (3.7.6) converts the problem of assessing a prior on the $m \times k$ dimensional joint distribution of the βs into the problem of assessing hyper-parameters, $\bar{\Delta}, A, v, V$. A combination of the data and these hyper-parameters (plus the functional forms of the distributions) will influence the posterior on Δ and V_β. The Bayes estimators of the $\{\beta_i\}$ are of a shrinkage variety and will exhibit less variation than least squares estimates computed equation by equation. The amount of shrinkage will be dictated both by the prior hyperparameters and the data. If we assess a tight prior on a "small" covariance matrix by setting v large and V to a small location value, then there will be a great deal of shrinkage. In addition, if the data suggests little variation in the β vectors from equation to equation, then the Bayes estimator will "adapt" to a posterior centered around a small value of V_β and the Bayes estimates of each beta vector will be shrunk even if our prior is relatively uninformative. As discussed in Chapter 5, there are many subtle aspects of this normal hierarchical model. We will defer a full discussion of the prior used in this model until Chapter 5.

Carl Morris was the first to observe that the Gibbs samplers were ideal for analysis of hierarchical models. The key observation is that given Δ and V_β, the $\{\beta_i, \sigma_i^2\}$ are independent with a prior that is the product of a normal prior on β_i and the inverse of a χ^2 prior. This can be analyzed via a Gibbs sampler by drawing $\beta_i | \sigma_i^2$ and then $\sigma_i^2 | \beta_i$. Once the $\{\beta_i\}$ are drawn, they are sufficient for V_β and Δ. Given $\{\beta_i\}$, V_β and Δ can be drawn using the algorithm for the MRM given in Chapter 2, Appendix 1. Thus, a Gibbs sampler for this model can be constructed by first drawing the regression parameters, $\{\beta_i, \sigma_i^2\}$, given the parameters of the first stage prior, Δ, V_β and then drawing the prior parameters conditional on $\{\beta_i, \sigma_i^2\}$. In the definition below, we will use τ_i to denote σ_i^2 to reduce notational clutter.

GIBBS SAMPLER FOR HIERARCHICAL LINEAR MODEL
Start with $\{\tau_i^0\}$, Δ^0, V_β^0
Draw

$$\beta_i^1 \left| y_i, X_i, (\Delta^0)^t z_i, V_\beta^0, \tau_i^0 \right.$$

and

$$\tau_i^1 \left| y_i, X_i, \beta_i^1, v_i, s_{0,i}^2 \quad i = 1, \ldots, m \right.$$

Draw

$$V_\beta^1 \left| \{\beta_i^1\}, v, V, Z, \bar{\Delta}, A \right.$$

and

$$\Delta^1 \left| \{\beta_i\}, V_\beta^1, Z, A, \bar{\Delta} \right.$$

Repeat as necessary.

The draw of $\{\beta_i, \sigma_i^2\}$ is conducted with a prior that specifies that β_i and σ_i^2 are a priori independent rather than dependent as in the natural conjugate prior, $\beta_i \sim N\left(\Delta' z_i, V_\beta\right)$ and $\sigma_i^2 = v_i s_{0,i}^2 / \chi_{v_i}^2$.

$$\beta_i^1 \left| y_i, X_i, \Delta^{0'} z_i, V_\beta^0, \tau_i^0 \sim N\left(\tilde{\beta}, \left(\tilde{X}_i' \tilde{X}_i + \left(V_\beta^0\right)^{-1}\right)^{-1}\right) \right.$$

$$\tilde{\beta} = \left(\tilde{X}_i' \tilde{X}_i + \left(V_\beta^0\right)^{-1}\right)^{-1} \left(\tilde{X}_i' \tilde{y}_i + \left(V_\beta^0\right)^{-1} \Delta^{0'} z_i\right)$$

(3.7.7)

$\tilde{y}_i = y_i / \sigma_i$ and $\tilde{X}_i = X_i / \sigma_i$.

R   The R code for this sampler is in `rhierLinearModel` that is available in our package, `bayesm`. This code makes use of the function to draw from the posterior of a MRM given at the end of Chapter 2. Even though the Gibbs sampler requires looping, the bulk (about 80%) of the computing time is devoted to drawing the regression coefficient vectors.

To illustrate the functioning of this sampler, consider a very typical problem in promotional response modeling. Most Consumer Packaged Goods (CPG) manufacturers have data on the pricing and promotional activities of many of their "key" accounts. A "key" account is a combination of a retailer and market area. For example, Safeway-Denver is a key account that receives attention from P&G's salesforce. The data is usually weekly data for one to two years. P&G may define several hundred key accounts. In order to allocate funds and saleforce effort over these accounts, it would be useful to understand how the customers in these accounts respond to various types of promotional activity. We can use data on sales and measure of price and promotional activity to estimate a simple sales response model for each of these accounts.

As an example, consider the sliced cheese product manufactured by Borden.
R This dataset can be loaded from *bayesm* using the command data (cheese). Data is available on some 88 key accounts for an average of 65 weeks. Weekly observations are recorded of unit volume (number of units sold in all stores in this account during each week), price in $, and a measure of display activity. Displays are a form of in-store advertising that usually consists of special signage or display of the merchandise in a prominent location. The measure of display activity is a "percent of ACV on display." Kraft would like to see retailers use the optimal combination of display and pricing to promote and sell this product.

One straightforward approach would be to run 88 separate regressions each with about 60 observations. However, the independent variables do not always have very much variation at the account variable. For example, two of the accounts have no display activity in this period. Even ignoring the fact that the display coefficient is not estimable for two of the 88 regressions, the least squares coefficients are not very usable as shown in Figure 3.6. These are coefficients from a regression of ln(Volume) on Display and ln(Price). Some of the display coefficients are absurdly large. To interpret the coefficient, recall that the Display variable reaches a maximum of 1 which means 100% display coverage. Thus, a display coefficient of 5 implies a multiple of sales volume of e^5 or almost a 150 fold increase. Clearly, these coefficients have been influenced by sampling error and, perhaps, outlying observations.

Figure 3.6 plots the least squares coefficients against the posterior means obtained from the Linear Hierarchical model Gibbs sampler.[17]
The prior settings are

$$v_i = 3, \quad s_{0,i}^2 = \mathrm{Var}\,(y_i), \quad v = k+3, \quad V = v \times 0.1 I_k, \quad \bar{\Delta} = 0, \quad A = 0.01$$

These prior settings represent a proper but very diffuse prior. Even so, the posterior means display a strong shrinkage effect. The absurdly large values of the least squares coefficients are shrunk in toward more reasonable values. This shrinkage stems from two forces: 1. the

---

[17] Here there are no "z" variables, so Z is simply a vector of ones and Δ is a common mean vector for the coefficients.

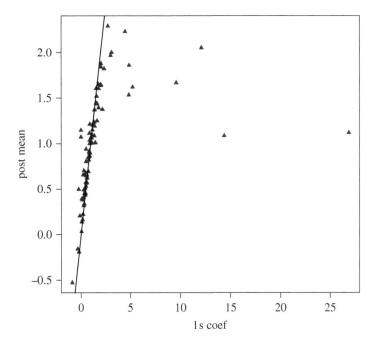

Figure 3.6   Linear hierarchical models: failure of least squares estimates

adaptive nature of the Hierarchical model that adapts V_β to the observed variation in the data and 2. the normal first stage prior that has very thin tails. It should be noted that, for most accounts, the least squares and posterior means are similar. This means that the sample information dominated the prior for these accounts (note that the prior is centered over 0 for the mean of the regression coefficients). The prior is set to a very diffuse setting so this is not very surprising. There are several coefficients for which both the posterior mean and the least squares estimates are negative. A negative coefficient would imply the displays depressed sales. This might be something that one would want to rule out a priori. Unfortunately, the conditional conjugate priors employed in this model are not capable of imposing a sign restriction on the regression coefficients. We will consider this in Chapter 5.

To illustrate how the prior settings affect the degree of shrinkage, we compute posterior means for three different values of v, $k + 3$, $k + 0.5\bar{n}$, $k + 2\bar{n}$. $\bar{n}$ is the average number of observations in each data set. The three values of v tighten down the IW prior on V_β. The V location matrix has already been set to a small value. This means that if v is large, our prior on V_β is highly informative and located over small values, inducing a great deal of shrinkage. Figure 3.7 plots the least squares coefficients vs the posterior means for three levels of v and for each of the three coefficients in the model. The small value of v is black, the medium dark grey and the high value light grey. The shrinkage is rather dramatic for v representing prior information obtained from a sample roughly twice the size of each regression sample. Although the IW and Wishart distributions only have one tightness parameter, this does not imply that the degree of shrinkage is the same for each coefficient. The degree of shrinkage depends on the amount of sample

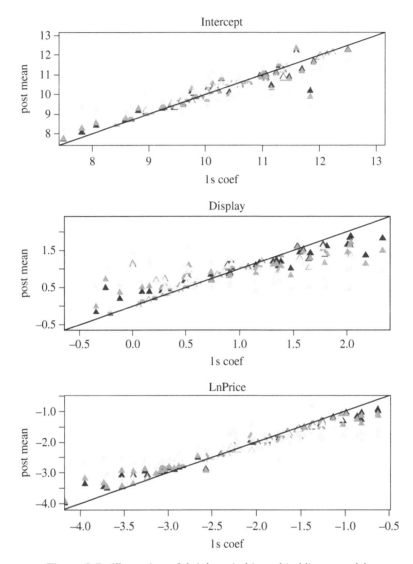

Figure 3.7   Illustration of shrinkage in hierarchical linear models

information available for each coefficient. The display coefficient is most difficult to estimate as displays are relatively rare compared to price changes. The intercept is the easiest to estimate and displays the least shrinkage.

3.8 DATA AUGMENTATION AND A PROBIT EXAMPLE

The examples that we have seen so far show that the Gibbs Sampler is extremely well-suited sets of linear models for which the conditional distributions are known and for which standard methods can be used to make direct draws from these conditional distributions. We enlarge the set of models that can be analyzed by requiring only conditional and not full conjugacy. This allows for analysis of systems of linear models which, heretofore, required approximate methods. However, the Gibbs Sampler can be applied to a much wider class of models, once the principal of data augmentation is introduced. The idea of data augmentation has its origin in the literature on missing values and the EM algorithm. The idea that missing values are unobserved and, therefore, should properly be considered as part of the "parameter" vector comes naturally to a Bayesian. In the EM algorithm for missing data models, the missing data is replaced with its expectation conditional on the observed data and the "complete" data likelihood is maximized over the "parameters." To a Bayesian, it is much more natural to compute the joint posterior of the missing values and the "parameters" and simply margin down to the parameters if this is all that is of interest. The idea of data augmentation extends to any situation in which there are unobservable constructs ((Tanner and Wong [1987]) were the first to point this out). For example, many distributions can be written as mixtures of other distributions so that data augmentation can be used to form a Gibbs sampler for these problems. Of particular relevance for marketing problems, is the use of latent variables to formulate models with discrete lumps of probability.

To illustrate the usefulness of the data augmentation concept for discrete dependent variables models, consider the latent variable formulation of the binary Probit model.

$$z_i = x_i'\beta + \varepsilon_i \quad \varepsilon_i \sim N(0,1)$$
$$y_i = \begin{cases} 0 \text{ if } z_i < 0 \\ 1 \text{ otherwise} \end{cases} \tag{3.8.1}$$

We observe (X, y). If this model is used to represent the choice between two alternatives, then z has the interpretation as the difference in utility between the two alternatives. We choose alternative "A" if it is more attractive than "B." We only partially observe latent utility in the sense that only x is observed. Other influences on utility are represented by the "error" term. Given that the latent structure in (3.8.1) is a standard regression model, we can use a normal prior for β, $\beta \sim N(\bar{\beta}, A^{-1})$.

Data augmentation proceeds by considering the entire vector of n z values as part of the parameter vector, $\theta' = (z, \beta)$. Given the normal prior for β, the model is complete in the sense that (3.8.1) specifies the joint distribution[18] of z and β.

$$p(z, \beta \,|\, X) = p(z \,|\, \beta, X) \, p(\beta)$$

We note that this is a highly correlated distribution, particularly if the prior on β is diffuse. We can write the directed graph for this model in (3.8.2).

$$\beta \rightarrow z \rightarrow y \tag{3.8.2}$$

This directed graph immediately reveals that β, y are independent conditional on z.

---

[18] If you regard the z values as meaningful objects for inference, then this is a prior distribution. If not, then the augmented parameters are simply devices by which one gets at the posterior distribution of β.

The posterior distribution of θ can easily be computed by using a Gibbs sampler.[19]

$$z\,|\,\beta\,,X,y \tag{3.8.3}$$

$$\beta\,|z,X \tag{3.8.4}$$

This Gibbs sampler recognizes that θ separates into two natural groups or "blocks." Given β and the data, the z's are independent truncated univariate normal distributions. Given z, inference on β is just a Bayes linear regression analysis with a normal prior and no scale parameter (note that z is sufficient for β and we don't need to add y to the conditioning arguments (3.8.4).

GIBBS SAMPLER FOR BINARY PROBIT
Start with β_0
Draw z from (3.8.3) by making n independent draws from $trun_{(a_i,b_i)}\,N\left(-x_i'\beta_0,1\right)$ with $a_i = 0$ if $y_i = 1, -\infty$ otherwise; $b_i = 0$ if $y_i = 0$, ∞ otherwise.
Draw β_1 from (3.8.4) using standard normal theory

$$\beta\,|y,X \sim N\left(\tilde{\beta},\left(X'X+A\right)^{-1}\right); \quad \tilde{\beta}=\left(X'X+A\right)^{-1}\left(X'y+A\bar{\beta}\right)$$

Repeat as necessary.

The dependence between draws of β comes entirely through the z vector. This will be useful Gibbs sampler to the extent to which the latent variables and β are not too highly correlated. R code for this binary probit sampler is available as the function,
R `rbprobitGibbs`, in our R package, *bayesm*.
Figure 3.8 shows the results of running this Gibbs Sampler with a simulated dataset with $n = 100$, two regressors which are uniform on $(0,1)$ and independent.
The left hand side shows histograms of the marginal posterior distribution of each parameter. The vertical line marks the "true" value of the parameter underlying the simulated data. The right hand panels show the autocorrelation functions. The marginal posterior distributions of the model parameters are very normal and the ACFs for this data are very reasonable even though there is not much information in this data on the betas.
One might be tempted to conclude that all of this machinery is an elaborate way of producing results from asymptotic theory (although one would have to verify that the posterior covariance is close to the asymptotic covariance). However, the parameters of this model are not as directly interpretable and relevant as in the linear regression context. In marketing applications, we are often interested in predicted probabilities for given

---

[19] Albert and Chib [1993] were the first to propose this sampler (see also, Chib [1992], for a closely related model).

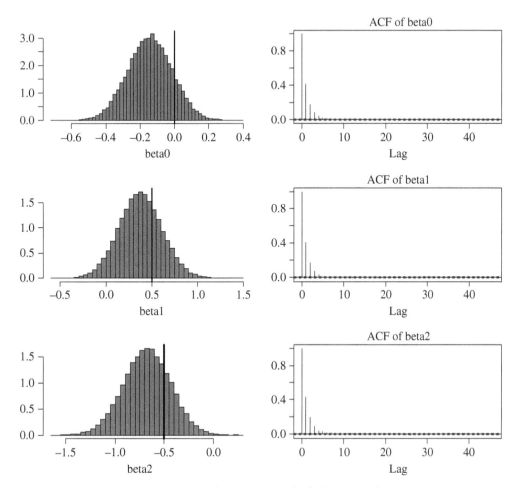

Figure 3.8   Gibbs sampler results for binary probit

values of x. Figure 3.9 shows the posterior distribution of the probability $y = 1$ for various x vectors, i.e., the posterior distribution of $\Phi(x'\beta)$. Since probabilities are bounded and there is posterior uncertainty, these distributions are very non-normal. Superimposed on the histogram are normal densities evaluated at the posterior mean and variance.[20] Even this small problem and highly regular model provide a powerful motivation to eliminate asymptotic approximations.

Finally, we should note that this model can easily be investigated by importance sampling performed on the posterior obtained by integrating out z.

$$p\left(\beta \mid X, y\right) \propto p\left(\beta\right) \prod_{i} \Phi\left(x_{i}'\beta\right)^{y_{i}} \left(1 - \Phi\left(x_{i}'\beta\right)\right)^{1-y_{i}} \qquad (3.8.5)$$

---

[20] Asymptotic theory would be even worse since it would keep the normal distribution assumption but insert asymptotic estimates of the mean and variance.

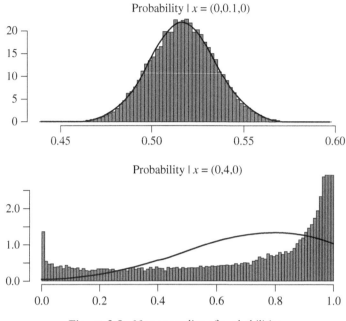

Figure 3.9 Nonnormality of probabilities

In fact, per unit of computing time, a direct approach via importance sampling will surely yield more information than the binary Gibbs Sampler due to the autocorrelation in the draw sequence. The binary probit Gibbs Sampler is of interest primarily because it suggests that data augmentation strategies can be useful for the multinomial probit and other problems for which the likelihood over the model parameters is difficult to evaluate.

3.9 MIXTURES OF NORMALS

Finite mixtures of multivariate normal distributions can provide a very flexible model for multivariate data. Mixtures of normals can accommodate thick tailed and skewed distributions. However, in the multivariate case, the possibilities are even broader. For example, we can create a joint distribution with "banana" shaped contours by arranging closely spaced normal distributions along a curve. There is a sense that with enough mixture components, one can approximate any multivariate distribution in the same sense that you can build any shaped hill by piling up small mounds of gravel.

The basic mixture of normals model can be written

$$y_i \sim N\left(\mu_{ind_i}, \Sigma_{ind_i}\right)$$
$$ind_i \sim Multinomial\left(pvec\right)$$

(3.9.1)

Here y_i is a p-dimensional vector and $pvec$ is a vector of K mixture probabilities. This model is referred to as a mixture of normals with K components. Equation (3.9.1) is a direct model for simulation from a mixture of normals (see `rmixture` in our R package, bayesm). First, we draw a multinomial distributed indicator of which component is "active" and then we draw a multivariate normal vector from this component. This representation of the model also suggests the basis of a Gibbs Sampler by augmenting the parameters with the vector of n indicators.

Priors for the mixture of normals model can be taken in convenient conditionally conjugate forms

$$pvec \sim Dirichlet(\alpha)$$
$$\mu_j \sim N\left(\bar{\mu}, \Sigma_k \otimes a_\mu^{-1}\right)$$
$$\Sigma_j \sim IW(v, V) \tag{3.9.2}$$
$$k = 1, \dots, K$$

In (3.9.2), the joint prior on the normal component parameters is independent conditional on p and in the form of the natural conjugate prior for multivariate regression (see Section 2.8.5). The DAG for the mixture of normals model can be written as

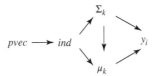

Given this DAG, we can easily write down the Gibbs Sampler as consisting of the following sets of conditionals.

$$ind \big| pvec, \{\mu_k, \Sigma_k\}, Y$$
$$pvec \big| ind$$
$$\{\mu_k, \Sigma_k\} \big| ind, Y \tag{3.9.3}$$
$$k = 1, \dots, K$$

Y is the $n \times p$ matrix of multivariate observations. This sampler was introduced by Diebolt and Robert [1994]. The key idea is that once the indicators are drawn, the observations are classified by normal component and then one can proceed with K independent conjugate draws of the normal component parameters.

The draw of the indicators is a multinomial draw based on the likelihood ratios with p as the prior probability of membership in each component.

$$ind_i \sim multinomial(\pi_i); \quad \pi' = (\pi_{i,1}, \dots, \pi_{i,K})$$
$$\pi_{i,k} = \frac{pvec_k \varphi\left(y_i \big| \mu_k, \Sigma_k\right)}{\sum_m pvec_m \varphi\left(y_i \big| \mu_m, \Sigma_m\right)} \tag{3.9.4}$$

Here $\varphi(\bullet)$ is the multivariate normal density.

The draw of $pvec$ given the indicators is a Dirichlet draw

$$pvec \sim Dirichlet(\tilde{\alpha})$$
$$\tilde{\alpha}_k = n_k + \alpha_k \tag{3.9.5}$$
$$n_k = \sum_{i=1}^{n} I\left(ind_i = k\right)$$

The draw of each $\left(\mu_k, \Sigma_k\right)$ can be made using the algorithm to draw from the MRM as detailed in Section 2.8.5. For each subgroup of observations, we have an MRM model of the form

$$Y_k = \iota\mu_k' + U; \quad U = \begin{bmatrix} u_1' \\ \vdots \\ u_{n_k}' \end{bmatrix}; \; u_i \sim N\left(0, \Sigma_k\right) \tag{3.9.6}$$

Here Y_k is the submatrix of Y that consists of the n_k rows where $ind_i = k$. The results of Chapter 2 simplify to the following draws:

$$\Sigma_k | Y_k, \upsilon, V \sim IW\left(\upsilon + n_k, V + S\right)$$

$$\mu_k | Y_k, \Sigma_k, \bar\mu, a_\mu \sim N\left(\tilde\mu_k, \frac{1}{(n_k + a_\mu)}\Sigma_k\right) \tag{3.9.7}$$

where

$$S = \left(Y_k - \iota\tilde\mu_k'\right)'\left(Y_k - \iota\tilde\mu_k'\right)$$

$$\tilde\mu_k = \left(n_k + a_\mu\right)^{-1}\left(n_k\bar y_k^* + a_\mu\bar\mu\right) \tag{3.9.8}$$

$$\bar y_k^* = \left(Y_k'\iota/n_k\right)'$$

R This Gibbs sampler for mixtures of normals is available as function, `rnmixGibbs`, in *bayesm*.

3.9.1 Identification in Normal Mixtures

It is well known that the likelihood for the normal mixture model can have up to $K!$ symmetric modes. This is due to what is referred to as the label-switching "problem." We can simply interchange or permute the labels for each of the components and have the same value of the likelihood. For example, consider the mixture of two univariate normal distributions. There are two equal height posterior modes (assuming the priors are identical). We can simply interchange the labels, calling mode 1 "2" and mode 2 "1" and leave the likelihood unchanged. This means that the marginal posteriors of the mean parameters will often have two modes. As we know from the examples considered in Chapter 3, this can affect many MCMC algorithms. It is possible that the algorithm may only investigate one of the modes or some subset of modes, leaving others completely untouched. This will occur when there is strong separation or classification information in the "data." However, when differences between components are small, there can be a good deal of switching from mode to mode. This switching from mode to mode by MCMC methods is often what researchers in this area mean by "label-switching."

Figure 3.10 illustrates the label-switching problem by considering the problem of inference about a mixture of two normals. Component "1" is a $N(1, 1)$ random variable and component "2" is $\sim N(2, 1)$. The mixture probability is 0.5. This is an example where there is little distance between the modes for each component. The mixture is a symmetric and unimodal density centered at 1.5. If we have a modest number of observations, the Gibbs Sampler which "flip" the component labels so that what had been labeled a draw of

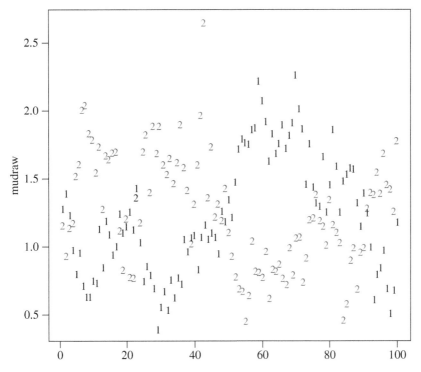

Figure 3.10   Illustration of label switching for mixture of normals model

the mean for the first component is now the mean of the second component $(\mu_1 \leftrightarrow \mu_2)$. Figure 3.10 plots the Gibbs Sampler output for μ_1 (labeled "1") and μ_2 (labeled "2"). We can see several label switches, including one around draw number 50.

This means that we cannot simply look at the marginal distribution of μ_1 from our Gibbs Sampler output. This parameter is not identified. However, the joint density is identified.

$$p(y) = p, \ \varphi(y|\mu_1, \sigma_1) + (1-p), \ \varphi(y|\mu_2, \sigma_2) \qquad (3.9.9)$$

For any given value of y, we can compute the posterior distribution of the density at this value. If we use a grid of y values, we can compute the posterior distribution of the entire density of the data. In particular, we can average the posterior to obtain the posterior mean as an estimator of the density. Figure 3.11 shows 10 posterior draws of the fitted density in (3.9.9). The solid density is the true density.

Much of the early work with mixture models attempted to solve the identification problem for individual component parameters by using various a priori ordering restrictions. For example, some advocate ordering the components by prior probability.

$$pvec_1 > pvec_2 > \cdots > pvec_K \qquad (3.9.10)$$

Unfortunately, (c.f. Stephens [2000]) imposing this restriction does not necessarily remove the identification problem. Other proposals include ordering by the normal mixture component parameters. Obviously, it may be difficult to define an ordering in

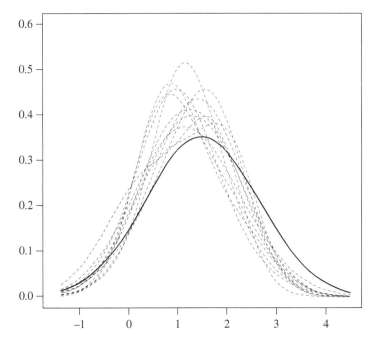

Figure 3.11   Posterior draws of density: example of univariate mixture of two normals

the case of mixtures of *multivariate* normal distributions. Moreover, even in the cases of univariate normals, Stephens has noted that identification cannot always be achieved. Choosing the right way to divide the parameter space so as to insure that only one mode remains can be somewhat of an art form that may be close to impossible in high dimensional problems (see Fruhwirth-Schnatter [2001] for a suggested method that may work in low dimensional situations).

Instead of imposing what can been termed "artificial" identification constraints, Stephens [2000] advocates postprocessing via relabeling of MCMC draws so as to minimize some sort of statistical criterion such as divergence of the estimated marginal posteriors from some unimodal distribution. This means that post-simulation optimization methods must be used to achieve a relabelling in the spirit of clustering algorithms.

There is no guarantee that such post processing will necessarily uncover the true structure of the data and there is still "art" in the choice of objective function and tuning of the re-labelling.

The label-switching phenomenon is only a problem to the extent to which the investigator wishes to attach meaning to specific normal components. If one interprets each normal component as a sub-population, then we might want to make inferences about the means and covariances for that component. For example, we might think of the population of respondents to a questionnaire as comprised of distinct groups each with some heterogeneity within group but where this may be small relative to across group heterogeneity. In this situation, label-switching becomes an issue of identification for the parameters of interest. The only way to "solve" this identification problem is by use of prior information. However, if we regard the mixture of normals as a flexible approximation to some unknown joint distribution, then the label-switching "problem" is not relevant. Ultimately, the data identify the joint distribution and we don't attach substantive importance to the mixture component parameters. We should recognize that this runs somewhat counter to the deeply ingrained tradition of identifying consumer segments. We do not really think of segments of homogeneous consumers as a reality but merely a convenient abstraction for the purpose of marketing strategy discussions. The empirical evidence, to date, overwhelmingly favors the view that there is a continuum of consumer tastes.

Thus, we view the object of interest as the joint density of the parameters. This density and any possible functions defined on it such as moments or probabilities of regions are identified without a priori restrictions or ad hoc post-processing of MCMC draws. Once you adopt this point of view, the identification and label-switching literature becomes irrelevant and, instead, you are faced with the problem of summarizing a fitted multivariate density function. Lower dimensional summaries of the mixture of normals density are required (as Fruhwirth-Schnatter [2004] point out any function of this density will also be immune to the label-switching identification issues). One possible summary would be the univariate marginal densities for each component of θ. However, this does not capture co-movement of different elements. It is certainly possible to compute posterior distributions of covariance but, as pointed out above, these lose interpretability for multi-modal and non-normal distributions.

3.9.2 Performance of the Unconstrained Gibbs Sampler

The Gibbs sampler outlined in (3.9.3) is referred to as the "unconstrained" Gibbs sampler in the sense that no prior constraints have been imposed to achieve identification of the mixture component parameters. As such, this sampler may exhibit label-switching. As pointed out above, any function that is invariant to label-switching such as the estimated mixture density will not be affected by this problem. There also may be algorithmic advantages to not imposing identification constraints. As pointed out by Fruhwirth-Schnatter [2001], identification constraints hamper mixing in single-move constrained Gibbs samplers. For example, if we imposed the constraint that the mixture probabilities must be ordered, then we must draw from a prior distribution restricted to a portion of the parameter space in which the ordering is imposed. A standard way to do this is to draw each of the K probabilities, one by one, given the others. The ordering constrains mean that as we draw the kth probability, it must lie between the $k - 1$th and $k + 1$th probabilities. This may leave little room for navigation. Thus, the unconstrained Gibbs sampler will often mix better than constrained samplers. Fruhwirth-Schnatter introduces as "random permutation" sample to promote better navigation of all modes in the unidentified parameters space of the mixture component parameters. This may improve mixing in the unidentified space but will not improve mixing for the identified quantities such as the estimated density and associated functions. The fact that the unconstrained Gibbs sampler may not navigate all or even more than one of the $K!$ symmetric modes does not mean that it doesn't mix well in the identified space. As Gilks [1997] comments "I am not convinced by the … desire to produce a unique labeling of the groups. It is unnecessary for valid Bayesian inference concerning identifiable quantities; it worsens mixing in the MCMC algorithm; it is difficult to achieve in any meaningful way, especially in high dimension, and it is therefore of dubious explanatory or predictive value" (p. 771).

Our experience is that the unconstrained Gibbs Sampler works very well, even for multivariate data and with a large number of components. To illustrate this, consider five dimensional data simulated from a three component normal mixture, $n = 500$.

$$\mu_1 = \begin{pmatrix} 1 \\ 2 \\ 3 \\ 4 \\ 5 \end{pmatrix}; \quad \mu_2 = 2\mu_1; \quad \mu_3 = 3\mu_1; \quad \Sigma_k = \begin{bmatrix} 1 & 0.5 & \cdots & 0.5 \\ 0.5 & 1 & \ddots & \vdots \\ \vdots & \ddots & \ddots & 0.5 \\ 0.5 & \cdots & 0.5 & 1 \end{bmatrix} \quad (3.9.11)$$

$$pvec = \begin{pmatrix} 1/2 \\ 1/3 \\ 1/6 \end{pmatrix}$$

We started the unconstrained Gibbs sampler with nine normal components. Figure 3.12 represents the distribution of the indicator variable across the nine components for each of the 400 first draws. The width of each horizontal line is proportional to the frequency with which that component number occurs in the draw of the indicator variable. The sampler starts out with an initial value of the indicator vector that is split evenly among the nine components. The sampler quickly shuts downs a number of the components. By 320 or so draws, the sampler is visiting only three components with frequency corresponding to the mixture probabilities that were used to simulate the data.

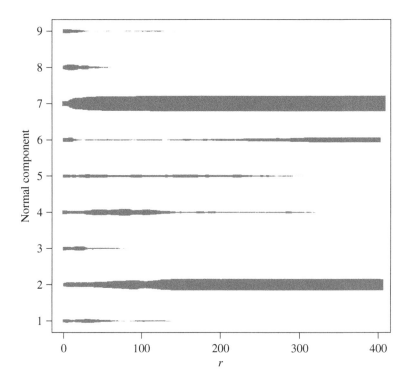

Figure 3.12   Plot of frequency of normal components: multivariate normal mixture example

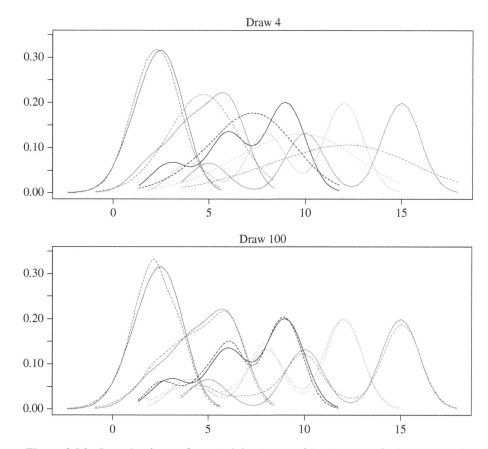

Figure 3.13   Posterior draws of marginal densities: multivariate normal mixture example

Figure 3.13 shows each of the five marginal distributions for particular draws in the MCMC run shown in Figure 3.13. The solid lines are the "true" marginal distributions implied by the normal mixture given in (3.9.11). The top panel shows the draw of the five marginal distributions for the 4th draw of the sampler. The marginals vary in the extent to which they show "separation" in modes. The marginal of the 5th component has the most pronounced separation. After only four draws, the sampler has not yet "burned-in" to capture the multimodality in the marginal distribution of the 3rd through 5th components. However, as shown in the bottom panel, by 100 draws the sampler has found the rough shape of the marginal distributions. Variation from draw to draw after the 100th draw simply reflects posterior uncertainty.

3.10 METROPOLIS ALGORITHMS

The Gibbs sampler is enormously useful, particularly for models built up from hierarchies of relatively standard distributions. However, there are many problems for which the conditional distributions are not of a known form that is easy to simulate from. For this reason, it is useful to have a more "general purpose" tool. By "general purpose," we mean a tool that can be applied to, at least in principle, to any posterior distribution. Of course, in practice, there will be problems for which the general purpose algorithm will produce a poorly performing MCMC sampler. This means that the general purpose algorithm means an approach to generating a candidate MCMC method for virtually any problem. The performance of this candidate will not be assured and must still be investigated.

The Metropolis class of algorithms is a general purpose approach to producing Markov chain samplers.[21] The idea of the Metropolis approach is to generate a Markov Chain with the posterior, $\pi()$ as its invariant distribution by appropriate modifications to a related Markov chain that is relatively easy to simulate from. Viewed in this way, Metropolis algorithms are similar in spirit to the accept/reject method of iid sampling. Accept/reject methods sample from a proposal distribution and then reject draws to modify the proposal distribution to achieve the desired target distribution. The proper invariant distribution is achieved by constructing a new chain that is time reversible with respect to π.

We will start with a discrete state space version of the Metropolis algorithm that will illustrate the essential workings of the algorithm. We have a transition matrix, Q, which we want to modify to insure that the resultant chain has a stationary distribution given by the vector π. The dimension of the state space is d.

DISCRETE METROPOLIS ALGORITHM

Start in state i, $\theta_0 = \theta^i$
Draw state j with probability given by $q_{ik}, k = 1, \ldots, d$ (multinomial draw)
Compute $\alpha = \min \left\{ 1, \frac{\pi_j q_{ji}}{\pi_i q_{ij}} \right\}$
With prob α, $\theta_1 = \theta^j$ (move) ELSE $\theta_1 = \theta^i$ (stay)
Repeat, as necessary

Note that only the ratios, π_j / π_i, are required and we only need to know posterior distribution up to a constant. The unique aspect of this algorithm is the possibility that it will not move on a given iteration of the chain. With probability $1 - \alpha$, the chain will repeat the value from the rth to $(r + 1)$st iteration. These repeats are to insure that new chain is reversible. If $\pi_i q_{ij} > \pi_j q_{ji}$ then there will be "too many" transitions from state i to state j and not enough reverse transitions from state j to i. For this reason, the Metropolis chain only accepts the α fraction of the transitions from i to j and all of the moves from j to i.

This algorithm is constructed to be time reversible with respect to π. To see this, recall that time reversibility requires $\pi_i p_{ij} = \pi_j p_{ji}$. The transition probability matrix for the chain defined by the Metropolis algorithm is defined by

$$p_{ij} = q_{ij} \alpha \left(i, j \right)$$

---

[21] Although the Metropolis method was developed for discrete distributions and discussed in the statistics literature as early as 1970 (Hastings [1970]), the popularity of this method is due in large part to Tierney [1994] and Chib and Greenberg [1995b] who provided a tutorial in the method as well as many useful suggestions for implementation.

Therefore,

$$\pi_i p_{ij} = \pi_i q_{ij} \min \left\{ 1, \frac{\pi_j q_{ji}}{\pi_i q_{ij}} \right\} = \min \left\{ \pi_i q_{ij}, \pi_j q_{ji} \right\}$$

$$\pi_j p_{ji} = \pi_j q_{ji} \min \left\{ 1, \frac{\pi_i q_{ij}}{\pi_j q_{ji}} \right\} = \min \left\{ \pi_j q_{ji}, \pi_i q_{ij} \right\}$$

and the condition for time reversibility is satisfied.

The continuous version of the Metropolis algorithm has exactly the same formulation as the discrete case except that the analysis of reversibility and convergence is slightly more complex. In particular, the transition kernel for the Metropolis does not have a standard density but rather has a mixture of discrete and continuous components. The continuous state space Metropolis algorithm starts with a proposal transition Kernel defined by the transition function $q(\theta, \vartheta)$. Given θ, $q(\theta, \bullet)$ is a density. The continuous state space version of the Metropolis algorithm is as follows.

CONTINUOUS STATE SPACE METROPOLIS
Start at θ_0.
Draw $\vartheta \sim q(\theta_0, \bullet)$
Compute $\alpha(\theta, \vartheta) = \min \left\{ 1, \frac{\pi(\vartheta) q(\vartheta, \theta)}{\pi(\theta) q(\theta, \vartheta)} \right\}$
With prob α, $\theta_1 = \vartheta$ ELSE $\theta_1 = \theta_0$
Repeat as necessary

To see that the continuous version of the Metropolis algorithm has π as its invariant distribution,[22] we first define the Kernel. Recall that the Kernel provides the probability that the chain will advance to a set A given that it is currently at point θ.

$$K(\theta, A) = \int_A p(\theta, \vartheta) \, d\vartheta + r(\theta) \delta_A(\theta) \tag{3.10.1}$$

$$p(\theta, \vartheta) = \alpha(\theta, \vartheta) q(\theta, \vartheta) \tag{3.10.2}$$

$$\delta_A(\theta) = \begin{cases} 1 & \theta \in A \\ 0 & \text{otherwise} \end{cases}$$

p defined in (3.10.2) is the transition function for the Metropolis chain. The probability that the chain will move away from θ is given by

$$\int_\Theta p(\theta, \vartheta) \, d\vartheta = \int_\Theta \alpha(\theta, \vartheta) \, q(\theta, \vartheta) \, d\vartheta$$

Since $K(\theta, \Theta) = 1$, this implies that $r(\theta)$ is the probability that the chain will stay at θ. Since there is a possibility for the chain to repeat the value of θ and, therefore, given θ, we can't provide a standard density representation for the distribution of ϑ since there is a mass point at θ. The conditional distribution defined by the Metropolis kernel can be interpreted as a mixture of a mass point at θ and a continuous density $p(\theta,\vartheta)/(1-r(\theta))$.

The Metropolis transition function p satisfies the "detailed balance" condition

$$\pi(\theta)\, p(\theta, \vartheta) = \pi(\vartheta)\, p(\vartheta, \theta) \tag{3.10.3}$$

or $\pi(\theta)\, \alpha(\theta, \vartheta)\, q(\theta, \vartheta) = \pi(\vartheta)\, \alpha(\vartheta, \theta)\, q(\vartheta, \theta)$ which is true by the construction of the α function. Equation (3.10.3) ensures that the Metropolis chain will be time reversible. We can now show that (3.10.3) implies that π is the invariant distribution of the chain. Recall that, for a continuous state chain, π is the invariant distribution if $\int_A \pi(\theta)\, d\theta = \int_\Theta \pi(\theta)\, K(\theta, A)\, d\theta$

$$\int_\Theta \pi(\theta)\, K(\theta, A)\, d\theta = \int_\Theta \pi(\theta) \left[\int_A p(\theta, \vartheta)\, d\vartheta + r(\theta)\, \delta_A(\theta) \right] d\theta$$

$$= \int_\Theta \int_A \pi(\theta)\, p(\theta, \vartheta)\, d\vartheta d\theta + \int_\Theta \pi(\theta)\, r(\theta)\, \delta_A(\theta)\, d\theta$$

Interchanging the order of integration and applying the detailed balanced equation,

$$= \int_A \int_\Theta \pi(\vartheta)\, p(\vartheta, \theta)\, d\theta d\vartheta + \int_\Theta \pi(\theta)\, r(\theta)\, \delta_A(\theta)\, d\theta$$

$$= \int_A \pi(\vartheta) \left[\int_\Theta p(\vartheta, \theta)\, d\theta \right] d\vartheta + \int_A \pi(\theta)\, r(\theta)\, d\theta$$

The integral of the Metropolis function in the first term above is simply the probability of moving.

$$= \int_A \pi(\vartheta)\, [1 - r(\vartheta)]\, d\vartheta + \int_A \pi(\theta)\, r(\theta)\, d\theta$$

$$= \int_A \pi(\vartheta)\, d\vartheta$$

Convergence of the Metropolis algorithm is assured by positivity of the proposal transition function $q(\theta, \vartheta)$[23] which assures that the chain is irreducible (see Robert and Casella [2004], Section 7.3.2). The challenge is to choose a "proposal" or candidate distribution that is relatively easy to evaluate and simulate from and yet produces a Metropolis chain with acceptable convergence properties. There are a wide variety of different styles of proposal densities. We will review some of the most useful.

3.10.1 Independence Metropolis Chains

Importance Sampling relies on having a reasonable approximation to π. Usually, the importance function is based on an asymptotic approximation to the posterior with fattened tails. This idea can be embedded in a Metropolis chain by taking q to be independent of the current value of the chain, $q(\theta, \vartheta) = q_{imp}(\vartheta)$ and based on the same sort of importance function ideas. We denote the independent Metropolis transition density by q_{imp} to draw the close analogy with importance sampling and with the criterion for a useful importance function.

---

[23] An additional condition is required to insure that the Metropolis chain is aperiodic. This condition requires that there be a non-zero probability of repeating.

INDEPENDENCE METROPOLIS
Start with θ_0
Draw $\vartheta \sim q_{imp}$
Compute $\alpha = \min \left\{ 1, \frac{\pi(\vartheta)q_{imp}(\theta)}{\pi(\theta)q_{imp}(\vartheta)} \right\}$
With prob α, $\theta_1 = \vartheta$, ELSE $\theta_1 = \theta_0$
Repeat, as necessary

If q is an excellent approximation to π, then most draws will be accepted since the ratio in the α computation will be close to one. This means that we will have a chain with almost no autocorrelation. To understand how the chain handles discrepancies between q and π, rewrite the ratio in the α computation as

$$\frac{\pi(\vartheta)q_{imp}(\theta)}{\pi(\theta)q_{imp}(\vartheta)} = \frac{\pi(\vartheta)\big/q_{imp}(\vartheta)}{\pi(\theta)\big/q_{imp}(\theta)}$$

If π has more relative mass at ϑ than at θ, the chain moves to ϑ with probability one to build up mass at ϑ. On the other hand, if π has less relative mass at ϑ than at θ, then there is a positive probability that the chain will repeat θ that builds up mass at that point and introduces dependence in the sequence of chain draws. This is really the opposite of accept/reject sampling in which the proposal distribution is "whittled down" to obtain π by rejecting draws.

It is important that the q proposal distribution has fatter tails than the target distribution for the same intuition as applies to importance sampling. If the target distribution has fatter tails than the proposal, the chain will wander off into the tails and then start repeating values to build up mass. If the proposal distribution dominates the target distribution in the sense that we can find a finite number M such that $\pi \leq M q_{imp}$ for all $\theta \in \Theta$, then the independence Metropolis chain is ergodic in an especially strong sense in that the distance between the repeated Kernel and π can be bounded for any starting point θ_0 (see Robert and Casella, theorem 7.8). It should be noted that if a normal prior is used to form the posterior target density, then the conditions for uniform ergodicity are met unless the likelihood is unbounded.

For problems of low dimension with very regular likelihoods, the independence Metropolis chain can be very useful. However, for higher dimensional problems, the same problems that plague importance sampling apply to the independence Metropolis chain. Even if the dominance condition holds, the proposal distribution can miss where the target distribution has substantial mass and can give rise to misleading results. For this reason, independence Metropolis chains are primarily useful in Hybrid samplers (see Section 3.10), in which the independence chain is imbedded inside of a Gibbs Sampler.

3.10.2 Random Walk Metropolis Chains

A particularly appealing Metropolis algorithm can be devised by using a random walk to generate proposal values.

$$\vartheta = \theta + \varepsilon \tag{3.10.4}$$

This proposal corresponds to the proposal transition function, $q(\theta, \vartheta) = q_\varepsilon (\vartheta - \theta)$ which depends only on the choice of the density of the increments. This proposal function is symmetric, $q(\theta, \varphi) = q(\varphi, \theta)$. A "natural" choice might be to make the increments normally distributed.[24]

GAUSSIAN RANDOM WALK METROPOLIS
Start with θ_0
Draw $\vartheta = \theta + \varepsilon$, $\varepsilon \sim N\left(0, s^2 \Sigma\right)$
Compute $\alpha = \min\left\{1, \frac{\pi(\vartheta)}{\pi(\theta)}\right\}$
With prob α, $\theta_1 = \vartheta$, ELSE $\theta_1 = \theta_0$
Repeat, as necessary

The simplicity of this algorithm gives it great appeal. It can be implemented for virtually any model and it does not appear to require the same in-depth a priori knowledge of π as the independence Metropolis does. In addition, there is an intuitive argument that the random walk algorithm should work better than the independence Metropolis in the sense that it might roam more freely in the parameter space (due to the drifting behavior of a random walk) and "automatically" seek out areas where π has high mass and then navigate those areas. From a theoretical point of view, however, strong properties such as uniform ergodicity cannot be proven for the RW Metropolis without further restrictions on the tail behavior of π. Jarner and Tweedie [2001] show that exponential tails of π is a necessary condition for geometric ergodicity[25] irrespective of the tail behavior of the proposal density q. While there is research into the relative tail behavior of the target and proposal density, there are few definitive answers. For example, it is debatable whether it is desirable to have the proposal (increment) density have thicker tails than the target. There are some results suggestive that this can be helpful in obtaining faster convergence but only for special cases (see Robert and Casella [2004], theorem 6.3.6, which applies only to log-concave targets and Jarner and Tweedie which only applies to univariate problems). In almost all practical applications, a Gaussian Random Walk is used. Again, these theoretical discussions are not terribly relevant to practice since we can always use a normal prior that will provide exponential behavior for the target density.

The problem with the Random Walk Metropolis is that it must be tuned by choosing the increment covariance matrix. This is often simplified by choosing Σ to be either I or the asymptotic covariance matrix. Computation of the asymptotic covariance matrix can be problematic for models with non-regular likelihoods. We must also choose the scaling factor, s. The scaling factor should be chosen to insure that the chain navigates the area where π has high mass while also producing a chain that is as informative as possible regarding the posterior summaries of interest. These goals cannot be met without some prior knowledge of π, which removes some of the superficial appeal of the random walk Metropolis. Even so, the random walk Metropolis has found widespread application and appears to work well for many problems.

---

[24] Interestingly enough, the original paper by Hastings includes examples in which the increments are uniformly distributed.

[25] Roughly speaking, geometric ergodicity means that the distance between the n-step kernel and the invariant distribution decreases to the power of n.

3.10.3 Scaling of the Random Walk Metropolis

The scaling of the Random Walk Metropolis is critical to its successful use. If we scale back on the variance of the increments by taking small values of the scaling factor, s, then we will almost always "move" or accept the proposed draw. This will mean that the algorithm will produce a chain that behaves (locally, at least) like a random walk. The chain will exhibit extremely high autocorrelation and, therefore, can provide extremely noisy estimates of the relevant posterior quantities. Given that small values of the scale factor will result in a slowly moving chain that can fail to properly explore regions of high target mass, we might take the opposite approach and take scale factors considerably larger than one. This can result in a chain proposed moves which are far away from the region of posterior mass (overshoot the target) and a high rejection rate. In the extreme case, over-scaled RW Metropolis chains can get stuck.

"Optimal" scaling of the Random Walk Metropolis is a balancing act in which the scaling factor chosen to optimize some performance criterion. A reasonable criterion is to measure information obtained about a posterior quantity per unit of computing time. Since we are using averages of a function of the draws to approximate the integral of that function with respect to the target posterior, the reciprocal of the variance of the sample average can be used as a measure of information. This clearly depends on the particular function used. Given a choice of function (such as the identity function for posterior means), we must be able to calculate the variance of the sample average of this function where the sample comes from a stationary but autocorrelated process. Let $\mu = E_\pi \left[g\left(\theta\right)\right]$, then we estimate μ with $\hat{\mu} = \frac{1}{R}\sum_r g\left(\theta^r\right) = \frac{1}{R}\sum_r g^r$.

$$\text{var}\left(\hat{\mu}\right) = \frac{1}{R^2}\left[\begin{array}{l} \text{var}\left(g^1\right) + \text{cov}\left(g^1, g^2\right) + \cdots + \text{cov}\left(g^1, g^R\right) + \\ \text{cov}\left(g^2, g^1\right) + \text{var}\left(g^2\right) + \cdots + \text{var}\left(g^R\right) \end{array}\right] \tag{3.10.5}$$

Since $\left\{g^r\right\}$ is a stationary process, we write (3.10.5) using the autocorrelations of the process.

$$\text{var}\left(\hat{\mu}\right) = \frac{\text{var}\left(g\right)}{R}\left[1 + 2\sum_{j=1}^{R-1}\left(\frac{R-j}{R}\right)\rho_j\right] = \frac{\text{var}\left(g\right)}{R}f_R \tag{3.10.6}$$

The "price" of autocorrelation in the MCMC draws is represented by the factor f_R, which is the multiple by which the variance is increased over the estimate based on an iid sample. Some use the reciprocal of f as a measure of the *Relative Numerical Efficiency* of the sampler.

In practice, we can use the sample moments of the g process to estimate (3.10.6). Some guidance is required for the choice of the number of autocorrelations to include in the computation of f. This is subject of a considerable literature in time series. We shouldn't take this formula literally and use all $R - 1$ computable autocorrelations. This would create a noisy and inconsistent estimate of f. There are two strategies in the time series literature. Some put a "taper" or declining weights on the sample autocorrelations used in estimating f. Others simply advocate truncating the sum at some point m.

$$\hat{f}_R = 1 + \sum_{j=1}^{m}\left(\frac{m+1-j}{m+1}\right)\hat{\rho}_j \tag{3.10.7}$$

Some guidance is required in the choice of m. In order for the estimator to be consistent, we must promise to increase m as the sample size (R) increases (albeit at a slower rate than R). This still does not provide guidance in the choice of m for a fixed sample size. Many MCMC applications will typically involve 10,000 or more draws. The autocorrelation structure of many chains used in practice is complicated and can have significant autocorrelation out to lags of order 100 or more. The exponential decline in autocorrelation associated with linear time series models is often not present in the MCMC chains. For these reasons, m should be chosen to be at least 100.

Given a measure of the information content, we can consider the problem of optimal scaling of the RW Metropolis chain. That is, we can optimize the choice of s to maximize relative numerical efficiency or to minimize f. Clearly, the optimal choice of s depends on the target distribution. Gelman et al. [1996]) and Roberts and Rosenthal [2001] provide some results on optimal scaling but only for target distributions that are products of identical normal densities. They also consider the interesting asymptotic experiment of increasing the dimension of the parameter vector. This analysis presents both the optimal scaling factor as well as the acceptance rate of the optimally scaled RW chain. For the case of a target distribution consisting of only one normal univariate density, the optimal scaling factor is $s = 2.38$. As the dimension increases, an asymptotic result has the scaling reduced at the rate of the square root of the dimension, $s = 2.3/\sqrt{d}$. Corresponding to these scaling results are implied optimal acceptance rates which are around 0.23.[26] In practice, it is impossible to determine how differences between the target for our problem and an iid normal distribution will translate into differences in optimal scaling. It is true that we are scaling the asymptotic covariance matrix so that if the target density is close to a normal with this covariance matrix then we can expect the Roberts and Rosenthal results to apply. However, if our target density differs markedly from the asymptotic normal approximation, it is possible that the optimal scaling is quite different from this rule of thumb. We recommend that, where possible, shorter runs of the RW chain be used to help tune or choose the scaling constant to maximize numerical relative efficiency for parameters of interest. Of course, this assumes that the researcher has already determined the "burn-in" period necessary to dissipate initial conditions. We will provide some guidelines on selection of burn-in period in Section 3.11. Contrary to current practice, choice of scaling should not be made on the basis of the acceptance rate of the chain but rather on the measure of numerical efficiency which is the more directly relevant quantity.

3.11 METROPOLIS ALGORITHMS ILLUSTRATED WITH THE MULTINOMIAL LOGIT MODEL

As discussed in Chapter 2, the Multinomial Logit Model is arguably the most frequently applied model in marketing applications. Individual data on product purchase often has the property that individuals are seldom observed to purchase more than one product of a

---

[26] Gelman et al. (1996) refer to optimal acceptance rates of closer to 0.5 for $d = 1$. This is because this paper only considered the first order autocorrelation. This has led to the incorrect interpretation that the optimal acceptance rate declines from around 0.5–0.23 or so as d increases.

specific type on one purchase occasion. The logit model has also been applied in a variety of forms to aggregate market share data. The MNL model has a very regular log-concave likelihood but it is not in a form that is easily summarized. Moments of functions of the parameter vector are not computable using analytic methods. In addition, the natural conjugate prior is not easily interpretable so that it is desirable to have methods which would work with standard priors such as the normal prior. If we assess a standard normal prior, we can write the posterior as

$$
\pi\left(\beta \,|\, X, y\right) \propto \ell\left(\beta \,|\, X, y\right) \pi\left(\beta\right)
$$
$$
\pi\left(\beta\right) \propto |A|^{1/2} \exp\left\{-\tfrac{1}{2}\left(\beta - \bar\beta\right)' A\left(\beta - \bar\beta\right)\right\} \tag{3.11.1}
$$

The likelihood in (3.11.1) is just the product of the probabilities of the observed choices or discrete outcomes over the n observations.

$$
\ell\left(\beta \,|\, X, y\right) = \prod_{i=1}^{n} \Pr\left(y_i = j \,|\, X_i, \beta\right)
$$
$$
\Pr\left(y_i = j \,|\, X_i, \beta\right) = \frac{\exp\left(x'_{i,j}\beta\right)}{\sum_{j=1}^{J} \exp\left(x'_{i,j}\beta\right)} \tag{3.11.2}
$$

y is a vector with the choices $(1, \ldots, J)$ and X is an $nJ \times k$ matrix of the values of the x variables for each alternative on each observation.

Experience with the MNL likelihood is that the asymptotic normal approximation is excellent. This suggests that Metropolis algorithms based on the asymptotic approximation will perform extremely well. We also note that the MNL likelihood has exponential tails (even without the normal prior) and this should provide very favorable theoretical convergence properties. We implement both an independence and RW Metropolis algorithm for the MNL model. Both Metropolis variants use the asymptotic normal approximation.

$$
\pi\left(\beta \,|\, X, y\right) \propto |H|^{1/2} \exp\left\{\tfrac{1}{2}\left(\beta - \hat\beta\right)' H\left(\beta - \hat\beta\right)\right\} \tag{3.11.3}
$$

We have a number of choices for $\hat\beta$, H. We can simply use the MLE for $\hat\beta$ or we could find the posterior mode (preferable for truly informative priors) at about the same computational cost. H could be minus the actual Hessian of the likelihood (alternatively, the posterior) evaluated at $\hat\beta$ or the sample information matrix. Alternatively, we can use expected sample information which can be computed for the MNL.

$$
H = -E\left[\frac{\partial^2 \log \ell}{\partial\beta\partial\beta'}\right] = \sum_i X_i A_i X'_i
$$
$$
X = \begin{bmatrix} X_1 \\ \vdots \\ X_n \end{bmatrix}; \quad A_i = \mathrm{Diag}\left(p_i\right) - p_i p'_i \tag{3.11.4}
$$

p_i is a J vector of the probabilities for each alternative for observation i.

We illustrate the functioning of Metropolis algorithms using a small sample of simulated data, chosen to put the asymptotic approximations to a severe test. $N = 100$, $J = 3$, and $k = 4$ (two intercepts and two independent variables that are produced by iid unif(0,1) draws). We set the beta vector to $(-2.5, 1.0, 0.7, -0.7)$. The first two elements of β are intercepts for alternatives 2 and 3 that are expressed (in the usual manner) relative to alternative 1 which is set to 0. These parameter settings imply that the probability of alternative 2 will be very small. In our simulated sample, we observed alternative 1 26 times, alternative 2 only 2 times and alternative 3 72 times.

To implement the independence Metropolis, we use a MVst candidate sampling distribution. That is, we draw candidate parameter vectors using $\beta \sim MVst\left(v, \hat{\beta}, H^{-1}\right)$. The only "tuning" required is the choice of v. Too small values of v will be inefficient in the sense of producing such fat tails that we will reject draws more often than for smaller values of v. In addition, very small values of v such as 4 or less produce a distribution which has fat tails but is also very "peaked" without the "shoulders" of the normal distribution. This also implies that the Metropolis algorithm would suffer inefficiencies from repeating draws to build-up mass on the shoulders of the peaked t distribution. Clearly, one could tune the independence Metropolis by picking v so as to maximize a numerical efficiency estimate or minimize f in (3.10.7). Using the mean of the parameters as the function whose numerical efficiency should be assessed, we find that numerical efficiency is relatively flat in the range of $5-15$. Very small values of v result in only slightly reduced numerical efficiency. All results reported here are for $v = 6$. `rmnlIndepMetrop` provides the R implementation of this algorithm (available in the package, `bayesm`).

R

The Random Walk Metropolis must be scaled in order to function efficiently. In particular, we propose β values using the equation

$$\beta_{cand} = \beta_{old} + v \quad v \sim N\left(0, s^2 H^{-1}\right) \tag{3.11.5}$$

According to the Roberts and Rosenthal guidelines, s values close to $2.93/\sqrt{d} = 2.93/2$ should work well. Given the accuracy of the normal approximation, we might expect the Roberts and Rosenthal guidelines to work very well since we can always transform the asymptotic normal approximation into a product of identical normal densities. However, the result is "asymptotic" in d. Figure 3.14 shows numerical efficiencies as measured by the square root of f as a function of the scaling constant s. We use the square root of f as we are interested in minimizing the numerical standard error and the square root of f is the multiple of the iid standard error. Each curve in Figure 3.14 shows numerical efficiency for each of the four parameters and "tuning" runs of 20,000 iterations. We consider only the estimation of the posterior mean in Figure 3.14. The small inset figure shows the acceptance rate as a function of s. As s increases, we expect the acceptance rate to decline as the chain navigates more freely. However, the numerical efficiency is ultimately the more relevant criterion. Numerical efficiency is maximized at around 1.25 which is close to the Roberts and Rosenthal value of 1.47. It is worth noting that minimum numerical efficiency of the RW Metropolis is around 4.

In comparison to the RW, the independence Metropolis functions much more efficiency with an acceptance rate of 0.70 and a numerical efficiency of 1.44 or barely less than iid sampling. Figure 3.15 shows the estimated posterior distribution of β_1 constructed from 50,000 draws of the independence and RW Metropolis chains. To the

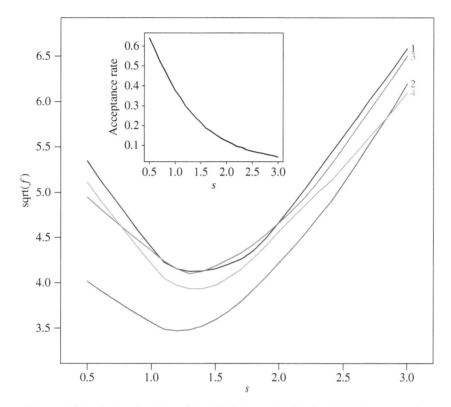

Figure 3.14   Optimal scaling of the RW Metropolis for the MNL logit example

right of the distribution are the corresponding ACFs. Differences between the ACFs of the RW and Independence chains result in a numerical efficiency ratio of 3:1. The light grey line on the distributions is the diffuse normal prior. The solid density curve is the asymptotic normal approximation to the posterior. Even in this extreme case, the actual posterior is only slightly skewed to the left from the asymptotic approximation.

For highly correlated chain output, there is a practice of "thinning" the output by selecting only every mth draw. The hope is that the "thinned" output will be approximately iid so that standard formulas can be used to compute numerical standard errors. Not only is this practice unnecessary in the sense that the investigator is literally throwing away information, but it can be misleading. In the case of the RW chain here, keeping every 10th observation (a very standard practice) will still produce a chain with non-trivial autocorrelation. The only possible reason to thin output is for convenience in storage. Given that correct numerical standard errors are trivial to compute, it seems odd that this practice continues.

3.12 HYBRID MCMC METHODS

In practice, many problems are solved with a "Gibbs-style" strategy for constructing the Markov Chain. In particular, hierarchical models have a structure that allows for an

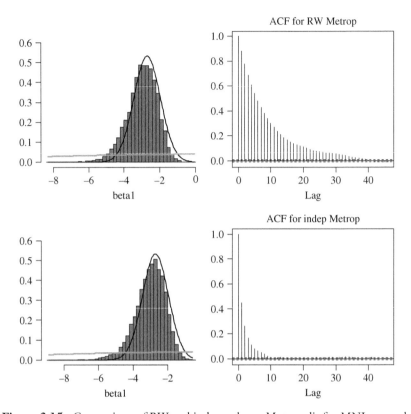

Figure 3.15   Comparison of RW and independence Metropolis for MNL example

efficient strategy which involves various conditional draws. This is because of conditional independence assumptions which are employed in constructing hierarchical models. However, in many cases, direct draws from the requisite conditional distribution are not available and some combination of a Gibbs style chain and Metropolis steps must be used. For example, consider the hierarchical structure given below.

$$\theta_2 \quad \rightarrow \quad \theta_1 \quad \rightarrow \quad Data \tag{3.12.1}$$

The Gibbs sampler for (3.12.1) would consist of

$$\theta_2 \,|\, \theta_1 \tag{3.12.2}$$

and

$$\theta_1 \,|\, \theta_2, Data \tag{3.12.3}$$

If conjugate priors are used, the draw defined by (3.12.2) is often a direct 1 for 1 draw from the conditional distribution. However, for many models the draw in (3.12.3) can be difficult to achieve. One possible strategy would be to break the draw each of the elements of the θ_1 vector one by one using Griddy Gibbs, sliced sampling or Adaptive

Rejection Metropolis Sampling (ARMS, see Gilks et al. [1995]). However, in many real problems, the dimension of θ_1 will be large so that this is not practical.

A useful idea is to replace the "Gibbs" draw in (3.12.3) with a Metropolis step. This algorithm is called by some a "Hybrid" chain or "Metropolis within Gibbs." To construct the Metropolis chain all this is required is to evaluate the conditional density in (3.12.3) up to a normalizing constant. Of course, the joint density evaluated at the draw of θ_2 is proportional to the conditional density and can be easily evaluated.

$$p\left(\theta_1 \,|\, \theta_2, Data\right) \propto p\left(Data \,|\, \theta_1, \theta_2\right) p\left(\theta_1, \theta_2\right)$$

If we implement a Metropolis step with the conditional distribution, $\theta_1 \,|\, \theta_2, Data$, as its invariant distribution, then it can be shown that the hybrid chain has the full posterior as its equilibrium distribution. To see this, we first return to a discussion of the Gibbs Sampler and develop the notion of the composition of two Markov chains.

The standard "two-stage" Gibbs sampler can be written:

$$\theta_1 \,|\, \theta_2, Data \tag{3.12.4}$$

and

$$\theta_2 \,|\, \theta_1, Data \tag{3.12.5}$$

The Markov Chain that (3.12.4) and (3.12.5) represent can be thought of as the combination of two chains, each one of which updates one of the components of θ. If we denote K_1^G as the kernel of a chain formed by the conditional distribution in (3.12.4) and K_2^G as the kernel of the chain formed by (3.12.5), then the two-stage Gibbs Sampler is the composition of these two chains:

$$[K^G = K_1^G \circ K_2^G] \tag{3.12.6}$$

The Kernel of the composed chain is

$$K(\theta, A) = \int_A p(\theta, \varphi)\, d\varphi$$
$$p(\theta, \varphi) = p\left((\theta_1, \theta_2), (\varphi_1, \varphi_2)\right) = \pi_{1|2}\left(\varphi_1|\theta_2\right) \pi_{2|1}\left(\varphi_2|\theta_1\right) \tag{3.12.7}$$

While each of the sub-chains are reducible (i.e., they only navigate on the sub-space formed by holding one component of θ fixed), the full chain is irreducible as we have seen before. What is less obvious is that the full posterior is the invariant distribution of both K_1^G and K_2^G, viewed separately. To see this, consider K_1^G. If we start with a draw from the joint posterior, we must then show that the one iteration of K_1^G reproduces the joint posterior. The joint posterior factors into the conditional and marginal densities.

$$\pi\left(\theta_1, \theta_2 | Data\right) = \pi_2\left(\theta_2 | Data\right) \pi_{1|2}\left(\theta_1 | \theta_2, Data\right)$$

This means that the "initial" value is a draw that can always be viewed as obtained by first drawing from the appropriate marginal and then drawing from the conditional distribution.

$$\theta_2^0 \sim \pi_2 \text{ and } \theta_1^0 \sim \pi_{1|2}$$

To show that K_1^G produces a draw from the joint posterior, we observe that

$$\theta_1^1 \sim \pi_{1|2}\left(\theta_2^0\right)$$

The notation, $\pi_{1|2}\left(\theta_2^0\right)$ means the conditional distribution of $\theta_1|\theta_2 = \theta_2^0$. Thus, the draw from K_1^G is also a draw from the joint.

$$\left(\theta_1^1, \theta_2^0\right) \sim \pi_{1|2}\,\pi_2 = \pi_{1,2}$$

The same argument applies to K_1^G and we can see that the full two-stage Gibbs Sampler is irreducible and has the joint posterior as its variant distribution.

The hybrid chain formed by substituting a Metropolis step for first step in the two-stage Gibbs sampler is also the composition of two reducible chains, each having the joint posterior as its invariant distribution. $K^H = K_1^M \circ K_2^G$. This follows directly by exactly the same argument that we applied to the Gibbs sampler. There is one additional important point. The Metropolis step will update the value of θ_1 given the previous iteration value of θ_2. This means that the values of θ_2 from the previous iteration can be used in the computation of candidate or proposal values of θ_1 in the Metropolis step. As long as we obey the proper conditioning and use only values of the last iteration, we will still have a Markov chain with the correct invariant distribution. This means that we can "automatically" adjust the Metropolis candidate sampling density depending on the last value of θ_2. We should also note that this is not the same as "adaptive" schemes that use information from a past subsequence to adjust the Metropolis step. These schemes are not Markovian and require additional analysis to establish that they have the proper invariant distribution.

3.13 DIAGNOSTICS

The theoretical properties of MCMC methods are quite appealing. For many algorithms and problems, especially with normal priors, it is easy to establish ergodicity and even stronger results such as geometric rates of convergence and uniform ergodicity. The problem with these rates of convergence results is that they do not specify the constants that govern the actual rate of convergence. In practice, we can produce an MCMC sampler which has desirable theoretical properties but poor performance given our finite computer resources. In addition, errors in formulating the model or the MCMC algorithm can be very important and difficult to diagnose.

The performance of MCMC algorithms is related to the speed at which the chain navigates the state space. Highly autocorrelated samplers require long periods of time to navigate the parameter space fully enough to properly represent the level of uncertainty in the posterior. A related problem is the dissipation of initial conditions. In practice, we start our samplers from reasonably arbitrary points and hope that they dissipate the effect of these initials after a burn-in period. Draws after the burn-in period are viewed as draws from the stationary distribution. Clearly the speed of navigation and dissipation of initial conditions are related. Near iid MCMC samplers will both dissipate initial conditions and rapidly navigate the parameter space. However, in practice we see examples of samplers that dissipate the initial conditions rapidly but are slow to navigate regions of high posterior mass. For example, if we consider the bivariate normal Gibbs sampler, high (much higher than 0.95) correlation will create a narrow ridge which the Gibbs sampler may require many iterations to navigate. However, given the thin tails of the normal distribution, if we start the sampler at a point far from this ridge, the sampler will quickly move to the ridge and dissipate the initial condition. The classic "witch's hat" examples are examples where a Gibbs sampler can get stuck in one mode of a very peaked distribution and fail to break out to the other mode even though this is theoretically possible. In our experience with models relevant to micro data and marketing applications, these situations are rarely encountered. More common are situations in which the MCMC sampler navigates slowly with a high degree of autocorrelation. In these situations, it may take days of computing to properly navigate the posterior.

Unfortunately, there is a sense in which there can be no powerful diagnostic or "test" for failure of convergence without substantial prior knowledge of the posterior. If the asymptotic approximation to the posterior is poor and/or if the parameter space is of high dimensions, then we are engaging in MCMC sampling in order to learn about the posterior. This means that we can never be very sure that we have gotten the right result. Many of the proposals for convergence diagnostics utilize a sub-sequence of an MCMC run in order to assess convergence. This requires confidence that the sampler has already navigated the relevant regions of the parameter space and the issue is more one of the information content of the sampler sequence and not whether or not navigation is complete. The information content of the sequence can easily be gauged by computing numerical standard errors for quantities of interest using the formula in (3.10.7). This doesn't address the convergence question. Other proposals in the convergence diagnostics areas involve starting the sampler from a number of initial conditions (Gelman and Rubin [1992])) to check for dissipation of initial conditions. Again, choice of these initial conditions requires information about the posterior. In problems where the parameter space is more than a few dimensions, it may be impractical to choose a comprehensive set of initial conditions. Others have suggested using multiple "parallel" runs of the MCMC sampler to compare the distribution obtained within one run with the cross-section simulation of many runs. This is not practical in the large scale problems considered here.

In practice, we rely on sequence or time series plots of MCMC output as well as computed autocorrelation and associated standard errors to monitor convergence. The assumption is that a slowly navigating chain (one with near random walk behavior) must be monitored very closely for possible convergence failure. These highly autocorrelated chains should be run as long as possible to investigate the sensitivity of the estimation of the posterior variability to run length. In many situations, the MCMC sampler may exhibit dependence of an extremely long lasting or persistence variety, often with autocorrelations that do not damp off quickly. In these situations, more attention should be focused on monitoring convergence. We have found that one of the most effective ways to assess convergence properties is to conduct sampling experiments that have been calibrated to match the characteristics of the actual data under analysis. While large scale sampling experiments may not be possible due to the size of the parameter space or due to computing limitations, we have found that a small scale experiment can be very useful.

In most situations, slowly navigating chains are produced by high correlation between groups of parameters. The classic example of this is the normal Gibbs sampler. Other examples include latent variables added for data augmentation. Here high correlation between the latent variables and the model "parameters" can produce slowly navigating chains. There are two possible solutions to high correlation between two subsets of parameters. If the two subsets can be "blocked" or combined into one iid draw, then the autocorrelation can be reduced substantially. However, care must be exercised in blocking as there are examples where the blocked sampler gets stuck. If one set of parameters is not of direct interest and can be integrated out (sometimes called "collapsing"), then the autocorrelation can also be reduced (see Liu [2001] for a theoretical comparison of collapsed and blocked samplers). We will see an example of this technique in the case study on scale usage heterogeneity. Of course, whether these are practical strategies depends on the model and priors and our ability to perform the integration. Finally, we have seen that, with a good proposal density, the Independence Metropolis can outperform the Random Walk.

One neglected, but important, source of concern are errors made in the formulation and coding of an MCMC method. One can easily formulate an MCMC sampler with the wrong posterior for the model under consideration. In addition, errors can be made in the coding of the densities used in Metropolis methods or in the draws from the conditional distributions. While simulation experiments will often detect the presence of these errors, Geweke [2004] has proposed an additional useful check on the formulation of MCMC samplers. The idea is to simulate from the joint distribution of the observables and the parameters in two independent ways. One way will use the MCMC method to be tested and the other will simply require drawing from the prior. Draws made in these two ways can then be compared to see if they are similar.

To draw from the joint distribution of the observations and parameters, we simply need the ability to simulate from the prior and the model.

$$p(y, \theta) = p(\theta)p(y|\theta) \tag{3.13.1}$$

IID draws from this joint density can be achieved by drawing from the prior and then drawing $y|\theta$ from a model simulator. Code to simulate from the prior and the model is not usually needed to implement the MCMC method.[27] This is the sense that coding errors in the MCMC method could be independent of coding errors in the direct draws from (3.13.1). To indirectly draw from the joint distribution of the observables and θ, we can use the MCMC method to draw from the posterior in conjunction with the model simulator. These two draws can be used to construct a hybrid chain with the joint distribution of (y, θ) as the invariant distribution.

EXTENDED SAMPLER
Start from some value θ^0
Draw $y^1 | \theta^0$ using the model simulator

Draw $\theta^1 | y^1, \theta^0$ using the MCMC method to be tested
Repeat as necessary

---

[27] For conditionally conjugate models, draws from the prior are usually achieved by using the same functions required to draw from the conditional posteriors (e.g., normal and Wishart draws). Thus, the Geweke method will not have "power" to detect errors in these simulation routines. However, simulating from the model is truly an independent coding task.

Denote the sequence of iid draws as $\{y^r, \theta^r\}$ and the sequence from the hybrid sampler as $\{\hat{y}^r, \hat{\theta}^r\}$. Geweke suggests that we compare $\bar{g} = \frac{1}{R}\sum_r g\left(y^r, \theta^r\right)$ with $\bar{\hat{g}} = \frac{1}{R}\sum_r g\left(\hat{y}^r, \hat{\theta}^r\right)$ as a diagnostic check on the MCMC algorithm. Differences between $\bar{g}$ and $\bar{\hat{g}}$ could represent conceptual errors in the MCMC algorithm which result in a different invariant distribution than the posterior of this model and prior or coding errors in implementing various aspects of the MCMC algorithm. Clearly, the power of this diagnostic to detect errors depends on the scope of the g function. There will be a trade-off between power for certain types of errors vs ability to perform omnibus duty against a wide spectrum of errors. In particular, g does not have to be a function of the observables at all. In many contexts, investigators use very diffuse, but proper priors, these prior hyperparameter settings will result in low power for the Geweke method. In order to obtain better power, it is useful to tighten up the priors by using hyper-parameter settings very different from those used in analysis of data. Finally, we prefer to compare the entire distribution rather than the particular moments implied by the g function (at the risk of focusing on univariate marginals). For example, qqplots using elements of (y, θ) drawn by the two methods can be very useful.

4

Unit-Level Models and Discrete Demand

Abstract

This chapter reviews models for discrete data. Much of the disaggregate data collected in marketing has discrete aspects to the quantities of goods purchased. Sections 4.1–4.3 review the latent variable approach to formulating models with discrete dependent variables, while Section 4.4 derives models based on a formal theory of utility maximization. Those interested in Multinomial Probit or Multivariate Probit models should focus on Sections 4.2 and 4.3. Section 4.2.1 provides material on understanding the difference between various Gibbs samplers proposed for these models and can be omitted by those seeking a more general appreciation. Section 4.4 forges a link between statistical and economic models and introduces demand models which can be used for more formal economic questions such as welfare analysis and policy simulation.

We define "unit level" as the lowest level of aggregation available in a data set. For example, retail scanner data is available at many levels of aggregation. The researcher might only have regional or market-level aggregate data. Standard regression models can suffice for this sort of highly aggregated data. However, as the level of aggregation declines to consumer level, sales response becomes more and more discrete. There are a larger number of zeroes in this data and often only a few integer-valued points of support. If, for example, we examine the prescribing behavior of a physician over a short period of time, this will be a count variable. Consumers often choose to purchase only a small number of items from a large set of alternatives. The goal of this chapter is to investigate models appropriate for disaggregate data. The common characteristic of these models will be the ability to attach lumps of probability to specific outcomes. It should also be emphasized that even if the goal is to analyze only highly aggregate data, the researcher could properly view this data as arising from individual level decisions aggregated up to form the data observed. Thus, individual-level demand models and models for the distribution of consumer preferences (the focus of Chapter 5) are important even if the researcher only has access to aggregate data.

4.1 LATENT VARIABLE MODELS

We will take the point of view that discrete-dependent variables arise via partial observation of an underlying continuous variable. For example, the binary probit model discussed in Chapter 3 was formulated as a binary outcome, which is simply the sign (0 if negative, 1 if positive) of the dependent variable in a normal regression model. There are three advantages of a latent variable formulation: 1. The latent variable formulation is very flexible and is capable of generating virtual any sort of outcome that has a discrete component; 2. The latent variable formulation allows for easy formulation of MCMC algorithms using data augmentation; and 3. In some situations, the latent variable can be given a random utility interpretation, which relates latent variable models to the formal econometric specification of demand models based on utility maximization.

We can start with a normal regression model as the model for the latent variables.

$$z_i = X_i \delta + v_i \tag{4.1.1}$$

We consider the case in which z_i can be either a scalar or a vector with corresponding changes to the dimensionality of both X_i and v_i.

The outcome variable, y_i, is modeled as a function of the latent variable z_i.

$$y_i = f(z_i) \tag{4.1.2}$$

In order to create discrete y, the f function must be constant over some regions of the space in which z resides. We distinguish between the cases in which z is univariate and multivariate. For the case of univariate z, standard models such as the binary logit/probit and ordered probit models can be obtained. The binary probit model assumes $f(z) = I(z > 0)$ and v has a normal distribution. The binary logit assumes the same f function and the extreme value type I with constant scale distribution for v.

Ordered models can be obtained by taking the f function to be an indicator function of which of several intervals z resides in.

$$f(z) = \sum_{c=1}^{C+1} c \times I(\gamma_{c-1} < z \le \gamma_c) \tag{4.1.3}$$

y is multinomial with values 1, ... , C and $\gamma_0 = -\infty$ and $\gamma_{C+1} = \infty$. The ordinal nature of y is enforced by the assumption of a unidimensional latent variable. As $x_i' \delta$ increases, the probability of obtaining higher values of y increases. By allowing for arbitrary cut-off points, the ordered model has flexibility to allow for different probabilities of each integer value of y conditional on x. For convenience in implementation of MCMC algorithms, we typically use an ordered probit model in which the latent variable is conditionally normal. As with the binary probit/logit models there is little difference between ordered probits and other models obtained by other distributional assumptions on v.

In other situations, y is multinomial but we do not wish to restrict the multinomial probabilities to be monotone in $x_i \delta$. In these cases, the underlying latent structure must be multivariate in order to provide sufficient flexibility to accommodate general patterns of the influence of the x variables on the probability that y takes on various values. That is, if we increase x, this may increase the probability of some value of y and decrease the probability of others in no particular pattern. Consider the multinomial case in which y can assume one of p values associated with the integers $1, \ldots, p$. In this case, we assume that z has a continuous, p dimensional distribution. A convenient choice of f is the indicator of the component with the maximum value of z.

$$f(z) = \sum_{j=1}^{p} j \times I\left(\max(z) = z_j\right) \qquad (4.1.4)$$

The multinomial distribution is appropriate for choice situations in which one of p alternatives is chosen. However, in some situations, multiple choices are a possible outcome. For example, we may observe consumers purchasing more than one variety of a product class on the same purchase occasion. In a survey context, a popular style of question is the "*m* choose *k*" or *m* alternatives are given and the questionnaire requests that the respondent "check all that apply." For these situations, a multivariate discrete choice model is appropriate. We represent this situation by allowing y to be p dimensional with each component having only two values. $y = (y_1, \ldots, y_p)$; $y_i = 0, 1$. For this model, we must also use a multivariate latent variable and this becomes the direct generalization of the "univariate" binary model.

$$f : R^p \rightarrow R^p; \quad f(z) = \begin{pmatrix} I(z_1 > 0) \\ \vdots \\ I(z_p > 0) \end{pmatrix} \qquad (4.1.5)$$

In some contexts, the response variable might reasonably be viewed as consisting of both discrete and continuous components. For example, we may observe both product choice and quantity. On many occasions, the consumer does not purchase the product but on some occasions a variable quantity is purchased. The quantity purchased is also discrete in the sense that most products are available in discrete units. However, we might assume that a continuous variable is a reasonable approximation to the quantity demanded conditional on positive demand. These situations can also be accommodated in the latent variable framework. For example, we can specify a simple Tobit model as

$$f(z) = z \times I(z > 0) \qquad (4.1.6)$$

However, we will defer a formal discussion of the mixed discrete continuous models until the introduction of a utility maximization model and a random utility framework.

Each of the models introduced in (4.1.1)–(4.1.5) has associated identification problems. These problems stem from the invariance of the $f(\bullet)$ functions to location or scale transformations of z, i.e., $f(cz) = f(z)$ or $f(z + k) = f(z)$. In the binary models, there is a scaling problem for the error, v. Changing the variance of v will not change likelihood as multiplication of z by a positive constant in (4.1.2) will not alter y. Typically, this is solved by imposing a scale restriction such as $Var(v) = 1$. In the ordered model in (4.1.3), there is both a scaling and location problem. Even if we fix the scale of v, we can add a constant to all of the cut-offs and subtract the same constant from the intercept in δ and leave the likelihood unchanged. For this reason, investigators typically set the first cutoff to zero, $\gamma_1 = 0$. In the multinomial model (4.1.4), there is both a location and scale problem. If we add a constant to every component of z or if we scale the entire z vector by a positive constant, then the likelihood is unchanged. We will discuss identification of multinomial models in Section 4.2. For the multivariate model in (4.1.5), we recognize that we can scale each of the components of z by a *different* positive amount and leave the likelihood unchanged. This means that only the correlation matrix of z is identified.

4.2 MULTINOMIAL PROBIT MODEL

If latent utility is conditionally normal and we observe outcome value i (of p) if max(z) = z_i, then we have specified the Multinomial Probit model.

$$y_i = f\left(z_i\right)$$
$$f\left(z_i\right) = \sum_{j=1}^{p} j \times I\left(\max\left(z_i\right) = z_{ij}\right) \qquad (4.2.1)$$
$$z_i = X_i \delta + v_i \quad v_i \sim iid\ N\left(0, \Omega\right)$$

If the multinomial outcome variable y represents consumer choice among p mutually exclusive alternatives, then X matrix consists of information about the attributes of each of the choice alternatives as well as covariates, which represent the characteristics of the consumer making choices.

The general X would have the structure $X_i = \left[\left(1, d_i'\right) \otimes I_p, A_i\right]$ where d is a vec-
R tor of "consumer" characteristics and A is a matrix of choice attributes. createX in our package, bayesm, can be used to create the X array with this structure. For example, we might have information on the price of each alternative as well as the income of each consumer making choices. Obviously, we can only identify the effects of consumer covariates with a sample taken across consumers. In most marketing contexts, we obtain a sample of consumers in panel form. We will consider modeling differences between consumers in Chapter 5. In a modern hierarchical approach, covariates are included in the distribution of parameters across consumers and not directly in the X matrix above. For expositional purposes, we will consider the case in which X contains only alternative specific information. In typical applications, δ contains alternative-specific intercepts as well as marketing mix variables for each alternative. It is common to assume that the coefficient on the marketing mix variables is the same across choice alternatives. The random utility derivation of multinomial choice models provides a rationale for this, which we will develop in Section 4.3. The linear model for the latent variable in (4.2.1) is a SUR model, written in a somewhat nonstandard form, and with restrictions that some of the regression coefficients are the same across regressions. For example, with a set of intercepts and one marketing mix variable, $X_i = \left[I_p \quad m_i\right]$, where I is a $p \times p$ identity matrix and m is a vector with p values of one marketing mix variable for each of p alternatives.

As indicated in Section 4.1, the model in (4.2.1) is not identified. In particular, if we add a scalar random variable to each of the p latent regressions, the likelihood remains unchanged. That is, $Var\left(z_i | X_i, \delta\right) = Var\left(z_i + u | X_i, \delta\right)$ where u is a scalar random variable. This is true only if Ω is unrestricted. If, for example, Ω is diagonal, then adding a scalar random variable will change the covariance structure by adding correlation between choice alternatives and the elements of the diagonal of Ω will be identified. In the case of unrestricted Ω, it is common practice to subtract the pth equation from each of the first $(p-1)$ equations to obtain a differenced system. The differenced system can be written

$$w_i = X_i^d \beta + \varepsilon_i \quad \varepsilon_i \sim N(0, \Sigma)$$

$$w_{ij} = z_{ij} - z_{ip}; \ X_i = \begin{bmatrix} x'_{i,1} \\ \vdots \\ x'_{i,p} \end{bmatrix}; \ X_i^d = \begin{bmatrix} x'_{i,1} - x'_{i,p} \\ \vdots \\ x'_{i,p-1} - x'_{i,p} \end{bmatrix}; \ \varepsilon_{ij} = v_{ij} - v_{ip} \qquad (4.2.2)$$

$$y_i = f(w) = \sum_{j=1}^{p-1} j \times I\left(\max(w_i) = w_{ij} \text{ and } w_{ij} > 0\right) + p \times I(w < 0)$$

We also note that if δ contains intercepts, then the intercept corresponding to the pth choice alternative has been set to zero and β contains all of the other elements of δ.

The system in (4.2.2) is still not identified as it is possible to multiply w by a positive scalar and leave the likelihood of the observed data unchanged. In other words, $f(w) = f(cw)$. This identification problem is a normalization problem. We have to fix the scale of the latent variables in order to achieve identification. It is common to set $\sigma_{11} = 1$ to fix the scale of w. However, it is also possible to achieve identification by setting one of the components of the β vector to some fixed value. For example, if price were included as a covariate, we might fix the price coefficient at -1.0. Of course, this requires a prior knowledge of the sign of one of the elements of the β vector. In the case of models with price, we might feel justified in imposing this exact restriction on the sign of the price coefficient, but in other cases we might be reluctant to make such an imposition.

In classical approaches to the MNP, the model in (4.2.2) is reparameterized to the identified parameters $(\tilde{\beta}, \tilde{\Sigma})$ by setting $\sigma_{11} = 1$. In a Bayesian approach, it is not necessary to impose identification restrictions. With proper priors, we can define the posterior in the unidentified space, construct an MCMC method to navigate this space and then simply "margin down" or report the posterior distribution of the identified quantities. In many cases, it is easier to define an MCMC method on the unrestricted space. Moreover, it turns out that the unrestricted sampler will often have better mixing properties.

In Section 3.7, we saw how data augmentation can be used to propose a Gibbs sampler for the binary probit model. These same ideas can be applied to the MNP model. If β, Σ are a priori independent, the DAG for the MNP model is given by

$$\begin{array}{c} \Sigma \searrow \\ \quad w \to y \\ \beta \nearrow \end{array} \qquad (4.2.3)$$

McCulloch and Rossi [1994] propose a Gibbs sampler based on cycling through three conditional distributions.

$$w \mid \beta, \Sigma, y, X^d$$
$$\beta \mid \Sigma, w \qquad (4.2.4)$$
$$\Sigma \mid \beta, w$$

This sampler produces a chain with $w, \beta, \Sigma | y, X^d$ as its stationary distribution. We can simply marginalize out w if the posterior of β, Σ is desired. We note that w is sufficient for β, Σ and this is why it is not necessary to include y as a conditioning argument in the second and third distributions in (4.2.4). Since the underlying latent structure is a normal Multivariate regression model, we can use the standard theory developed in Chapter 2 to draw β and Σ. The only difficulty is the draw of w. Given β, Σ, the $\{w_i\}$ are independent, $p-1$ dimensional truncated normal random vectors. The regions of truncation are the R_{y_i}, as defined in (4.2.11). Direct draws from truncated multivariate normal random vectors are difficult to accomplish efficiently. The insight of McCulloch and Rossi [1994] is to recognize that one can define a Gibbs sampler by breaking each draw of w_i into a sequence of $p - 1$ univariate truncated normal draws by cycling through the w vector.

That is, we do not draw $w_i | \beta, \Sigma$ directly. Instead, we draw from each of the $(p - 1)$ truncated univariate normal distributions.

$$w_{ij} | w_{i,-j}, y_i, \beta, \Sigma \sim N\left(m_{ij}, \tau_{jj}^2\right)$$
$$\times \left[I\left(j = y_i\right) I\left(w_{ij} > \max\left(w_{i,-j}, 0\right)\right) + I\left(j \neq y_i\right) I\left(w_{ij} < \max\left(w_{i,-j}, 0\right)\right) \right]$$
$$m_{ij} = x_{ij}^{d'} \beta + F'\left(w_{i,-j} - X_{i,-j}^d \beta\right) \tag{4.2.5}$$
$$F = -\frac{1}{\sigma^{j,j}} \gamma_{j,-j}$$
$$\tau_{jj}^2 = 1/\sigma^{j,j}$$

$\sigma^{i,j}$ denotes the (i,j)th element of Σ^{-1} and

$$\Sigma^{-1} = \begin{bmatrix} \gamma_1' \\ \vdots \\ \gamma_{p-1}' \end{bmatrix} \tag{4.2.6}$$

The univariate distribution in (4.2.5) is a truncated normal distribution. $\gamma_{j,-j}$ refers to the jth row of Σ^{-1} with the jth element deleted. $X_{i,-j}^d$ is the matrix, X_i, with the jth column deleted. We st art with the first element $(j = 1)$ and "Gibbs thru" each observation, replacing elements of w, one by one, until the entire vector is updated. In bayesm, this is done in C code. To implement this sort of sampler in R, condMom can be useful to compute the right moments for each univariate truncated normal draw.

To implement full Gibbs sampler for the MNP model, we need to specify a prior over the model parameters, β, Σ. The model parameters are not identified without further restrictions due to the scaling problem for the MNP model. The identified parameters are obtained by normalizing with respect to one of the diagonal elements of Σ.

$$\tilde{\beta} = \beta/\sqrt{\sigma_{11}}; \ \tilde{\Sigma} = \Sigma/\sigma_{11} \tag{4.2.7}$$

For ease of interpretation, we will report the correlation matrix and vector of relative variances as the set of identified covariance parameters. One approach would put a prior on the full set of unidentified parameters as in McCulloch and Rossi [1994].

$$\beta \sim N\left(\bar{\beta}, A^{-1}\right) \ \Sigma \sim IW\left(v, V_0\right) \tag{4.2.8}$$

Equation (4.2.8) induces a prior on the identified parameters in (4.2.7). The induced prior on the identified parameters is not in standard form.[1] Imai and van Dyk [2005] suggest a very similar prior but with the advantage that the prior induced on the identified regression coefficients is more easily interpretable.

$$\beta|\Sigma \sim N\left(\sqrt{\sigma_{11}}\tilde{\beta}, \sigma_{11} A^{-1}\right); \ \Sigma \sim IW\left(v, V_0\right) \tag{4.2.9}$$

---

[1] McCulloch et al. [2000] derive these distributions. Assessment of these priors for all but the highly diffuse case may be difficult. Under the prior in (4.2.8), $\tilde{\beta}$ and $\tilde{\Sigma}$ are not independent. In addition, the prior on $\tilde{\beta}$ can be skewed for nonzero values of $\bar{\beta}$

With the prior in (4.2.9), $\tilde{\beta} \sim N(\bar{\beta}, A^{-1})$. This prior specification makes assessment of the prior more straightforward. It also allows for the use of an improper prior on the identified regression coefficients.

Another approach would be to put a prior directly on the set of identified parameters and define a Gibbs sampler on the space $(w, \tilde{\beta}, \tilde{\Sigma})$ as in McCulloch et al. [2000]. This requires a prior on the set of covariance matrices with 1,1 element set equal to 1. This is done via a reparameterization of Σ.

$$\Sigma = \begin{bmatrix} \sigma_{11} & \gamma' \\ \gamma & \Phi + \gamma\gamma' \end{bmatrix} \tag{4.2.10}$$

This reparameterization is suggested by considering the conditional distribution of $\varepsilon_{-1} | \varepsilon_1 \sim N(\gamma\varepsilon_1, \Phi)$. This is a multivariate regression of ε_{-1} on ε_1 and the standard conjugate prior can be used for γ, Φ (as in Section 2.8.5). σ_{11} can be set to 1 in order to achieve identification. This creates a prior over covariance matrices with the 1,1 element fixed at 1. However, the Gibbs sampler suggested by McCulloch et al. [2000] (hereafter termed the ID MNP sampler) navigates the identified parameter space more slowly than the sampler of McCulloch and Rossi [1994]. For this reason, we cover the approach of McCulloch and Rossi [1994] here.

Given the prior in (4.2.8), we can define the McCulloch and Rossi [1994] sampler, which we term the NID, for nonidentified sampler, as follows:

NID MNP Gibbs Sampler
Start with initial values, w_0, β_0, Σ_0
Draw $w_1 | \beta_0, \Sigma_0$ using (4.2.5)
Draw $\beta_1 | w_1, \Sigma_0 \sim N(\tilde{\beta}, V)$

$$V = \left(X^{d*'} X^{d*} + A \right)^{-1} \quad \tilde{\beta} = V \left(X^{d*'} w^* + A\bar{\beta} \right)$$
$$\Sigma_0^{-1} = C'C$$
$$X_i^{d*} = C'X_i^d \quad w_i^* = C'w_i$$
$$X^d = \begin{bmatrix} X_1^d \\ \vdots \\ X_n^d \end{bmatrix}$$

Draw $\Sigma_1 | w_1, \beta_1$ *using* $\Sigma^{-1} | w, \beta \sim W \left(v + n, (V_0 + S)^{-1} \right)$

$$S = \sum_{i=1}^n \varepsilon_i \varepsilon_i'$$
$$\varepsilon_i = w_i - X_i^d \beta$$

Repeat as necessary

R   This sampler is implemented in `rmnpGibbs` in our package, `bayesm`.

To illustrate the functioning of the NID sampler, consider a simulated example from McCulloch and Rossi [1994].

$N = 1600$, $p = 6$. $X \sim \text{iid} \, \text{Unif}(-2, 2)$. $\beta = 2$.
$\Sigma = \text{diag}(\sigma) \left(\rho\iota\iota' + (1-\rho) I_{p-1} \right) \text{diag}(\sigma)$. $\rho = 0.5$ and $\sigma' = (1, 2, 3, 4, 5)^{0.5}$.

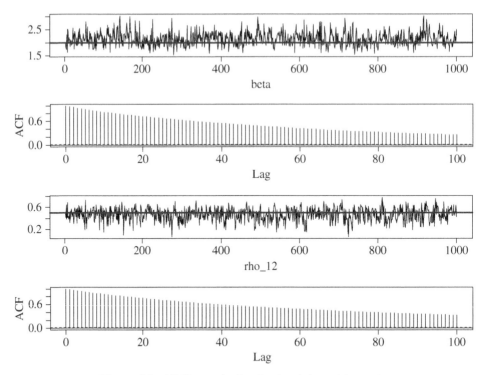

Figure 4.1   MNP sampler for simulated data with $p = 6$.

Figure 4.1 shows the MCMC trace plot and ACFs for $\tilde{\beta}$ and ρ_{12}. These are achieved by normalizing the draws of the unidentified parameters appropriately. For example, to construct an MCMC estimate of the posterior distribution of $\tilde{\beta}$, we simply post-process the draws of β by dividing by the (1, 1) element of Σ. Figure 4.1 shows the trace plots for the "thinned" draw sequence obtained by extracting every 100th draw from a draw sequence of 100,000. However, the ACFs are computed using every draw. β was started at 0 and Σ at the identity matrix. Very diffuse priors were used, $\beta \sim N(0, 100)$ and $\Sigma \sim IW\left(v = (p-1) + 2, vI_{p-1}\right)$. The MCMC sampler dissipates the initial conditions very rapidly but exhibits very high autocorrelation. One way to gauge the extent of this auto correlation is to compute the f factor or relative numerical efficiency, which is the ratio of the variance of the mean of the MCMC draws to the variance assuming an iid sample. For $\tilde{\beta}$, $f = 110$ and, for ρ_{12}, $f = 130$. This means that our "sample" of 100,000 has the information content of approximately 800 $\left(\approx 100{,}000 / \sqrt{f}\right)$ iid draws from the posterior. The posterior mean of $\tilde{\beta}$ is estimated to be 2.14 with a numerical standard error of 0.0084 and the posterior mean of is 0.39 with a numerical standard error of 0.005. These numerical standard errors must be viewed relative to the posterior standard deviations of these quantities. If the posterior standard deviation is large, then it means that we cannot make precise inferences about these parameters and we may be willing to tolerate larger numerical standard errors. In this example, the estimated posterior standard deviations for both $\tilde{\beta}$ and ρ_{12} are about 30 times the size of the numerical standard errors.

4.2.1 Understanding the Autocorrelation Properties of the MNP Gibbs Sampler

MNP Gibbs sampler exhibits higher autocorrelation than MCMC examples considered in Chapter 3. As with all Gibbs samplers, high autocorrelation is created by dependence among the random variables being sampled. It appears that the introduction of the latent variables via data augmentation has created the problem. There must be high dependence, between w, β, and Σ. To verify and explore this intuition, consider the binary probit model that is a special case of the MNP model with $p = 2$. For this case, we can easily identify two samplers: 1. the NID sampler[2] and 2. the ID sampler of Albert and Chib (see also Chapter 3). Both samplers include both w and β, but the Albert and Chib sampler sets $\Sigma = 1$ and navigates in the identified parameter space.

It is easier to develop an intuition for the high dependence between w and the model parameters in the case of the ID sampler as there is one fewer model parameter. Consider a simple case where there is only one X variable and this variable takes on only two values $(0,1)$. In this case, all observations fall into a 2×2 contingency table defined by X and Y. The latent variables are drawn from four different truncated distributions corresponding to cells of this table.

	$Y = 0$	$Y = 1$
$X = 0$	$w_{0,0}$	$w_{0,1}$
$X = 1$	$w_{1,0}$	$w_{1,1}$

To see why β and w can be very highly autocorrelated, consider the case depicted in Figure 4.2. If $X = 0$, then w will be sampled from the normal distribution centered at zero, which is depicted by the dotted density. For $X = 0$ and $Y = 1$, w will be sampled from the positive half normal and, for $X = 0$ and $Y = 0$, from the negative half normal. If β changes, these draws of w will remain unchanged in distribution. However, consider what happens to the draws of w for $X = 1$. These draws are made from truncations of the normal distributions depicted by solid densities in Figure 4.2. If β increases, then the draws of w will come from a normal distribution centered at a higher mean and truncated either from above by 0 or below by 0. These draws will be larger in expectation. This sets up a situation in which if β increases, w draws tend to increase for $X = 1$ and remain unchanged for $X = 0$. In turn, the posterior mean of β will increase as the least

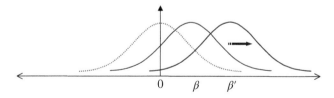

Figure 4.2   Understanding the correlation between latent variable and binary logit parameters

---

[2] For $p = 2$, our NID sampler is identical to that proposed by Van Dyk and Meng [2001].

squares estimate of $\beta\,|w$ will simply be the difference in mean w for $X = 1$ vs $X = 0$. The only force working against the strong dependence of w and β is the variability in the distribution of $w\,|X = 1, \beta$.

If $Var\left(x'\beta\right)$ is large relative to $Var(\varepsilon)$, then the dependence between β and w will be high. In the ID sampler, $Var(\varepsilon)$ is fixed at 1. However, in the NID sampler the variance of ε is drawn as a parameter. Large values of $Var(\varepsilon)$ will reduce the dependence between w and β. The NID sampler can draw large values of $Var(\varepsilon)$, particularly for very diffuse priors. Van Dyk and Meng [2001] observe that it would be optimal to set an improper prior for Σ in terms of the properties of the NID sampler. However, both the NID and ID samplers can exhibit high dependence between draws. Both samplers will perform poorly when $Var\left(x'\beta\right)$ is large. This may be disconcerting to the reader – when the X variables are highly predictive of choice, the performance of both samplers degrades.

To illustrate the dependence between β and w, we consider some simple simulated examples. X is one column that is created by taking draws from a Bernoulli(0.5) distribution. $N = 100$, $\Sigma = 1$. We will simulate data corresponding to an increasing sequence of β values. The intuition is that as β increases, we should observe an increase in dependence between w_i draws for $X = 1$, $Y = 1$, and β. Figure 4.3 plots the sample mean of the w draws for $X = 1$, $Y = 1$ against the β draw for the same ID MCMC iteration.

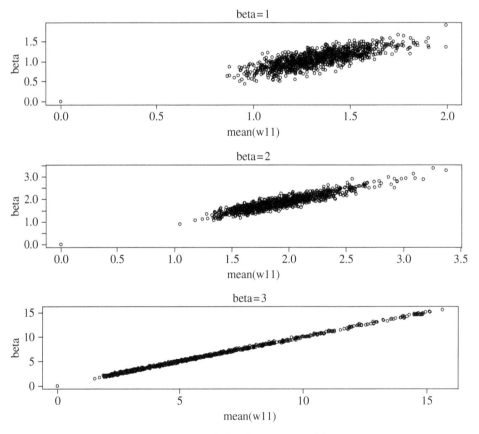

Figure 4.3   Correlation between β and latents

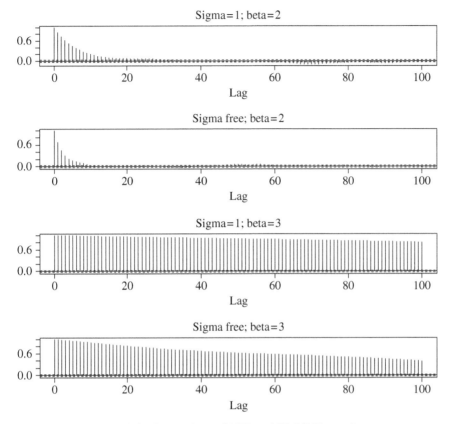

Figure 4.4   Comparison of NID and ID MNP samplers

Four simulated data sets with $\beta = (1, 2, 3)$ were created. This correlation exceeds 0.99 for $\beta = 3$. Figure 4.4 provides the ACFs for the draw sequence of β for each of the simulated datasets. The dependence between w and β results in extremely high autocorrelation for large values of β. It should be pointed out that a β value of 3 or larger is fairly extreme as this implies that the $\text{Prob}(Y = 1 \,|\, X = 1) > 0.99^3$.

However, the NID sampler performs much better as illustrated in Figure 4.4. The top two ACFs show the ID and NID samplers for data simulated with $\beta = 2$. The NID sampler exhibits much lower autocorrelation. For data simulated with $\beta = 3$, the autocorrelation for the ID sampler becomes very extreme. In that case, the ID sampler output is some 15 times less informative than an iid sample. The NID sampler is highly autocorrelated but with output approximately twice as informative as the ID sampler. These results for the binary probit model suggest that the ID sampler for the MNP model proposed by McCulloch et al. [2000] can exhibit very high autocorrelation. Nobile [1998] points this out. The intuition developed here is that navigating in the unidentified parameter space can result in a chain with better mixing properties in the space of identified parameters. Nobile [1998] proposes a modification of the McCulloch and Rossi [1994] algorithm, which includes a Metropolis step that increases

---

[3] It should be noted that $Prob(Y = 0 | X = 0) = .5$. This means that we are not in the situation in which the observations can be separated and the posterior with flat priors becomes improper.

the variability of the σ draws. Nobile observes improvement over the McCulloch and Rossi NID sampler under extreme conditions. As VanDyk and Meng [2001] point out, any latent variable model that margins out to the right likelihood can be used as the basis for a data augmentation MCMC method.[4] It is entirely possible that even better performing chains can be constructed using different data augmentation schemes.

4.2.2 The Likelihood for the MNP Model

The likelihood for the MNP model is simply the product of the relevant probabilities of the outcomes. Each of these probabilities involves integrating a normal distribution over a $p-1$ dimensional cone

$$\ell(\beta, \Sigma) = \prod_{i=1}^{n} \mathrm{Pr}\left(y_i \left| X_i^d, \beta, \Sigma \right.\right)$$

$$\mathrm{Pr}\left(y_i \left| X_i^d, \beta, \Sigma \right.\right) = \int_{R_{y_i}} \varphi\left(w \left| X_i^d \beta, \Sigma \right.\right) dw \qquad (4.2.11)$$

$$R_{y_i} = \begin{cases} \left\{ w : w_{y_i} > \max\left(w_{-y_i}, 0\right) \right\} & \text{if } y_i < p \\ \{w : w < 0\} & \text{if } y_i = p \end{cases}$$

Here w_{-j} denotes all elements in the w vector except the jth element (a $(p-2)$ dimensional vector) and $\varphi(\bullet)$ is the multivariate normal density function. Figure 4.5 shows

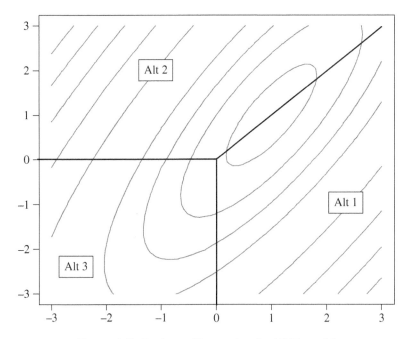

Figure 4.5 Regions of integration for MNP model

---

[4] In the sense that $p(y|\beta, \Sigma) = \int p(y, w|\beta, \Sigma)\, dw = \int p(y|w)\, p(w|\beta, \Sigma)\, dw$ for many latent models.

this situation for a three choice example. The contours correspond to a bivariate normal distribution centered at a mean determined by the X values for the ith observation. Three regions are shown on the figure, corresponding to each of the three choice alternatives. If $y_i = 1$ then we must have $w_1 > \max(w_2, 0)$. If $y_i = 2$, then $w_2 > \max(w_1, 0)$. If $y_i = 3$, then $w_1, w_2 < 0$. Direct evaluation of the likelihood is only computationally feasible if these integrals can be evaluated accurately. The GHK method discussed in Chapter 2 can be used to evaluate the likelihood in (4.2.11). The GHK method is designed to compute integral of a normal distribution over a rectangle defined by vectors of lower and upper truncation points. In order to use this method, we have to re-express the region defined in (4.2.11) as a rectangular region. To do so, we must define a matrix, specific to each choice alternative, that can be used to transform the w vector to a rectangular region.

For $j = 1, \ldots, p - 1$, define[5]

$$A_j = \begin{bmatrix} -I_{j-1} & 1 & 0 \\ & \vdots & \\ 0 & 1 & -I_{(p-1)-j} \end{bmatrix}$$

The condition $w_j > \max(w_{-j})$ and $w_j > 0$ is equivalent to $A_j w > 0$. We can then re-express the inequalities by applying A_j to both sides of the latent variable equation

$$A_j w = A_j \mu + A_j \varepsilon > 0$$
$$A_j \varepsilon > -A_j \mu$$
$$\text{where } \mu = X^d \beta$$

If we define $u = A_j \varepsilon$, then we can express choice probability as a multivariate normal integral truncated from below.

$$u > -A_j \mu \quad u \sim N\left(0, A_j \Sigma A_j'\right)$$
$$\Pr\left(y = j \mid X, \beta, \Sigma\right) = \int_{-A_j X \beta}^{\infty} \varphi\left(u \mid 0, A_j \Sigma A_j'\right) du \tag{4.2.12}$$

For the case of $j = p$, the region is already defined in terms of ε

$$\varepsilon < -\mu$$
$$\Pr\left(y = p \mid X, \beta, \Sigma\right) = \int_{-\infty}^{-X\beta} \varphi\left(u \mid 0, \Sigma\right) du \tag{4.2.13}$$

R The function, llmnp, in bayesm, implements this approach.

We could use the likelihood evaluated via GHK to implement a Metropolis Chain for the MNP without data augmentation. However, this would require a good proposal function for both β, Σ. Experience with Metropolis methods for covariance matrices suggests that these methods are only useful for small dimensional problems.

---

[5] Note that if $j = 1$ or $j = p - 1$, then we simply omit the requisite I matrix (i.e., I_0 means nothing)!

4.3 MULTIVARIATE PROBIT MODEL

The Multivariate Probit model is specified by assuming the same multivariate regression model as for the MNP model but with a different censoring mechanism. We observe the sign of the components of the underlying p-dimensional multivariate regression model.

$$w_i = X_i\beta + \varepsilon_i \quad \varepsilon_i \sim N(0, \Sigma)$$
$$y_{ij} = \begin{cases} 1 \ if \ w_{ij} > 0 \\ 0 \ \text{otherwise} \end{cases} \tag{4.3.1}$$

Here choice alternatives are not mutually exclusive as in the MNP model. The multivariate probit model has been applied to purchase of products in two different categories (Manchanda et al. [1999]) or to surveys with pick j of p questions (Edwards and Allenby [2003]). In the econometrics literature, the multivariate probit has been applied to a binary phenomenon that is observed over adjacent time periods (e.g., labor force participation observed for individual workers).

The identification problem in the Multivariate Probit can be different from the identification problem for the MNP depending on the structure of the X array. Consider the general case that includes intercepts for each of the p choice alternatives and covariates that are allowed to have different coefficients for each p choices.

$$X_i = \left(z_i' \otimes I_p\right) \tag{4.3.2}$$

z is a $d \times 1$ vector of observations on covariates. Thus, X is a $p \times k$ matrix with $k = p \times d$.

$$\beta = \begin{bmatrix} \beta_1 \\ \vdots \\ \beta_d \end{bmatrix} \tag{4.3.3}$$

β_i, $i = 1, \ldots, d$ are p dimensional coefficient vectors. The identification problem arises from the fact that we can scale each of the p means for w with a different scaling constant without changing the observed data. This implies that only the correlation matrix of Σ is identified and that transformation from the unidentified to the identified parameters $((\beta, \Sigma) \to (\tilde{\beta}, R))$ is defined by

$$\tilde{B} = \Lambda B$$
$$\tilde{\beta} = vec\left(\tilde{B}\right) \tag{4.3.4}$$
$$R = \Lambda \Sigma \Lambda$$

where

$$\begin{cases} B = \begin{bmatrix} \beta_1, \ldots, \beta_d \end{bmatrix} \\ \Lambda = \begin{bmatrix} 1/\sqrt{\sigma_{11}} & & \\ & \ddots & \\ & & 1/\sqrt{\sigma_{pp}} \end{bmatrix} \end{cases}$$

However, if the coefficients on a given covariate are restricted to be equal across all p choices, then there are fewer unidentified parameters. Then we cannot scale each equation by a *different* positive constant. This brings us back in to the same situation as in the MNP model where we must normalize by one of the diagonal elements of Σ. For example, we might have an attribute like price of the p alternatives under consideration. We might want to restrict the price attribute to have the same effect on w for each alternative. This amounts to the restriction that $\beta_{j1} = \beta_{j2} = \cdots \beta_{jp}$ for covariate j.

To construct a MCMC algorithm for the multivariate probit model, we can use data augmentation just as in the MNP model by adding w to the parameter space. We simply "Gibbs thru" the w vector using the appropriate conditional univariate normal distribution but with an upper (lower) truncation of 0 depending on the value of y.

$$w_{ij} \,\big|\, w_{i,-j}, y_i, \beta, \Sigma \sim N\left(m_{ij}, \tau_{jj}^2\right) \left[I\left(y_{ij} = 1\right) I\left(w_{ij} > 0\right) + I\left(y_{ij} = 0\right) I\left(w_{ij} < 0\right)\right]$$

$$m_{ij} = x'_{ij}\beta + F'\left(w_{i,-j} - X_{i,-j}\beta\right)$$

$$F = -\frac{1}{\sigma_{jj}}\gamma_{j,-j}$$

$$\tau_{jj}^2 = 1/\sigma^{jj}$$

$$(4.3.5)$$

Here the vector y is an np vector of indicator variables.

We must make a choice of whether to navigate in the unidentified (β, Σ) space or the identified $(\tilde{\beta}, R)$ space. The unidentified parameter space is larger by p dimensions than the identified space. The intuition developed from the MNP model and generalized by VanDyk and Meng [2001] is that navigating in the higher dimensional unidentified parameter space with diffuse priors will produce a chain with superior mixing properties to a chain defined on the identified space. An additional complication will be the method for drawing valid R matrices. The algorithm of Chib and Greenberg [1998] or Barnard et al. [2000] can be used to draw R. However, given the additional complication and computational cost of these methods and the fact that we expect these chains to have inferior mixing properties, we recommend using a more straightforward Gibbs sampler on the unidentified parameter space (see Edwards and Allenby [2003] for details).

NID Multivariate Probit Gibbs Sampler
Start with initial values, w_0, β_0, Σ_0
Draw $w^1 \big| \beta_0, \Sigma_0, y$ using (4.3.5)
Draw $\beta_1 \big| w_1, \Sigma_0 \sim N\left(\tilde{\beta}, V\right)$

$$V = \left(X^{*\prime}X^* + A\right)^{-1} \quad \tilde{\beta} = V\left(X^{*\prime}w^* + A\bar{\beta}\right)$$
$$\Sigma_0^{-1} = C'C$$
$$X_i^* = C'X_i \quad w_i^* = C'w_i$$
$$X = \begin{bmatrix} X_1 \\ \vdots \\ X_n \end{bmatrix}$$

Draw $\Sigma_1 \big| w_1, \beta_1$ using $\Sigma^{-1} \big| w, \beta \sim W\left(v + n, \left(V_0 + S\right)^{-1}\right)$

$$S = \sum_{i=1}^n \varepsilon_i \varepsilon_i'$$
$$\varepsilon_i = w_i - X_i^d \beta$$

Repeat as necessary

R This algorithm has been implemented in the function, rmvpGibbs, in *bayesm*.

To illustrate this sampler, we consider a data example from Edwards and Allenby
R [2003]. This dataset is available in *bayesm* and can be loaded using the R command,
data(Scotch). 2218 respondents were given a list of 21 scotch brands and asked to indicate whether or not they drink these brands on a regular basis. The interest in this example
is in understanding the correlation structure underlying brand choice. Correlation in the
latent variable can be viewed as a measure of similarity of two brands. In this example,
$X_i = I_{21}$ so that the β vector is simply a vector of intercepts or means for the latent
variable w.

Figure 4.6 shows MCMC traces (every 20th draw from sequence of 20,000) and
ACFs for two of the elements of β corresponding to popular blended whiskeys (note: the
brand chosen most often is Chivas Regal so all other brands have a smaller intercepts).
These plots show quick dissipation of the initial condition ($\beta_0 = 0$ and $\Sigma_0 = I$). The
autocorrelations of the intercept draws is small with a numerical efficiency roughly one
half of that of an iid sample. Figure 4.7 shows MCMC traces (again every 20th draw
from a sequence of 20,000) and ACFs for two of the correlations. These correlations are
between two single malts and between the most popular blended whiskey (Chivas Regal)
and a single malt. The single malts exhibit high correlation showing similarity in product
branding and taste. There is a negative correlation between the single malt and the
blended whiskey showing some divergence in consumer preference. The autocorrelations

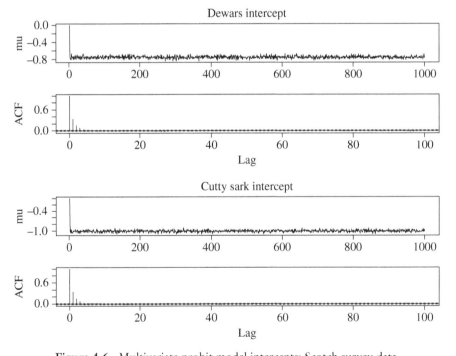

Figure 4.6 Multivariate probit model intercepts: Scotch survey data

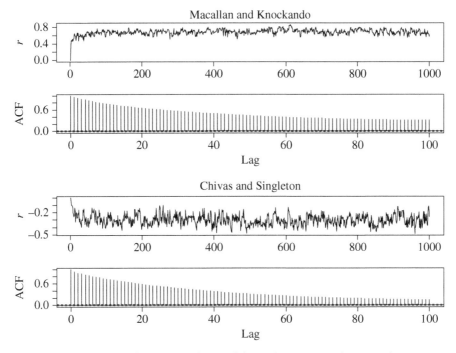

Figure 4.7   Multivariate probit model correlations: Scotch survey data

for the correlation parameters are higher than the intercept or mean parameters, but still less than for some of the MNP examples considered in Section 4.2 (numerical efficiency here is 1/9th of a random sample as compared to 1/11th for the MNP examples). The intuition for the better performance of the NID sampler for the Multivariate Probit (in comparison to the NID MNP sampler) is that there is a higher dimensional unidentified parameter space improving the mixing characteristics of the sampler in the identified parameter space.

In our analysis of the scotch data, we used a relatively diffuse but proper prior: $\beta \sim N\left(0, 100 I_p\right)$ and $\Sigma \sim IW\left(p+2, (p+2) I_p\right)$. Our analysis of the binary probit model developed an intuition that the diffusion of the prior on Σ would affect the performance of the sampler. In particular, as emphasized by Meng and Van Dyk [1999] and Van Dyk and Meng [2001], mixing should be maximized by allowing the prior on Σ to be improper. However, as McCulloch et al. [2000] have pointed out, improper priors on Σ can be extraordinarily informative on functions of Σ such as correlations. The implied marginal prior on each correlation will be U-shaped with substantial mass near -1 and 1.

We reran the analysis with an improper prior on Σ and show the results in Figure 4.8. As expected, the improper prior produces better mixing, reducing the autocorrelations substantially. Numerical efficiency is now at 1/7th of an iid sample. However, the improper prior on Σ is extraordinarily informative on the correlations changing both the location and tightness of the posterior distributions of the correlations. Figure 4.9 shows the prior distribution of a correlation for the barely proper case, $\Sigma \sim IW\left(p, p I_p\right)$. This "u-shaped" distribution puts high mass near high positive or negative correlations. The improper prior can be viewed as the limit of proper priors as the diffusion increases

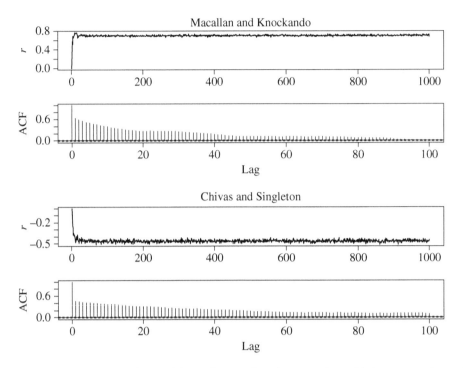

Figure 4.8 Multivariate probit model correlations: Scotch survey data with improper prior on Σ

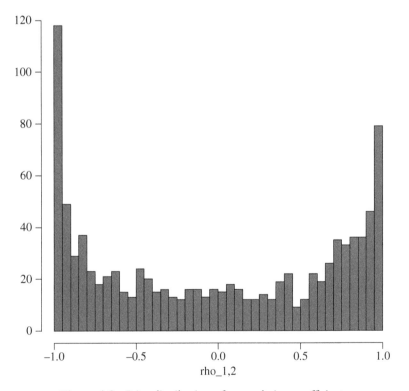

Figure 4.9 Prior distribution of a correlation coefficient

and, therefore, will be even more informative than the barely proper space, putting high mass on extreme values of correlation.

Thus, improper priors are very dangerous for this analysis even though they might receive attention due to mixing considerations. The problem here is that there is a conflict between prior settings, which promote maximum mixing and substantive informative considerations. It would be desirable to produce a MCMC sampler with a prior that separates "tuning" from substantive informativeness.

4.4 DEMAND THEORY AND MODELS INVOLVING DISCRETE CHOICE

The multinomial and multivariate choice models developed in Sections 4.2 and 4.3 are simply statistical models that use a latent variable device to link discrete outcomes to covariate information. Since the outcomes in many marketing situations are the result of customer decisions, we should relate these statistical models to models of optimizing behavior and consumer demand theory. If we assume that consumers are maximizing utility subject to a budget constraint, then our modeling efforts focus on choice of the utility function and assumptions regarding the distribution of unobservables.

In much of standard demand theory (c.f. Deaton and Muellbauer [1980]), utility functions are chosen so that only interior solutions are obtained. That is, consumers faced with a set of goods would purchase all goods in various amounts. For aggregate data or for highly aggregated goods (such as food and housing), these utility functional forms are appropriate. However, at the level of the individual customer, few of the possible alternative goods are chosen. When faced with a set of similar products, customers are often observed to only purchase one product. For these situations, utility functions must be chosen, which allow for corner solutions. This means that marginal utility must remain finite as consumption approaches zero (at least for some of the goods). In other words, the indifference curves cannot be tangent to the axes.

For situations in which consumers are only observed to choose one alternative, a linear utility function is a reasonable place to start.

$$\max_x U(x) = \psi' x \\ s.t. \; p'x = E \tag{4.4.1}$$

For this functional form, marginal utility is constant and the optimal solution is to choose the alternative with the highest ratio of marginal utility to price (ψ_j / p_j) and consume $x = E/p_j$ units. In this situation, if we observe p and E, then consumer choice will be deterministic. The standard random utility approach is to assume that there are factors affecting the marginal utility, which are not observed by the researcher but are known to the consumer. These factors could be time varying (i.e., consumer inventory or promotional activities) or represent taste differences between consumers. Since marginal utility must be positive, it is natural to consider a multiplicative specification for inclusion of these omitted unobservables.

$$\psi = \bar{\psi} \, e^{\varepsilon} \tag{4.4.2}$$

If we take logs, allow ψ to be a function of observable covariates such as consumer characteristics and product attributes, and subtract log price, we obtain a model similar to the standard multinomial models.

$$\ln \psi - \ln p = \bar{\psi} - \ln p + \varepsilon = \tilde{X}\beta - \ln p + \varepsilon \tag{4.4.3}$$

$\tilde{X}$ is a $p \times k$ matrix containing the relevant covariates. Letting $z = \ln \psi - \ln p$ and $X = [\tilde{X} \ \ln p]$, we have the latent variable formulation of a multinomial choice model in (4.2.1). Alternative j is chosen if $z_j = \max_i (z_i)$.

To complete the model, we must specify the distribution of the latent component of marginal utility. First, consider iid errors with a given scale, σ. In (4.4.3), σ is identified as the price coefficient is set to -1. Alternatively, we could introduce a price coefficient. If we allow a free price coefficient, we must fix the scale of the marginal utility errors to avoid the scale invariance identification problem. This means that we can interpret a price coefficient as $1/\sigma$. As the variance of the marginal utility errors grows, the influence of price on the probabilities of choice declines since the factors other than price are "larger" in scale. If we allow ε to have a multivariate normal distribution, we obtain the standard MNP model. The standard multinomial logit model can be derived by assuming that errors are iid extreme value type I. Demand theory provides some guidance as to the sort of covariates that might be included as well as the restriction that the price coefficient is negative and the same across choice alternatives.

4.4.1 A Nonhomothetic Choice Model

The restrictive properties of the MN Logit model can be relaxed either by specifying a richer class of error distributions (as in the MNP model that avoids the IIA property) or by specifying a more flexible utility structure. The linear structure assumes constant marginal utility and no interactions (e.g., the marginal utility of alternative i is not affected by the consumption of other alternatives). One could argue that for choice data, constant marginal utility is appropriate in that we would need quantity information to estimate the curvature of the utility function. However, even with only access to choice information, the constant marginal utility assumption implies that price changes can only have substitution and not income effects (i.e., the utility function is homothetic). In the homothetic world, as greater amounts are allocated to expenditure on the products under study, we continue to purchase the same brand. In many marketing applications, there are products of differing quality levels. We observe consumers trading up to high-quality models as expenditure increases or during sales. The opposite phenomena of "trading down" to lower quality brands as expenditure decreases or as lower quality brands are discounted does not occur as frequently. This phenomenon has been dubbed "asymmetric switching." The standard homothetic logit or probit models cannot exhibit asymmetric switching.

A nonhomothetic choice model can be specified by retaining the linear structure of utility but assuming that marginal utility is a function of the overall level of attainable utility. This retains the assumption of no interactions in utility but allows for a nontrivial income effect that occurs either across consumers as consumers allocate a greater expenditure to the product category or as the category becomes "cheaper" due to price reductions on some of the items. One convenient specification that defines a valid utility function is given below.

$$\psi_i(u) = \exp\left\{\alpha_i - k_i u\right\} \qquad (4.4.4)$$

In (4.4.4), the marginal utility of the ith product is a function of the maximum attainable level of utility. This defines the utility function implicitly. k governs the rate at which marginal utility changes as the level of attainable utility increases. If $k_1 < k_2$, then, as the budget increases, consumers will tend to purchase product 1 more than 2. The utility function for this case has a set of linear but rotating indifference curves. In this sense, we can regard product 1 as a superior product that provides a higher marginal utility for consumers willing to allocate a greater real expenditure to the product category.

Since the indifference curves are linear, consumers will choose only one of the products in the category. As in the standard linear utility model, consumers choose the alternative that provides the highest level of attainable utility conditional on the level of expenditure for the category, E, and vector of prices, p. That is, consumers find the maximum of $\{u^1, u^2, \dots, u^p\}$. u^1 solves

$$u^i = \psi_i\left(u^i\right) E/_{p_i} \qquad (4.4.5)$$

Taking logs of both sides, we can write this as the implicit solution to an equation of the form $\ln(y) = C_1 - C_2 y$. If C_2 is positive, this equation has an implicit solution that can be easily calculated by Newton's method. To complete this model, we must specify a random utility error as well as an expenditure function. In Allenby and Rossi [1991], an expenditure function was specified as

$$\ln E = \gamma' z \qquad (4.4.6)$$

z is a vector covariates. In Allenby and Rossi, z includes a price index for the product category. This is a somewhat ad hoc solution that avoids the specification of a bivariate utility function over all other goods and this product category. In addition, one might be tempted to make the α and k parameters a function of demographic variables. However, since marginal utility is specified as an implicit nonlinear function of α and k, we would not be able to write the nonhomothetic model in form of (4.4.3).

If we add a standard extreme value type I error to the implicitly defined marginal utility, we obtain a logit model in which the probability of choice is given by

$$\Pr(i) = \frac{\exp\left(\tau v_i\right)}{\sum_j \exp\left(\tau v_j\right)} \qquad (4.4.7)$$

where $v_j = \alpha_j - k_j u^j - \ln p_j.\tau$ is the scale parameter of the extreme value distribution. Note that u is also a function of E, p_j, α_j and k_j. Thus, we can write $v_j = f\left(\alpha_j, k_j \middle| E, p_j\right)$

R `llnhlogit` in *bayesm* evaluates the log-likelihood for this model.

4.4.2 Demand for Discrete Quantities

The application of discrete choice models to packaged goods requires researchers to adjust their models to accommodate demand quantities. Packaged goods are often available in multiple sizes, and it is not appropriate to treat the various package sizes as independent alternatives since the same good is contained in each of the packages. While it is possible to estimate discrete choice models that allow for dependence among the alternatives, additional restrictions on the coefficients are needed so that parameter estimates from the model conform to economic theory. For example, a discrete choice model calibrated on soft drink purchases would need to impose ordinal restrictions on the intercepts so that the utility of 6-pack, 12-pack, and 18-pack offerings would reflect diminishing marginal returns to quantity.

Quantity can be incorporated into models of consumer demand by imbedding a utility function for a discrete choice model into a utility function that relates the product class to an outside good. For example, consider the Cobb–Douglas utility function:

$$\ln U(x, z) = \alpha_0 + \alpha_x \ln U(x) + \alpha_z \ln(z) \tag{4.4.8}$$

where $x = (x_1, \dots , x_K)$ is the vector of the amount of each alternative (i.e., brand) purchased, K represents the number of brands in the product class, z represents the amount of the outside good purchased, and $U(x)$ denotes a sub-utility function. The sub-utility function, $U(x)$, can be specified as the linear function in equation (4.4.1) or the nonhomothetic function in equation (4.4.4).

Maximizing (4.4.8) subject to a budget constraint leads to a vector of demand (x, z) that, in general, is a mixture of corner and interior solutions. However, as discussed in Allenby et al. [2004], "A Choice Model for Packaged Goods," the utility maximizing solution will always be a corner solution in x when per-unit price schedules are concave, for example, the cost per ounce of a 6-pack of is greater than the cost per ounce of a 12-pack. When this occurs, the choice probability of observing quantity x_i is:

$$\Pr(x_i) = \frac{\exp\left[\ln(\psi_i) + \ln(x_i) + (\alpha_z/\alpha_x)\ln(E - p_i(x_i))\right]}{\sum_{k=1}^{K} \exp\left[\ln(\psi_k) + \ln(x_k) + (\alpha_z/\alpha_x)\ln(E - p_k(x_k))\right]} \tag{4.4.9}$$

where ψ_i is the marginal utility of brand i, E is the budgetary allotment, $p_i(x_i)$ is the price of x_i units of brand i, and x_k is the quantity of brand k that maximizes equation (4.4.8).

4.4.3 Demand for Variety

In some product categories, consumers are observed to purchase a subset of products in the category. For example, consumers purchase multiple varieties of soft drinks or yogurts. The standard multinomial models have zero likelihood for this sort of consumer behavior as the choice options are regarded as mutually exclusive. On the other hand, many common utility specifications are designed to give rise to strictly interior solutions in which all products in the category are purchased. What is needed is a demand system that can give rise to a mixture of corner and interior solutions. This can be achieved by translating a utility function so that its indifference curves intersect the axes with finite slope. One simple additive structure is given by

$$\bar{U}(x) = \sum_j \psi_j (x_j + \gamma_j)^{\alpha_j} \tag{4.4.10}$$

The $\{\gamma_j\}$ parameters serve to translate an additive power utility to admit the possibility of corner solutions. The utility function also exhibits curvature or diminishing marginal utility that allows for the possibility of "wear-out" in the consumption of a particular variety. The utility in (4.4.10) is an additive, but nonlinear utility function. Equation (4.4.10) defines a valid utility function under the restrictions that $\psi_j > 0$ and $0 < \alpha_j \leq 1$.

This utility specification can accommodate a wide variety of situations, including the purchase of a large number of different varieties as well as purchases where only one variety is selected. If a particular variety has a high value of ψ_j and a value of α_j close to one, then we would expected to see purchases of large quantities of only one variety (high baseline preference and low satiation). On the other hand, small values of α imply a high satiation rate, we expect to see multiple varieties purchased if the ψ's are not too different.

To develop a statistical specification, we follow a standard random utility approach and introduce a multiplicative normal error into marginal utility:

$$\ln \left(U_j \right) = \ln \left(\bar{U}_j \right) + \varepsilon_j \quad \varepsilon \sim N\left(0, \Sigma \right) \tag{4.4.11}$$

where $\bar{U}_j$ is the derivative of the utility function in (4.4.10) with respect to x_j. We use a log-normal error term to enforce positivity of marginal utility. We specify a full covariance matrix for the random marginal utility errors. In some applications, it may be difficult to identify this covariance matrix. Further restrictions may be necessary. Even the assumption that Σ is the identity matrix is not necessarily too restrictive as we have specified a log-normal distribution of marginal utility errors that exhibits heteroskedasticity of a reasonable form. However, this is largely an empirical matter.

We derive the demand system for the set of goods under study conditional on the expenditure allocation to this set of goods. In the random utility approach, it is assumed that the consumer knows the value of ε and that this represents omitted factors, which influence marginal utility but are not observable to the data analyst. If we derive the optimal demand by maximizing utility subject to the budget constraint and conditional on the random utility error, we define a mapping from p, E, and ε to demand. Assuming a distribution for ε provides a basis for deriving the distribution of optimal demand, denoted x^*. There are two technical issues in deriving the distribution of demand: 1. Optimal demand is a nonlinear function of ε and requires use of change-of-variable calculus and 2. The possibility of corner solutions means that there are point masses in the distribution of demand and, thus, the distribution of demand will be a mixed discrete-continuous distribution. Computing the size of these point masses involves integrating the normal distribution of ε over rectangular regions of R^m.

To solve for optimal demand, we form the Lagrangian for the problem and derive the standard Kuhn–Tucker first order conditions. It is important to remember in the utility function specified in (4.4.11), $\bar{U}$ is only the deterministic part of utility (that observed by us) and that the consumer maximizes U, which includes the realization of the random utility errors. The Lagrangian is given by

$$U\left(x \right) - \lambda \left(p'x - E \right)$$

Differentiating the Lagrangian gives the standard Kuhn–Tucker first order conditions:

$$\bar{U}_j e^{\varepsilon_j} - \lambda p_j = 0 \qquad \text{if } x_j^* > 0$$
$$\bar{U}_j e^{\varepsilon_j} - \lambda p_j < 0 \qquad \text{if } x_j^* = 0$$

$x*$ is the vector of optimal demands for each of the m goods under consideration. Dividing by price and taking logs, the Kuhn–Tucker conditions can be re-written as:

$$V_j\left(x_j^* \mid p\right) + \epsilon_j = \ln \lambda \qquad \text{if } x_j^* > 0$$
$$V_j\left(x_j^* \mid p\right) + \epsilon_j < \ln \lambda \qquad \text{if } x_j^* = 0 \qquad (4.4.12)$$

where λ is Lagrange multiplier and $V_j\left(x_j^* \mid p\right) = \ln\left(\psi_j \alpha_j\left(x_j^* + \gamma_j\right)^{\alpha_j - 1}\right) - \ln\left(p_j\right)$ $j = 1, \dots, m$.

Optimal demand satisfies the Kuhn–Tucker conditions in (4.4.12) as well as the "adding-up" constraint that total $p'x* = E$. The "adding-up" constraint induces a singularity in the distribution of $x*$. To handle this singularity, we use the standard device of differencing the first order conditions with respect to one of the goods. Without loss of generality, we assume that the first good is always purchased (one of the m goods must be purchased since we assume that $E > 0$) and subtract condition, (4.4.12), for good 1 from the others. This reduces the dimensionality of the system of equations by one. Equation (4.4.12) is now equivalent to:

$$v_j = h_j\left(x^*, p\right) \quad \text{if } x_j^* > 0$$
$$v_j < h_j\left(x^*, p\right) \quad \text{if } x_j^* = 0 \qquad (4.4.13)$$

where $v_j = \epsilon_j - \epsilon_1$ and $h_j(x^*, p) = V_1 - V_j$ and $j = 2, \dots, m$.

The likelihood for $x^* = \left(x_1^*, \dots, x_m^*\right)$ can be constructed by utilizing the p.d.f. of $v = (v_2, \dots, v_m)'$, the Kuhn–Tucker conditions in (4.4.13), and the adding up constraint $p'x* = E$. $v = (v_2, \dots, v_m)' \sim N(0, \Omega)$. $\Omega = A\Sigma A'$ where $A = \begin{bmatrix} -\iota & I_{m-1} \end{bmatrix}$. Given that corner solutions will occur with non-zero probability, the distribution of optimal demand will have a mixed discrete-continuous distribution with lumps of probability corresponding to regions of ε that imply corner solutions. Thus, the likelihood function will have a density component corresponding to the goods with nonzero quantities and a mass function corresponding to the corners in which some of the goods will have zero optimal demand. The probability that n of the m goods are selected is equal to:

$$P\left(x_i^* > 0 \text{ and } x_j^* = 0; \ i = 2, \dots, n \text{ and } j = n+1, \dots m\right)$$
$$= \int_{-\infty}^{h_m} \cdots \int_{-\infty}^{h_{n+1}} \phi(h_2, \dots, h_n, v_{n+1}, \dots, v_m \mid 0, \ \Omega) |J| dv_{n+1} \cdots dv_m \qquad (4.4.14)$$

where $\phi(\cdot)$ is normal density, $h_j = h_j(x^*, p)$, and J is the Jacobian,

$$J_{ij} = \frac{\partial h_{i+1}(x^*; p)}{\partial x_{j+1}^*}$$
$$i, j = 1, \dots, n - 1.$$

We should note that the adding up constraint, $p'x = E$, makes this Jacobian nondiagonal as we can always express the demand for the "first" good with nonzero demand as a function of the other demands.

The intuition behind the likelihood function in (4.4.14) can be obtained from the Kuhn–Tucker conditions in (4.4.13). For goods with nonzero demand, the first condition in (4.4.13) means that optimal demand is an implicitly defined nonlinear function of ε given by h(). We use the change-of-variable theorem to derive the density of x^* (this generates the Jacobian term). For goods not purchased, the second Kuhn–Tucker condition defines a region of possible values of v, which are consistent with this specific corner solution. The probability that these goods have zero demand is calculated by integrating the normal distribution of v over the appropriate region.

If there are only corner solutions with one good chosen, our model collapses to a standard choice model. The probability that only good one is chosen is given by

$$P\left(x_j^* = 0, \ j = 2, \ \dots, m\right)$$
$$= \int_{-\infty}^{h_m} \cdots \int_{-\infty}^{h_2} \phi\left(v_2, \ \dots, v_m\right) \ dv_2 \cdots dv_m$$

Similarly, we can derive the distribution of demand for the case in which all goods are at an interior solution.

$$P\left(x_i^* > 0; \ i = 2, \ \dots, m\right)$$
$$= \varphi\left(h_2, \ \dots, h_m \mid 0, \Omega\right) |J|$$

The joint distribution of $(x_2^*, \dots, x_m^*)'$ in (4.4.14) can be evaluated by noting that it can be factored into discrete and continuous parts. In evaluating the likelihood we transform (7) to the product of two factors as follows. By partitioning $v = (v_2, \dots, v_m)'$ into $v_a = (v_2, \dots, v_n)'$ and $v_b = (v_{n+1}, \dots, v_m)'$ such that

$$\begin{bmatrix} v_a \\ v_b \end{bmatrix} \sim MVN \left(\begin{bmatrix} 0 \\ 0 \end{bmatrix}, \ \begin{bmatrix} \Omega_{aa} & \Omega_{ab} \\ \Omega_{ba} & \Omega_{bb} \end{bmatrix} \right)$$

v_a and $v_b|v_a$ are normally distributed, then $v_a \sim MVN(0, \Omega_{aa})$ and $v_b|v_a = h_a \sim MVN$ (μ, Σ) where $\mu = \Omega_{ba}\Omega_{aa}^{-1}h_a$, $\Sigma = \Omega_{bb} - \Omega_{ba}\Omega_{aa}^{-1}\Omega_{ab}$, and $h_a = (h_2, \dots, h_n)'$. Then, (4.4.14) can be rewritten as the product of two factors:

$$P\left(x_i^* > 0 \text{ and } x_j^* = 0; \ i = 2, \ \dots, n \text{ and } j = n+1, \ \dots, m\right)$$
$$= \phi_{v_a}\left(h_2, \ \dots, h_n \mid 0, \Omega_{aa}\right) |J|$$
$$\times \int_{-\infty}^{h_m} \cdots \int_{-\infty}^{h_{n+1}} \phi_{v_b|v_a}\left(v_{n+1}, \ \dots, v_m \mid \mu, \Sigma\right) \ dv_{n+1} \cdots dv_m \qquad (4.4.15)$$

We use the GHK simulator (Chapter 2, Section 10 or Keane [1994], Hajivassiliou et al. [1996]) to evaluate the multivariate normal integral in (4.4.15). In Kim et al. [2002], "Modeling Consumer Demand for Variety," we will apply a heterogeneous version of this model to data on purchase of yogurt varieties.

The additive utility model used does not include any interactions in the utility function. That is, the marginal utility of consumption of good i does not depend on the consumption level of other goods. In particular, additive utility specifications impose the restriction that all goods are substitutes, ruling out complementarity. Gentzkow [2007] includes utility interaction in a choice model that allows for the possibility that two goods are complements.[6] He applies this formulation to purchase data on the print and online versions of newspapers. His results suggest that there are complementarities between the print and online versions of newspapers.

---

[6] We note that with only two goods there can only be substitution in demand. However, Gentzkow includes the usual outside alternative.

5

Hierarchical Models for Heterogeneous Units

Abstract

This chapter provides a comprehensive treatment of hierarchical models. Hierarchical models are designed to measure differences between units using a particular prior structure. Choice of the form of the hierarchical model (i.e. the form of the prior) as well as the MCMC algorithm to conduct inference are important questions. We explore a new class of hybrid MCMC algorithms that are customized or tuned to the posteriors for individual units. We also implement a mixture of normals prior for the distribution of model coefficients across units. We illustrate these methods in the context of a panel of household purchase data and a base or unit-level multinomial logit model. Those interested in the main points without technical details are urged to concentrate on Sections 5.1, 5.2, 5.4, and 5.5.3.

One of the greatest challenges in marketing is to understand the diversity of preferences and sensitivities that exists in the market. Heterogeneity in preferences gives rise to differentiated product offerings, market segments, and market niches. Differing sensitivities are the basis for targeted communication programs and promotions. As consumer preferences and sensitivities become more diverse, it becomes less and less efficient to consider the market in the aggregate. Marketing practices that are designed to respond to consumer differences require an inference method and model capable of producing individual or unit level parameter estimates. Moreover, optimal decision-making requires not only point estimates of unit level parameters but also a characterization of the uncertainty in these estimates. In this chapter, we will show how Bayesian hierarchical approaches are ideal for these problems as it is possible to produce posterior distributions for a large numbers of unit-level parameters.

In contrast to this emphasis on individual differences, economists are often more interested in aggregate effects and regard heterogeneity as a statistical nuisance parameter problem which must be addressed but not emphasized. Econometricians frequently

Bayesian Statistics and Marketing, Second Edition. Peter E. Rossi, Greg M. Allenby, and Sanjog Misra.
© 2024 John Wiley & Sons Ltd. Published 2024 by John Wiley & Sons Ltd.

employ methods which do not allow for the estimation of individual-level parameters. For example, random coefficient models are often implemented through an unconditional likelihood approach in which only hyper-parameters are estimated. Furthermore, the models of heterogeneity considered in the econometrics literature often restrict heterogeneity to subsets of parameters such as model intercepts. In the marketing context, there is no reason to believe that differences should be confined to the intercepts and, as indicated above, differences in slope coefficients are critically important. Finally, economic policy evaluation is frequently based on estimated hyper-parameters which are measured with much greater certainty than individual-level parameters. This is in contrast to marketing policies which often attempt to respond to individual differences that are measured less precisely.

This new literature emphasizing unit-level parameters is made possible by the explosion in the availability of disaggregate data. Scanner data at the household and store level is now commonplace. In the pharmaceutical industry, physician-level prescription data is also available. This raises both modeling challenges as well as major opportunities for improved profitability through decentralized marketing decisions that exploit heterogeneity. This new data comes in panel structure in which N, the number of units is large relative to T, the length of the panel. Thus, we may have a large amount of data obtained by observing a large number of decision units. For a variety of reasons, it is unlikely that we will ever have a very large amount of information about any one decision unit. Data describing consumer preferences and sensitivities to variables such as price are typically obtained through surveys or household purchase histories which yield very limited individual-level information. For example, household purchases in most product categories often total less than 12 per year. Similarly, survey respondents become fatigued and irritable when questioned for more than 20 or 30 minutes. As a result, the amount of data available for drawing inferences about any specific consumer is very small, although there may exist many consumers in a particular study.

The classical fixed-effects approach to heterogeneity has some appeal since it delivers the individual unit-level parameter estimates and does not require the specification of any particular probability distribution of heterogeneity. However, the sparseness of individual-level data renders this approach impractical. In many situations, incomplete household level data causes a lack of identification at the unit level. In other cases, the parameters are identified in the unit-level likelihood but the fixed effects estimates are measured with huge uncertainty which is difficult to quantify using standard asymptotic methods.

From a Bayesian perspective, modeling panel data is about the choice of a prior over a high dimensional parameter space. The hierarchical approach is one convenient way of specifying the joint prior over unit-level parameters. Clearly, this prior will be informative and must be in order to produce reasonable inferences. However, it is reasonable to ask for flexibility in the form of this prior distribution. In this chapter, we will introduce hierarchical models for general unit level models and apply these ideas to a hierarchical logit setting. Recognizing the need for flexibility in the prior, we will expand the set of priors to include mixtures of normal distributions.

5.1 HETEROGENEITY AND PRIORS

A useful general structure for disaggregate data is a panel structure in which the units are regarded as independent conditional on unit level parameters (see Yang and Allenby

[2003]), "Modeling Interdependent Consumer Preferences," for an example which relaxes this assumption). Given a joint prior on the collection of unit level parameters, the posterior distribution can be written as follows:

$$p\left(\theta_1, \dots, \theta_m | y_1, \dots, y_m\right) \propto \left[\prod_i p\left(y_i | \theta_i\right)\right] \times p\left(\theta_1, \dots, \theta_m | \tau\right) \qquad (5.1.1)$$

The term in brackets is the conditional likelihood and the rightmost term is the joint prior with hyperparameter, τ. Note here we generically denote the data for the ith unit as y_i. In many instances, the amount of information available for many of the units is small. This means that the specification of the functional form and hyperparameter for the prior may be important in determining the inferences made for any one unit. A good example of this can be found in choice data sets in which consumers are observed to be choosing from a set of products. Many consumers ("units") do not choose all of the alternatives available during the course of observation. In this situation, most standard choice models don't have a bounded maximum likelihood estimate (the likelihood will asymptote in a certain direction in the parameter space). For these consumers, the prior is, in large part, determining the inferences made.

Assessment of the joint prior for $\left(\theta_1, \dots, \theta_m\right)$ is difficult due to the high dimension of the parameter space and, therefore, some sort of simplification of the form of the prior is required. One frequently employed simplification is to assume that, conditional on the hyperparameter, $\left(\theta_1, \dots, \theta_m\right)$ are a priori independent.

$$p\left(\theta_1, \dots, \theta_m | y_1, \dots, y_m\right) \propto \prod_i p\left(y_i | \theta_i\right) p\left(\theta_i | \tau\right) \qquad (5.1.2)$$

This means that inference for each unit can be conducted independently of all other units *conditional* on τ. This is the Bayesian analogue of fixed effects approaches in classical statistics.

The specification of the conditionally independent prior can be very important due to the scarcity of data for many of the units. Both the form of the prior and the values of the hyperparameters are important and can have pronounced effects on the unit-level inferences. For example, it is common to specify a normal prior, $\theta_i \sim N\left(\bar{\theta}, V_\theta\right)$. The normal form of this prior means that influence of the likelihood for each unit may be attenuated for likelihoods centered far away from the prior. That is, the thin tails of the normal distribution diminish the influence of outlying observations. In this sense, the specification of a normal form for the prior, whatever the values of the hyper-parameters, is far from innocuous.

Assessment of the prior hyperparameters can also be challenging in any applied situation. For the case of the normal prior, some relatively diffuse prior may be a reasonable default choice. Allenby and Rossi [1993] use a prior based on a scaled version of the pooled model information matrix. The prior covariance is scaled back to represent the expected information in one observation to insure a relatively diffuse prior. Use of this sort of normal prior will induce a phenomenon of "shrinkage" in which the Bayes estimates (posterior means) $\{\tilde{\theta}_i = E\left[\theta_i | data_i, prior\right]\}$ will be clustered more closely to the prior mean than the unit-level maximum likelihood estimates $\{\hat{\theta}_i\}$. For diffuse prior settings, the normal form of the prior will be responsible for the shrinkage effects. In particular, outliers will be "shrunk" dramatically toward the prior mean. For many applications, this is a very desirable feature of the normal form prior. We will "shrink" the outliers in toward the rest of the parameter estimates and leave the rest pretty much alone.

5.2 HIERARCHICAL MODELS

In general, however, it may be desirable to have the amount of shrinkage induced by the priors driven by information in the data. That is, we should "adapt" the level of shrinkage to the information in the data regarding the dispersion in $\{\theta_i\}$. If, for example, we observe that the $\{\theta_i\}$ are tightly distributed about some location or that there is very little information in each unit level likelihood, then we might want to increase the tightness of the prior so that the shrinkage effects are larger. This feature of "adaptive shrinkage" was the original motivation for work by Efron and Morris [1975] and others on empirical Bayes approaches in which prior parameters were estimated. These empirical Bayes approaches are an approximation to a full Bayes approach in which we specify a second stage prior on the hyper-parameters of the conditional independent prior. This specification is called a Hierarchical Bayes Model and consists of the unit level likelihood and two stages of priors:

Likelihood: $p\left(y_i|\theta_i\right)$
First-stage prior: $p\left(\theta_i|\tau\right)$
Second-stage prior: $p\left(\tau|h\right)$

The joint posterior for the hierarchical model is given by

$$p\left(\theta_1,\ldots,\theta_m,\tau|y_1,\ldots,y_m,h\right) \propto \left[\prod_i p\left(y_i|\theta_i\right)p\left(\theta_i|\tau\right)\right] \times p\left(\tau|h\right).$$

In the hierarchical model, the prior induced on the unit level parameters is not an independent prior. The unit level parameters are conditionally, but not unconditionally, a priori independent.

$$p\left(\theta_1,\ldots,\theta_m|h\right) = \int \prod_i p\left(\theta_i|\tau\right)p\left(\tau|h\right)d\tau$$

If, for example, the second-stage prior on τ is very diffuse, the marginal priors on the unit-level parameters, θ_i, will be highly dependent as each parameter has a large common component. Improper priors on the hyperparameters are extremely dangerous not only because of their extreme implications for some marginals of interest as we have seen in Chapter 4 but also because the posterior may not be proper. As Robert and Casella [2004] point out, it is possible to define an MCMC method for a hierarchical model which does not have any posterior as its invariant distribution for the case of improper priors.

The first-stage prior (or random effect distribution) is often taken to be a normal prior. Obviously, the normal distribution is a flexible distribution with easily interpretable parameters. In addition, we can increase the flexibility of this distribution using a mixture of normals approach as outlined in Section 5.5. We can easily incorporate observable features of each unit by using a Multivariate Regression specification.

$$\theta_i = \Delta' z_i + u_i \quad u_i \sim N\left(0, V_\theta\right)$$

or   (5.2.1)

$$\Theta = Z\Delta + U$$

Θ is an $m \times k$ matrix whose rows contain each of the unit level parameter vectors. Z is an $m \times n_z$ matrix of observations on the n_z covariates which describe differences between units with row z_i. $\theta_i \sim N\left(\bar{\theta}, V_\theta\right)$ is a special case of (5.2.1) where Z is a vector of ones with length equal to the number of units. Given the Θ array, draws of Δ and V_θ can be accomplished using either a Gibbs sampler or direct draws for the Multivariate Regression model as outlined in Section 2.12 and implemented in the *bayesm* function, **R** rmultireg.

The hierarchical model specifies that both prior and sample information will be used to make inferences about the common parameter, τ. For example, in normal prior, $\theta_i \sim N\left(\bar{\theta}, V_\theta\right)$, the common parameters provide the location and the spread of the distribution of θ_i. Thus, the posterior for the θ_i will reflect a level of shrinkage inferred from the data. It is important to remember, however, that the normal functional form will induce a great deal of shrinkage for outlying units even if the posterior of V_θ is centered on large values.

In classical approaches to these models, the first-stage prior is called a random effects model and is considered part of the likelihood. The random effects model is used to average the conditional likelihood to produce an unconditional likelihood which is a function of the common parameters alone.

$$\ell(\tau) = \prod_i \int p\left(y_i | \theta_i\right) p\left(\theta_i | \tau\right) d\theta_i$$

In the classic econometric literature, much is made of the distinction between random coefficient models and fixed effect models. Fixed effect models are considered "non-parametric" in the sense that there is no specified distribution for the θ_i parameters.[1] Random coefficient models are often considered more efficient but subject to specification error in the assumed random effects distribution, $p\left(\theta_i | \tau\right)$. In a Bayesian treatment, we see that the distinction between these two approaches is in the formulation of the joint prior on $\{\theta_1, \ldots, \theta_m\}$. A Bayesian "fixed effects" approach specifies independent priors over each of the unit level parameters while the "random effects" approach specifies a highly dependent joint prior.

The use of a hierarchical model for prediction also highlights the distinction between various priors. A hierarchical model assumes that each unit is a draw from a "super-population" or that the units are exchangeable (conditional, perhaps, on some vector of co-variates). This means that if we want to make a prediction regarding a new unit we can regard this new unit as drawn from the same population. Without

---

[1] Classical inference for fixed effects models faces a fundamental conundrum: more time series observations are required for application of asymptotic theory (which is needed for non-linear models). However, we invariably have a short panel. Various experiments in which both the number of cross-sectional units and the time dimension increase are unpersuasive. While we might accept asymptotics that allow only N to increase to infinity, we are unlikely ever to see T increase as well. But most importantly, we avoid this altogether in the Bayesian approach.

the hierarchical structure, all we know is that this new unit is different and have little guidance as to how to proceed.

5.3 INFERENCE FOR HIERARCHICAL MODELS

Hierarchical models for panel data structures are ideally suited for MCMC methods. In particular, a "Gibbs" style Markov chain can often be constructed by considering the basic two sets of conditionals:

$$(1) \; \theta_i | \tau, y_i$$

and

$$(2) \; \tau | \{\theta_i\}$$

The first set of conditionals exploit the fact that the θ_i are conditionally independent. The second set exploit the fact that $\{\theta_i\}$ are sufficient for τ. That is, once the $\{\theta_i\}$ are drawn from (1), these serve as "data" to the inferences regarding τ. If, for example, the first stage prior is normal, then standard natural conjugate priors can be used, and all draws can be done one-for-one and in logical blocks. This normal prior model is also the building block for other more complicated priors. The normal model is given by

$$\theta_i \sim N\left(\bar{\theta}, V_\theta\right)$$

$$\bar{\theta} \sim N\left(\bar{\bar{\theta}}, A^{-1}\right)$$

$$V_\theta \sim IW\left(v, V\right)$$

In the normal model, the $\{\theta_i\}$ drawn from (1) are treated as a multivariate normal sample and standard conditionally conjugate priors are used. It is worth noting that in many applications the second stage priors are set to be very diffuse ($A^{-1} = 100\,I$ or larger) and the Wishart is set to have expectation I with very small degrees of freedom such as $\dim(\theta) + 3$. As we often have a larger number of units in the analysis, the data seems to overwhelm these priors and we learn a great deal about τ, or in the case of the normal prior, $\left(\bar{\theta}, V_\theta\right)$.

Drawing the $\{\theta_i\}$ given the unit level data and τ is dependent on the unit level model. For linear models, as illustrated in Chapter 3, we can implement a Gibbs sampler by making direct draws from the appropriate conjugate distributions. However, in most marketing applications, there is no convenient conjugate prior or a convenient way of sampling from the conditional posteriors. For this reason, most rely on some sort of Metropolis algorithm to draw θ_i. As discussed in Chapter 3, there are two very useful variants of the Metropolis algorithm – independence and random walk. Both could be used to develop a general purpose drawing method for hierarchical models. In either case, the candidate draws require a candidate sampling density (as in the case of the independence Metropolis) or an increment density in the case of the RW Metropolis. The performance of these algorithms will depend critically on the selection of these densities.

In both the independence and random walk cases, the densities should be selected to capture the curvature and tail behavior of the conditional posterior

$$p\left(\theta_{i}|y_{i},\tau\right) \propto p\left(y_{i}|\theta_{i}\right) p\left(\theta_{i}|\tau\right) \tag{5.3.1}$$

This suggests that the Metropolis algorithm used to draw each θ_i should be customized for each cross-sectional unit. In the Metropolis literature, there is also a tradition of experimentation with the scaling of the covariance matrix of either the random walk increments or the independence chain candidate density. Clearly, it is not practical to experiment with scaling factors which are customized to each individual unit. In order to develop a practical Metropolis algorithm for hierarchical models, we must provide a method of customization to the unit level which does not require experimentation.

A further complication for a practical implementation is that individual level likelihoods may not have a maximum. For example, suppose the unit level model is a logit model and the unit does not choose all alternatives in the sample. Then there is no maximum likelihood estimator for this unit if we include intercepts for each choice alternative. The unit-level likelihood is increasing in any direction which moves the intercepts for alternatives never chosen to $-\infty$. Most common proposals for Metropolis sampling densities involve use of maximum likelihood estimators. One could argue that the prior in (5.3.1) avoids this problem. Proper priors as well as normal tails will usually insure that a maximum exists (Allenby and Rossi [1993] suggests an approximate Bayes estimator which uses this posterior mode). However, using the posterior mode (and associated Hessian) in a Metropolis step would require computation of the mode and Hessian at every MCMC step and for each cross-sectional unit.[2] This could render the Metropolis algorithm computationally impractical.

There is a folk literature on the choices of Metropolis proposal densities for hierarchical models. Some advocate using the same proposal for all units and base this proposal on the asymptotic normal approximation to the pooled likelihood. Obviously, the pooled likelihood is a mixture of unit level likelihoods so that it is possible that this proposal (even if scaled for the relative number of unit-level and total observations) has a location and curvature that is different from any single unit level likelihood. Another popular proposal is to use the current draw of the prior as the basis of a random walk chain. That is, if the first stage prior is normal, we use the current draw of the variance of the θ as the variance of the proposal distribution. This is clearly not a good idea as it does not adapt to the information content available for a specific unit. If all units have very little information, then the prior will dominate and this idea may work acceptably. However, there are units with a moderate to large amount of information, an RW chain will exhibit very high autocorrelation due to rejected draws which come from a prior which is much less tight than the unit level likelihood.

We propose a class of Metropolis algorithms which use candidate sampling distributions which are customized to the unit level likelihoods but are computationally practical in the sense that they do not require order R (the number of MCMC draws) optimizations but only require an initial set of optimizations. These candidate sampling distributions can either be used as the basis of an independence or RW Metropolis chain. In addition, our proposal does not require that each unit-level likelihood have a

---

[2] Note that the parameters of the prior in (5.3.1) will vary from MCMC step to MCMC step.

maximum. To handle the problem of non-existence of maxima, we use a "fractional" likelihood approach in which we modify the individual level likelihood (but only for the purpose of a Metropolis proposal density by multiplying by a likelihood with a defined maximum.

$$\ell_i^*(\theta) = \ell_i(\theta)^{(1-w)}\,\overline{\ell}(\theta)^{w\beta} \qquad (5.3.2)$$

$\overline{\ell}$ can be the pooled likelihood which almost certainly has a maximum. The β weight is designed to scale the pooled likelihood to the appropriate order so that it does not dominate the unit-level likelihood.

$$\beta = \frac{n_i}{N} \qquad (5.3.3)$$

n_i is the number of observations for the ith unit and N is the total number of observations in all units. w is a tuning constant which represents the weight of the scaled pooled likelihood relative to the individual likelihood. We only bring in the pooled likelihood for the purpose of "regularizing" the problem so we would typically set w to a small value such as 0.1. The pseudo-likelihood[3] in (5.3.2) can be maximized to obtain a location and scale, $\widehat{\theta}_i$ and H_i. These quantities can then be combined with the prior to form a Metropolis proposal distribution. In many cases, the prior will be in a normal form so that we can combine the prior and normal approximation to the pseudo unit level likelihood using standard theory for the Bayes linear model. This provides us with a proposal that is customized to the curvature and possible location of each unit-level likelihood. That is, if the prior is $N\left(\overline{\theta}, V_\theta\right)$, then our proposal will be based on a normal density with moments

$$\text{mean: } \mu_i^* = \left(H_i + V_\theta^{-1}\right)^{-1}\left(H_i\widehat{\theta}_i + V_\theta^{-1}\overline{\theta}\right)$$
$$\text{variance: } V_i^* = \left(H_i + V_\theta^{-1}\right)^{-1} \quad H_i = -\left.\frac{\partial^2\log\left(\ell_i^*\right)}{\partial\theta\partial\theta'}\right|_{\theta=\widehat{\theta}_i} \qquad (5.3.4)$$

For an independence Metropolis, we will use both the customized location as well as the curvature estimate for each unit. We note that these will be updated from draw to draw of the prior τ parameters as the chain progresses. However, each update will only use the current draw of τ and the proposal location and scale parameters. An RW chain will use only the scale parameters combined with τ. We will adopt the scaling proposal of Roberts and Rosenthal [2001] and set scaling to $2.93/\sqrt{\dim(\theta)}$. This provides us with two Metropolis algorithms which are automatically tuned. The independence chain might be regarded as somewhat higher risk than the RW chain as we require that both the curvature and location obtained by the approximate pseudo-likelihood procedure be correct. If the location is off, the Independence chain can be highly autocorrelated as it rejects "out-of-place" candidates. The RW chain will adapt to the location of the unit

---

[3] Computation of the pseudo likelihood estimates need only be performed once prior to initiation of the Metropolis algorithm. It should also be noted that for models without lower dimensional sufficient statistics, the evaluation of the pseudo-likelihood in (5.3.2) requires evaluation of the pooled likelihood as well. To reduce this computational burden we can use the asymptotic normal approximation to the pooled likelihood. As we are using this likelihood only for the purpose of "regularizing" our unit level likelihood, the quality of this approximation is not crucial.

level likelihoods but this could be at the price of higher autocorrelation. Thus, it is the risk-averse alternative.

5.4 A HIERARCHICAL MULTINOMIAL LOGIT EXAMPLE

To examine the performance of various proposed chains, we consider first the case of a hierarchical logit model. Each of the units is assumed to have an MNL likelihood and we specify a normal distribution of the logit parameters over units with mean $\Delta' z_i$ as in (5.2.1). The hierarchical logit model takes the form:

$$
\begin{aligned}
&\ell\left(\beta_i | y_i, X_i\right) \ [MNL], \\
&B = Z\Delta + U \quad u_i \sim N\left(0, V_\beta\right), \\
&vec\left(\Delta | V_\beta\right) \sim N\left(vec\left(\overline{\Delta}\right), V_\beta \otimes A^{-1}\right), \\
&V_\beta \sim IW\left(v, V\right).
\end{aligned}
\tag{5.4.1}
$$

u_i and β_i are the ith rows of B and U. The DAG for the model in (5.4.1) is given by

$$
\begin{array}{c}
V_\beta \searrow \\
\downarrow \quad \beta_i \rightarrow y_i \\
\Delta \nearrow
\end{array}
\tag{5.4.2}
$$

Given a draw of B, draws of Δ, V_β can be made using standard conjugate theory for the MRM. We can define three possible chains for drawing the β_i:

1. An Independence Metropolis with draws from a multivariate student t with location and scale given by (5.3.4). Note that both the location and scale will be influenced by the current draw of both Δ and V_β. Candidates will be drawn from an multivariate student t with mean $\beta^* = \left(H_i + \left(V_\beta^r\right)^{-1}\right)^{-1}\left(H_i\hat{\beta}_i + \left(V_\beta^r\right)^{-1}(\Delta^r)'z_i\right)$ and covariance proportional to $\left(H_i + \left(V_\beta^r\right)^{-1}\right)^{-1}$.

2. An RW Metropolis with increments having covariance $s^2 V_\beta^r$ where s is a scaling constant and V_β^r is the current draw of V_β.

3. An improved RW Metropolis with increments having covariance $s^2(H_i + (V_\beta^r)^{-1})^{-1}$ where H_i is the Hessian of the ith unit likelihood evaluated at the MLE for the fraction likelihood defined by multiplying the MNL unit likelihood by the pooled likelihood raised to the β power.

We will choose an "automatic" tuning scheme in which we set $s = 2.93 / \sqrt{\dim(\beta_i)}$.

We might expect the chain defined by (1) to perform well if our location estimates for the posteriors of each MNL given the current draw of Δ and V_β and are good following the intuition developed for the single logit model in Chapter 3. However, we must recognize that the normal approximation to the logit likelihood may break down for likelihoods with no defined maximum as we have if a unit does not choose from all alternatives available. The independence Metropolis chain will not adapt to the proper location unlike the RW chains.

The RW chains offer adaptability in location at the expense of possibly slower navigation. The RW chain defined by (2) which simply uses V_β for the covariance of increments is not expected to perform well for cases in which some units have a good deal of information and others very little. For units with little information, the unit-level conditional posteriors are dominated by the prior term (the unit likelihood is relatively flat) and the RW defined by (2) may have increments of approximately the right scale. However, if a unit has a more sharply defined likelihood, the increments proposed by the chain in (2) will be too large on average. This could create high autocorrelation due to the rejection of candidates and the consequent "stickiness" of the chain. The RW sampler defined by (3) does not suffer from this problem.

To investigate the properties of these three chains, we simulated data from a five choice hierarchical logit model with four intercepts and one X variable drawn as $unif(-1.5, 0)$ which is meant to approximate a log-price variable with a good deal of variation. 100 units were created: 50 with only five observations and 50 with 50 observations. $\beta_i \sim N(\mu, V_\beta)$ $\mu' = (1, -1, 0, 0, -3)$; V_β has diagonal elements all equal to 3 and with the [4,5] and [5,4] elements set to 1.5. Diffuse priors were used, A = 0.01, $v = 5 + 3$, $V = vI_5$.

Figure 5.1 shows draw sequences (every 20th draw) from the Independence chain (1) and the improved RW chain (3) for the [5,5] element of V_β. The dark horizontal line is the "true" parameter value. The independence chain takes an extraordinary number of draws to dissipate the initial conditions. It appears to take at least 15,000 draws to reach the stationary distribution. On the other hand, the improved RW chain mixes well and dissipates the initial condition in fewer than 500 iterations. Figure 5.2 compares the two RW chains for a unit with 50 observations. The RW chain with increments based on V_β

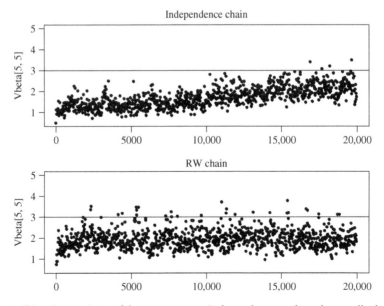

Figure 5.1 Comparison of draw sequences: independence and random-walk chains

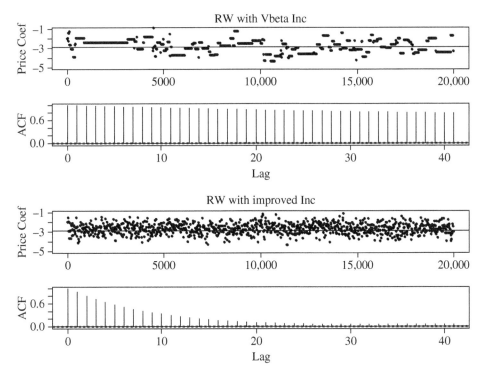

Figure 5.2 Comparison of two RW chains: draws of unit-level parameters

alone exhibits very poor mixing. Long runs of repeated values are shown in the figure as the chain rejects draws too far away from the mass of the posterior. The improved RW chain shows much better mixing. The numerical efficiency measure (see Section 3.9) for the improved RW chain is 3.83 vs 8.66 for the RW chain proposed in ii). For a unit with a small number of observations, both chains have comparable numerical efficiency (5 for the improved chain and 5.94 for the chain proposed in (2)). The improved RW chain is **R** implemented in the `bayesm` function, `rhierMnlRwMixture`.

We also consider an example using scanner panel data on purchases of margarine. This data set was originally analyzed in Allenby and Rossi [1991] and contains purchases on 10 brands of margarine and some 500 panelists. This dataset is available in `bayesm` and **R** can be loaded with the command, `data(margarine)`. Several of the brands have only very small share of purchases and, thus, for the purposes of illustration, we consider a subset of data on purchases of six brands: 1. Parkay stick, 2. Blue Bonnett Stick, 3. Fleischmanns stick, 4. House brand stick, 5. Generic stick, and 6. Shed Spread Tub. We also restricted attention to those households with five or more purchases. This gives us a data set with 313 households making a total of 3405 purchases. We have brand-specific intercepts and a log price variable in the hierarchy for a total of six unit-level logit coefficients. We also have information on various demographic characteristics of each household including household income and family size which form the Z matrix of observable determinants in the hierarchy. We use "standard" diffuse prior settings of $A = 0.01I$, $v = 6 + 3$, $V = vI$ and run the improved RW chain for 20,000 iterations. Figure 5.3 shows the posterior distributions of the price coefficient for selected households. In the

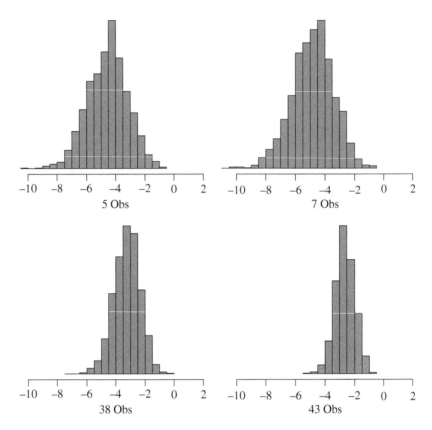

Figure 5.3   Posterior distribution of price coefficients: Selected households with small and large number of observations

top row, the posterior is displayed for two households with a relatively small amount of information. It should be emphasized that these households do not have defined maxima for their unit-level likelihoods. This does not mean that we can't learn something about their price sensitivity. The household level data plus the first-stage prior provide some limited information. In the bottom row of Figure 5.3, we display marginal posteriors for households with a larger number of observations. As might be expected, the posteriors sharpen up considerably.

Figure 5.4 shows the marginal posteriors for various functions of V_β. The top histogram shows the marginal posterior of the correlation between the house and generic brand intercepts. This is centered tightly over rather large values, suggesting that household preferences for house and generic brands are highly correlated as has been suggested in the literature on private label brands. In the bottom panel of Figure 5.4, the posterior distribution of the standard deviation of the price coefficient is displayed. We note that both quantities are nonlinear functions of V_β; it is inappropriate to apply these functions to the posterior mean, $E\left[V_\beta | data\right]$. Both posterior distributions exhibit substantial skewness and show, yet again, that asymptotic normal approximations to the posterior distribution of key parameters can be poor.

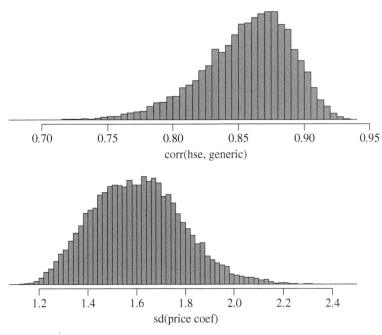

Figure 5.4   Posterior distribution of selected variance parameters

Table 5.1   Correlations and standard deviations of betas

Blue Bonnett	1.53	0.39	0.43	0.46	0.27	−0.07
	(0.13)	(0.13)	(0.10)	(0.10)	(0.13)	(0.14)
Fleischmanns		3.44	0.31	0.28	0.09	0.48
		(0.55)	(0.15)	(0.18)	(0.17)	(0.15)
House			2.5	0.86	0.49	−0.05
			(0.19)	(0.03)	(0.10)	(0.14)
Generic				3.0	0.55	−0.08
				(0.27)	(0.10)	(0.14)
Shed Spread Tub					3.0	0.05
					(0.33)	(0.15)
Price						1.6
						(0.18)

Diagonal contains standard deviations; off-diagonal the correlations.

Table 5.1 shows the posterior means (standard deviations) of all correlations in the off-diagonal and the standard deviations of each β on the diagonal. The posterior standard deviations of the households βs are very large. This shows tremendous heterogeneity between households in brand preference and price sensitivity. Table 5.2 shows that very little of this measured heterogeneity can be attributed to the household demographic attributes, log-income, and family size. Most of the elements of Δ displayed in Table 5.2 are very imprecisely measured, particularly for the effects of income. Larger families show some preference toward the house and generic brands and shy away from

Table 5.2   Posterior distribution of Δ

	Blue Bonnet Intercept	Fleischmanns intercept	House intercept	Generic intercept	Shed Spread intercept	log(price)
Intercept	−1.27	−3.37	−3.31	−4.96	0.03	−3.48
	(0.64)	(1.8)	(0.99)	(1.2)	(1.2)	(0.85)
log(Income)	0.07	0.80	0.02	−0.51	−0.62	−0.26
	(0.21)	(0.59)	(0.32)	(0.40)	(0.42)	(0.28)
Family size	−0.03	−0.70	0.24	0.55	0.06	0.08
	(0.10)	(0.28)	(0.14)	(0.18)	(0.20)	(0.12)

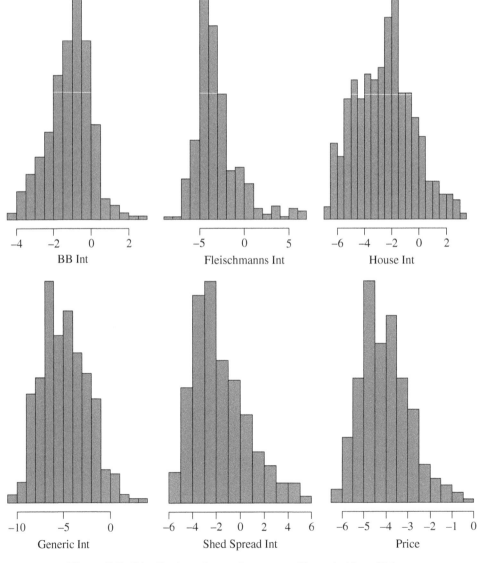

Figure 5.5   Distribution of posterior means of household coefficients

Fleischmanns. However, the general impression is of a weak relationship with these demographic variables.

Figure 5.5 displays the distributions of posterior means of coefficients across the 313 households. These distributions exhibit a good deal of skewness. In Section 5.7, we develop a diagnostic for our normal model of household heterogeneity which is based on comparing these distributions to the predictive distribution from our model. The predictive distribution will not be normal as we will integrate out the parameters of the first stage prior. However, for the settings of the hyper-parameters in this data analysis, the predictive distribution will be symmetric, albeit fatter tailed than the normal. This informal evidence suggests that the normal first stage prior may not be adequate. In the next section, we will allow for a more flexible family of priors based on mixtures of normals.

5.5 USING MIXTURES OF NORMALS

Much of the work in both marketing and in the general statistics literature has used the normal prior for the first-stage of the hierarchical model. The normal prior offers a great deal of flexibility and fits conveniently with a large Bayesian regression/multivariate analysis literature. The standard normal model can easily handle analysis of many units (Steenburgh et al. [2003]), and can include observable determinants of heterogeneity (see Allenby and Ginter [1995]; Rossi et al. [1996]). Typically, we might postulate that various demographic or market characteristics might explain differences in intercepts (brand preference) or slopes (marketing mix sensitivities). In linear models, this normal prior specification amounts to specifying a set of interactions between the explanatory variables in the model explaining y (see McCulloch and Rossi [1994] for further discussion of this point).

While the normal model is flexible, there are several drawbacks for marketing applications. As discussed above, the thin tails of the normal model tend to shrink outlying units greatly toward the center of the data. While this may be desirable in many applications, it is a drawback in discovering new structure in the data. For example, if the distribution of the unit-level parameters is bi-modal (something to be expected in models with brand intercepts) then a normal first-stage prior may shrink the unit-level estimates to such a degree as to mask the multi-modality (see below for further discussions of diagnostics). Fortunately, the normal model provides a building block for a mixture of normals extension of the first-stage prior. Mixtures of normal models provide a great deal of added flexibility. In particular, multiple modes are possible. Fatter tails than the normal can also be accommodated by mixing in normal components with large variance. It is well-known that the mixture of normals model provides a great deal of flexibility and that with enough components, virtually any multivariate density can be approximated. That is to say, we can "build-up" any distribution, no matter how non-normal, using many small normal components in much the same manner as it is possible to build a mountain will small piles of gravel. However, as a practical matter, we may not be able to identify significant deviations from a normal model of heterogeneity as we only observe the unit level parameters with considerable error. Intuition developed by direct application of the mixture of normals approach to estimation of densities for directly observed data may not carry over well to the use of mixture of normals in a hierarchical setting.

The mixture of normals model can also be viewed as a generalization of the popular finite mixture model. The finite mixture model views the prior as a discrete distribution with a set of mass points. This approach has been very popular in marketing due to the interpretation of each mixture point as representing a "segment" and to the ease of estimation. In addition, the finite mixture approach can be given the interpretation of a non-parametric method as in Heckman and Singer [1984]. Critics of the finite mixture approach have pointed to the implausibility of the existence of a small number of homogeneous segments as well as the fact that the finite mixture approach does not allow for extreme units whose parameters lie outside the convex hull of the support points. The mixture of normals approach avoids the drawbacks of the finite mixture model while incorporating many of the more desirable features.

The mixture of K multivariate normal models can be used as the basis of the heterogeneity distribution as follows:

$$\theta_i = \Delta' z_i + u_i$$

$$u_i \sim N\left(\mu_{ind_i}, \Sigma_{ind_i}\right) \tag{5.5.1}$$

$$ind_i \sim multinomial_K(pvec)$$

ind_i is an indicator latent variable for which component observation i is from. ind takes on values $1, \dots, K$. pvec is a vector of mixture probabilities of length K. In (5.5.1), the z vector of observable characteristics of the population does not include an intercept and has n_z elements. For this reason, we advise that z be centered so that the mean of θ given average z values will be entirely determined by the normal mixture component means. The moments of are θ given below[4]

$$E\left[\theta_i \middle| z_i = \overline{z}, p, \{\mu_k\}\right] = \overline{\mu} = \sum_{k=1}^{K} pvec_k \mu_k$$

$$Var\left(\theta_i \middle| z_i, p, \{\mu_k\}, \{\Sigma_k\}\right) = \sum_{k=1}^{K} pvec_k \Sigma_k + \sum_{k=1}^{K} pvec_k \left(\mu_k - \overline{\mu}\right)\left(\mu_k - \overline{\mu}\right)' \tag{5.5.2}$$

Of course, the variance loses much of its meaning and interpretability as we move farther from an elliptically symmetric distribution.

As in Section 3.9, priors for the mixture of normals model can be chosen in convenient conditionally conjugate forms.

$$vec(\Delta) = \delta \sim N\left(\overline{\delta}, A_\delta^{-1}\right)$$

$$pvec \sim Dirichlet(\alpha)$$

$$\mu_k \sim N\left(\overline{\mu}, \Sigma_k \otimes a_\mu^{-1}\right) \tag{5.5.3}$$

$$\Sigma_k \sim IW(v, V)$$

---

[4] The variance can be derived by using the identity $Var(\theta_i) = E\left[Var(\theta_i|ind)\right] + Var\left(E[\theta_i|ind]\right)$.

The DAG for this model can be written as

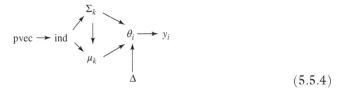

$$(5.5.4)$$

The K parameters in α determine the tightness of the prior on the mixture component probabilities as they are from a natural conjugate prior in which they can be interpreted as cell counts for the components from a previous sample of size, $n^* = \sum_k \alpha_k$. The priors on the mixture components are iid across components in (5.5.3). Much of the statistics literature on mixtures of normals has considered only univariate mixtures of normals or mixture of normals for low dimensional multivariate distributions. Unless there is only a small amount of data or a very large number of components, the priors for the mixture component parameters may not be very influential in this setting and, thus, may not require too careful consideration. However, in the case of hierarchical models and marketing applications, the mixture model may be applied to parameter vectors of relatively high dimension (such as Multinomial choice model parameters) and the priors will matter as the dimension of parameter space of the normal components may easily exceed 200 or 300.

5.5.1 A Hybrid Sampler

We can easily define an MCMC chain of a Gibbs style by alternating between the draws of individual unit-level parameters and mixture components.

$$\theta_i | ind_i, \Delta' z_i, \mu_{ind_i}, \Sigma_{ind_i} \qquad (5.5.5)$$

$$pvec, ind, \Delta, \{\mu_k\}, \{\Sigma_k\} | \{\Theta\}$$

$$\Theta = \begin{bmatrix} \theta'_1 \\ \vdots \\ \theta'_m \end{bmatrix}; \quad ind = \begin{bmatrix} ind_1 \\ \vdots \\ ind_m \end{bmatrix} \qquad (5.5.6)$$

Some advocate margining out the indicators of the mixture components and using a direct Metropolis step for (5.5.6). The argument here is that removal of these latent variables may improve the mixing of the chain. The likelihood function for the mixture of normals can be evaluated at very low cost. This is certainly possible for mixture of univariate normals in which one could stack up the means and log variances into a vector to be drawn either piecemeal or in one shot by a RW Metropolis. However, in the case of mixtures of multivariate normals, this would require using a Metropolis algorithm to navigate in a very high dimensional space of positive definite matrices. Experience with Metropolis algorithms for covariance structures has shown they are very difficult to tune for satisfactory performance for five and higher dimensions. For a mixture of normals, we can easily require a parameter space with as many as 10 covariance matrices, each one of which might have 20 or more parameters.

The draw of the hierarchical parameters in (5.5.6) can be broken down into a succession of conditional draws (note the prior parameters are suppressed in (5.5.7) to focus discussion on the nature of the draws).

$$
\begin{aligned}
& ind|pvec, Z, \Delta, \{\mu_k, \Sigma_k\}, \Theta \\
& pvec|ind \\
& \{\mu_k, \Sigma_k\}|ind, \Theta \\
& \Delta|ind, Z, \{\mu_k, \Sigma_k\}, \Theta
\end{aligned}
\tag{5.5.7}
$$

Here we view the $\{\theta_i\}$ or Θ as the "data" generated by a mixture of normals with mean driven by a multivariate regression with explanatory variables in the $m \times n_z$ matrix Z. First, we draw the indicators for each component, which provides a classification of the "observations" into one of each of the K components. Given the indicators, there are essentially K independent multivariate normal samples on which conjugate draws can be performed to update the $\{\mu_k, \Sigma_k\}$ parameters. We also must update our views on the mixture probabilities. Since the observable variables in Z affect the means of all components, the draw of Δ must be done by pooling across all observations, adjusting for heteroskedasticity.

The role of the z variables is to shift the mean of the normal mixture on the basis of observations. All of the normal mixture parameters should, therefore, be drawn on the "data" with this component of the mean removed.

$$
\Theta^* = \Theta - Z\Delta
\tag{5.5.8}
$$

As in 3.9, the draw of the indicators is a multinomial draw based on the likelihood ratios with p as the prior probability of membership in each component.

$$
ind_i \sim multinomial\,(\pi_i)\,;\; \pi_i' = (\pi_{i,1}, \dots, \pi_{i,K})
$$
$$
\pi_{i,k} = \frac{pvec_k\varphi\,(\theta_i^* | \mu_k, \Sigma_k)}{\sum_m pvec_m\varphi\,(\theta_i^* | \mu_k, \Sigma_k)}
\tag{5.5.9}
$$

Here $\varphi\,(\bullet)$ is the multivariate normal density.

The draw of $pvec$ given the indicators is a Dirichlet draw

$$
\begin{aligned}
& pvec \sim \text{Dirichlet}\,(\tilde{\alpha}) \\
& \tilde{\alpha}_k = n_k + \alpha_k \\
& n_k = \textstyle\sum_{i=1}^n I\,(ind_i = k)
\end{aligned}
\tag{5.5.10}
$$

The draw of each (μ_k, Σ_k) can be made using the algorithm to draw from the Multivariate regression model as detailed in Section 2.8.5. For each subgroup of observations, we have an MRM model of the form

$$
\Theta_k^* = \iota\mu_k' + U;\quad U = \begin{bmatrix} u_1' \\ \vdots \\ u_{n_k}' \end{bmatrix};\; u_i \sim N\,(0, \Sigma_k)
\tag{5.5.11}
$$

Here Θ_k^* is the submatrix of Θ^* that consists of the n_k rows where $ind_i = k$. We can use
R our function, rmultireg in *bayesm*, to achieve these draws.

The draw of Δ requires that we pool data from all K components into one regression
model. Since we are proceeding conditional on the component means and variances, we
can appropriately standardize the "data" and perform one draw from a standard Bayesian
regression model. To motivate the final draw result, let us first consider the kth component. We subset both the Θ and Z matrices to consider only those observations from
the kth component and subtract off the mean. Let Θ_k, Z_k be $n_k \times$ nvar and $n_k \times n_z$ arrays
consisting of only those observations for which $ind_i = k$. nvar is the dimension of the
parameter vectors $\{\theta_i\}$.

$$Y_k = \Theta_k - \iota \mu_k \tag{5.5.12}$$

We can write the model for these observations in the form

$$Y_k = Z_k \Delta + U_k \quad or \quad Y_k' = \Delta' Z_k' + U_k' \tag{5.5.13}$$

We will stack these nvar equations up to see how to standardize.

$$vec\left(Y_k'\right) = \left(Z_k \otimes I_{\text{nvar}}\right) vec\left(\Delta'\right) + vec\left(U_k'\right) \tag{5.5.14}$$

$Var\left(vec\left(U_k'\right)\right) = I_{n_k} \otimes \Sigma_k$ and $\Sigma_k = R_k' R_k$. Therefore, if we multiply thru by
$I_k \otimes \left(R_k^{-1}\right)'$, this will standardize the error variances in (5.5.14) to have an identity
covariance structure.

$$\left(I_{n_k} + \left(R_k^{-1}\right)'\right) vec\left(Y_k'\right) = \left(Z_k \otimes \left(R_k^{-1}\right)'\right) vec\left(\Delta'\right) + z_k$$
$$Var\left(z_k\right) = I_{n_k \times \text{nvar}} \tag{5.5.15}$$

We can stack up the K equations of the form of (5.5.15).

$$y = X\delta + z$$

$$y = \begin{bmatrix} I_{n_1} \otimes \left(R_1^{-1}\right)' vec\left(Y_1'\right) \\ \vdots \\ I_{n_K} \otimes \left(R_K^{-1}\right)' vec\left(Y_K'\right) \end{bmatrix} \tag{5.5.16}$$

$$X = \begin{bmatrix} Z_1' \otimes \left(R_1^{-1}\right)' \\ \vdots \\ Z_K' \otimes \left(R_K^{-1}\right)' \end{bmatrix}$$

$\delta = vec\left(\Delta'\right)$. Given our prior, $\delta \sim N\left(\bar{\delta}, \left(A_\delta\right)^{-1}\right)$, we can combine with (5.5.16) to compute the conditional posterior in the standard normal form.

$$\delta | y, X, \bar{\delta}, A_\delta \sim N\left(\left(X'X + A_\delta\right)^{-1}\left(X'y + A_\delta \bar{\delta}\right), \left(X'X + A_\delta\right)^{-1}\right) \tag{5.5.17}$$

The moments needed for (5.5.17) can be calculated efficiently as follows.

$$
\begin{aligned}
X'X &= \sum_{k=1}^{K} \left(Z_k' Z_k \otimes R_k^{-1} \left(R_k^{-1} \right)' \right) = \sum_{k=1}^{K} \left(Z_k' Z_k \otimes \Sigma_k \right) \\
X'y &= vec \left(\sum_{k=1}^{K} \Sigma_j^{-1} Y_k' Z_k \right)
\end{aligned}
\tag{5.5.18}
$$

5.5.2 Identification of the Number of Mixture Components

Given that it is possible to undertake posterior simulation of models with 10 or more components, there is some interest in determining the number of components from the data and priors. For mixtures of univariate normals, Richardson and Green [1997] propose an application of the reversible jump sampler that, in principle, allows for MCMC navigation of different size mixture models. The Richardson and Green sampler can "jump" up or down to mixture models of different sizes. In theory, one might be able to use the frequency with which the chain visits a given size component model as an estimate of the posterior probability of that size model. The reversible jump sampler requires a mapping from a lower dimensional mixture component model to a higher dimensional mixture component model.

The other approach to determining the number of mixture components is to attempt to compute the posterior probability of models with a fixed number of components on the basis of simulation output. That is to say, we run 1, 5, and 10 components models and attempt to compute the Bayes Factors for each model. Some have used asymptotic approximations to the Bayes Factors such as the Schwarz approximation. DiCiccio et al. [1997] provide a review of various methods which use simulation output and various asymptotic approximations. All asymptotic methods require finding the posterior mode either by simulation or numerical optimization. This may be particularly challenging in the case of the mixture of multivariate normals in which the likelihood exhibits multiple modes and the parameter space can be extremely high dimensional. In Chapter 6, we will review a number of these methods and return to the problem of computing Bayes Factors for high dimensional models with non-conjugate set-ups. Lenk and DeSarbo [2000] compute Bayes Factors for the number of mixture components in hierarchical generalized linear models.

Given that the normal mixture model is an approximation to some underlying joint density, the main goal in exploring models with different numbers of components is to insure the adequacy of the approximation. That is to say, we want to insure that we include enough components to capture the essential features of the data. The danger of including too many components is that our estimated densities may "overfit" the data. For example, the mixture approximation may build many small lobes on the joint density in an attempt to mimic the empirical distribution of the data in much the same way as kernel smoothing procedures produce lumpy or multimodal density estimates with a too small bandwidth selection. In a hierarchical setting, this is made all the more difficult by the fact that the "data" consist of unknown parameters and we are unable to inspect the empirical distribution. This means that prior views regarding the smoothness of the density are extremely important in obtaining sensible and useful density estimates. The fact that we are in hierarchical setting where the parameters are not observed directly may help us obtain smoother density estimates as the normal mixture will not be able to fit particular noise in the empirical distribution of the "data" as this distribution is only known with error. Thus, devoting a mixture component to accommodating a few outlying data points will not occur unless these outliers are determined very precisely.

"Testing" for the number of components or computing Bayes Factors is of greater interest to those who wish to attach substantive importance to each component, for example, to proclaim there are X number of subpopulations in the data. It is also possible to use posterior probabilities for model averaging. The purpose of model averaging, in this context, is to insure smoothness by averaging over models of varying numbers of components. Our view is that individual components are not very interpretable and are only included for flexibility. Adequate smoothness can be built in via the priors on mixture component parameters. Thus, we take a more informal approach where we investigate fits from models with varying numbers of components. With informative priors that insure adequate smoothness, addition of components that are not necessary to fit the patterns in the data will not change the fitted density.

5.5.3 Application to Hierarchical Models

The normal mixture model provides a natural generalization to the standard normal model of the distribution of heterogeneity. In this section, we will apply this model to a hierarchical MNL model. In the literature on mixtures of normals, investigators typically use very diffuse informative priors on the mixture component parameters $\{\mu_k, \Sigma_k\}$ and p. Improper priors or even very diffuse proper priors are dangerous in any Bayesian context and especially so in the case of marketing data. Typically, panel data on choice of products includes a subset of panelists who do not purchase from the complete set of products. This means that prior beliefs on the logit model parameters will be very important for this set of panelists. Diffuse but proper priors applied to these panelists will result in the inference that these panelists are essentially never willing to purchase these products under any setting of the model covariates if product or brand intercepts are included in the model. We will simply set the intercepts for products not purchased to large negative numbers. Given the logistic probability locus, this will result in zero probability of purchase for all intents and purposes. We do not find this plausible. The probabilities of purchase for these products may be low but is not zero. In a one component normal mixture, the other households inform the first stage prior so that we never obtain extreme estimates of intercepts for these panelists with incomplete purchase histories. The thin tails of the normal density as well as reasonable values of the covariance matrix keep us from making extreme inferences. However, in the case of more than one normal component, this can change. If, for example, there are a group of panelists who do not purchase product A, then the mixture model can assign these panelists to one component. Once this assignment is made, the mean product A intercept values for this component will drift off to very large negative numbers. This problem will be particularly acute when a reasonably large number of components are considered.

There are two ways to deal with this problem (note the option of deleting panelists with incomplete purchase records is not defensible!): 1. use models with very small numbers of components or 2. use informative priors on the means of each component. Given that we center the Z variables, the prior on $\{\mu_k\}$ reflects our views regarding intercepts.

Recall that we use a $N\left(\overline{\mu}, \Sigma_k \otimes A_{\mu}^{-1}\right)$ prior. In much of the work with mixtures, this A_{μ} is set to very small values (e.g., 0.01). Our view is that this admits implausible intercept values of -20 or $+20$. This, of course, is only meaningful if all X variables are on the same scale and location. For this reason, we advocate standardizing the X. variables. We then set a value of A_{μ} of $1/16$ or so rather than $1/100$.

The prior on Σ_k is also important. If we set a tight prior around a small value, then we may force the normal mixture to use a large number of components to fit the data. In addition, the natural conjugate prior links the location and scale so that tight priors over Σ will influence the range of plausible μ values. We will set the prior on Σ to be relatively diffuse by setting v to nvar $+3$ and $V = vI$.

We return to the margarine example discussed in Section 5.4. There are five brand intercepts and one price sensitivity parameter so that nvar $= 6$ and we are fitting a six dimensional distribution of β_i over households. We can combine the Gibbs sampler for normal mixtures with an RW Metropolis chain defined along the lines of 5.3 to draw the household level parameters. We should emphasize that we do have some 300 households but this is not the same as 300 direct observations on six dimensional data. With 300 direct observations, the normal mixture sampler works very well, recovering components with relative ease. However, in the hierarchical logit example, there is only a small amount of information about each household parameter vector. It will be much more difficult to recover complex structure with this effective sample size.

Figure 5.6 presents posterior means of the marginal densities, contrasting one and five component mixture models. For each MCMC draw, we have one fitted multivariate density and we can average these densities over the R draws. To obtain the posterior mean of a marginal density for a specific element of β, we average the marginal densities.

$$\overline{d}_j\left(\beta_j\right) = \frac{1}{R}\sum_{r=1}^{R}\sum_{k=1}^{K}pvec_k^r\phi_j\left(\beta_j\mid \mu_k^r, \Sigma_k^r\right) \tag{5.5.19}$$

We set down a grid of possible values for each element of β and then evaluate the posterior mean of the density in (5.5.19). These densities are shown in the figure for $K = 1$ and $K = 5$. For at least four of the six elements of β, we see pronounced deviations from normality. For Fleischmans, House and Generic intercepts, we obtain highly left skewed distributions with some left lobes. Recall that these are intercepts with the base brand set to Parkay stick. This means that these are relative preferences, holding price constant. For Fleischmanns and the House brands, there is a mass centered close to zero with a very thick left tail. We can interpret this as that there are a number of households who view the House and Fleischmanns brands as equivalent in quality to Parkay but that there are a number of other households who regard these brands as decidedly inferior.

The non-normality of the estimated first stage prior also has a strong influence on the estimates of household posterior means as illustrated in Figure 5.7. There fat tails of the five component normal mixture allow for more extreme estimates of both brand intercepts and price sensitivity. Thus, Figure 5.7 demonstrates that the observed differences in Figure 5.6 make a material difference even if one is only concerned with developing household estimates. However, one should be cautious before using these household estimates. The one component model provides very strong shrinkage of extreme estimates and should, therefore, be regarded as somewhat conservative.

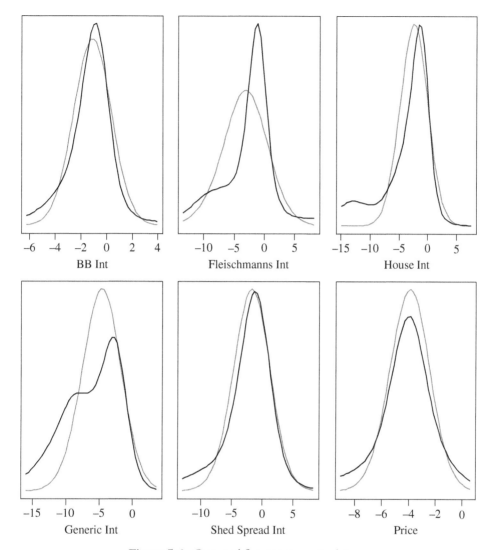

Figure 5.6 One- and five-component mixtures

Our fitting strategy for normal mixture models also included adding a large number of components to see if this makes a material difference in the estimated distribution of household level parameters. It should be pointed out that the marginal computation cost of increasing the number of normal components is rather trivial compared to the cost of the Metropolis draws of the individual household parameters. From a computational point of view, the mixture model is basically free and, therefore, can be used routinely. In Figure 5.8, we consider a 10-component mixture and compare this to the five component. Our view is that the differences in fitted densities in Figure 5.8 is rather small. Figure 5.9 compares the household posterior means for five- and ten-component models. There is quite close agreement between the estimates derived from 5 and 10 component models.

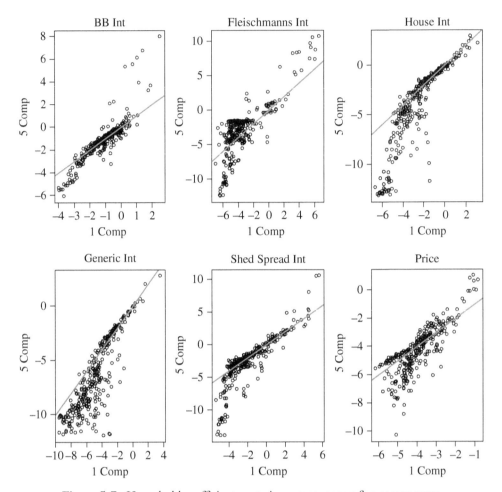

Figure 5.7   Household coefficient posterior means: one vs five components

5.6 FURTHER ELABORATIONS OF THE NORMAL MODEL OF HETEROGENEITY

In many situations, we have prior information on the signs of various coefficients in the base model. For example, price parameters are negative and advertising effects are positive. In a Bayesian approach, this sort of prior information can be included by modifying the first-stage prior. We replace the normal distribution with a distribution with restricted support, corresponding to the appropriate sign restrictions. For example, we can use a log-normal distribution for a parameter which is restricted via sign by the reparameterization, $\theta' = \ln(\theta)$. However, note that this change in the form of the prior can destroy some of the conjugate relationships which are exploited in Gibbs-sampler. However, if Metropolis-style methods are used to generate draws in the Markov chain, it

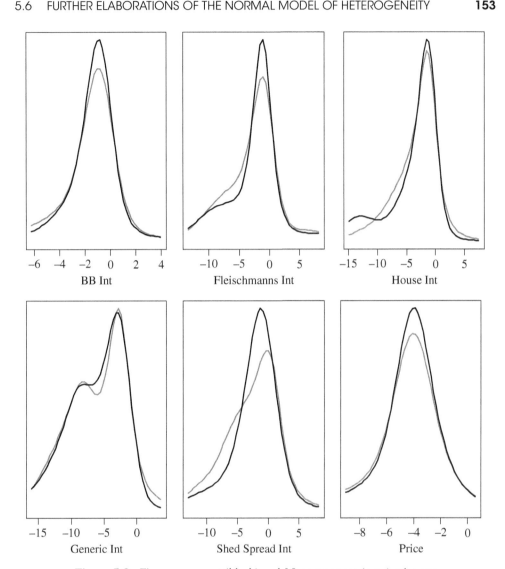

Figure 5.8 Five-component (black) and 10-component (gray) mixtures

is a simple matter to directly re-parameterize the likelihood function, by substituting $\exp(\theta')$ for θ, rather than rely on the heterogeneity distribution to impose the range restriction.

What is more important is to ask whether the log-normal prior is appropriate. The left tail of the log-normal distribution declines to zero, insuring a mode for the log-normal distribution at a strictly positive value. For situations in which we want to admit zero as a possible value for the parameter, this prior may not be appropriate. Boatwright et al. [1999] explore the use of truncated normal priors as an alternative to the log-normal

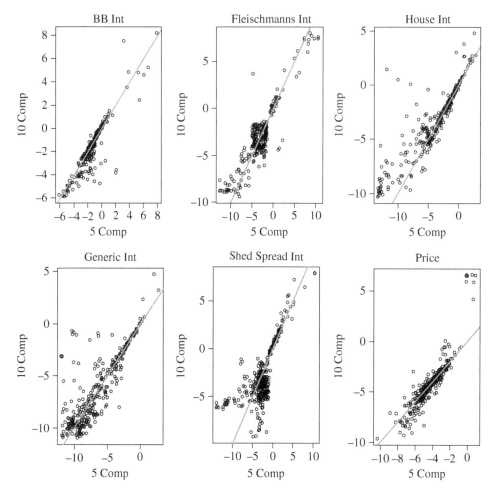

Figure 5.9 Household coefficient posterior means: five vs ten components

reparameterization approach. Truncated normal priors are much more flexible allowing for mass to be piled up at zero.

Bayesian models can also accommodate structural heterogeneity, or changes in the likelihood specification for a unit of analysis. The likelihood is specified as a mixture of likelihoods:

$$p(y_{it}|\{\theta_{ik}\}) = r_1 p_1(y_{it}|\theta_{i1}) + \cdots + r_K p_K(y_{it}|\theta_{iK}),$$

and estimation proceeds by appending indicator variables for the mixture component to the state space. Conditional on the indicator variables, the datum, y_{it}, is assigned to one of K likelihoods. The indicator variables, conditional on all other parameters, have a multinomial distribution with probabilities proportional to the number of observations assigned to the component and the probability that the datum arises from likelihood. Models of structural heterogeneity have been used to investigate intra-individual change in the decision process due to environmental changes (Yang and Allenby [2000]).

Finally, Bayesian methods have recently been used to relax the commonly made assumption that the unit parameters, θ_i, are iid draws from the distribution of heterogeneity. Ter Hofstede et al. [2002] employ a conditional Gaussian field specification to study spatial patterns in response coefficients:

$$p(\theta_i|\tau) = p(\theta_i|\{\theta_j : j \in S_i\}, V_\theta)$$

where S_i denote units that are spatially adjacent to unit i. Since the MCMC estimation algorithm employs full conditional distributions of the model parameters, the draw of θ_i involves using a local average for the mean of the mixing distribution. Yang et al. [2003] employ a simultaneous specification of the unit parameters to reflect the possible presence of interdependent effects due to the presence of social and information networks.

$$\theta = \rho W\theta + u$$
$$u \sim N\left(0, \sigma^2 I\right)$$

where W is a matrix that specifies the network, ρ is a coefficient that measures the influence of the network, and u is an innovation. We discuss this model at length in case study 2.

5.7 DIAGNOSTIC CHECKS OF THE FIRST STAGE PRIOR

In the hierarchical model, the prior is specified in a two stage process:

$$\theta \sim N\left(\bar{\theta}, V_\theta\right)$$
$$p\left(\bar{\theta}, V_\theta\right)$$

In the classical literature, the normal distribution of θ would be called the random effects model and would be considered part of the likelihood rather than part of the prior. Typically, very diffuse priors are used for the second stage. Thus, it is the first stage prior which is important and will always remain important as long as there are only a few observations available per household. Since the parameters of the first stage prior are inferred from the data, the main focus of concern should be on the form of this distribution.

In the econometric literature, the use of parametric distributions of heterogeneity (e.g., normal distributions) are often criticized on the grounds that their mis-specification leads to inconsistent estimates of the common model parameters (c.f. Heckman and Singer [1984]). For example, if the true distribution of household parameters were skewed or bimodal, our inferences based on a symmetric, unimodal normal prior could be misleading. One simple approach would be to plot the distribution of the posterior household means and compare this to the implied normal distribution evaluated at the Bayes estimates of the hyperparameters, $N\left(E\left[\bar{\theta}|data\right], E\left[V_\theta\right]\right)$. The posterior means are not constrained to follow the normal distribution since the normal distribution is only part of the prior and the posterior is influenced by the unit-level data. This simple approach is in the right spirit but could be misleading since we do not properly account for uncertainty in the unit-level parameter estimates.

Allenby and Rossi [1999] provide a diagnostic check of the assumption of normality in the first stage of the prior distribution that properly accounts for parameter uncertainty. To handle uncertainty in our knowledge of the common parameters of the normal distribution, we compute the predictive distribution of $\theta_{i'}$ for unit i' selected at random from the population of households with the random effects distribution. Using our data and model, we can define the predictive distribution of $\theta_{i'}$ as follows:

$$\theta_{i'} | data = \iint \phi\left(\theta | \bar{\theta}, V_\theta\right) p\left(\bar{\theta}, V_\theta | data\right) d\bar{\theta} \, dV_\theta$$

Here $\phi\left(\theta_{i'} | \bar{\theta}, V_\theta\right)$ is the normal prior distribution. We can use our MCMC draws of $\bar{\theta}, V_\theta$, coupled with draws from the normal prior to construct an estimate of this distribution. The diagnostic check is constructed by comparing the distribution of the unit-level posterior means to the predictive distribution based on the model, given above.

5.8 FINDINGS AND INFLUENCE ON MARKETING PRACTICE

The last ten years of work on heterogeneity in marketing has yielded several important findings. Researchers have explored a rather large set of first stage models with a normal distribution of heterogeneity across units. In particular, investigators have considered a first stage normal linear regression (Blattberg and George [1991]), a first stage logit model (Allenby and Lenk [1994, 1995]), a first stage probit (McCulloch and Rossi [1994]), a first stage Poisson (Neelameghan and Chintagunta [1999]), and a first stage generalized gamma distribution model (Allenby et al. [1999]). The major conclusion is that there is a substantial degree of heterogeneity across units in various marketing data sets. This finding of a large degree of heterogeneity holds out substantial promise for the study of preferences, both in terms of substantive and practical significance. There may be substantial heterogeneity bias in models that do not properly account for heterogeneity (Chang et al. [1999]), and there is large value in customizing marketing decisions to the unit level (see Rossi et al. [1996].

Yang et al. [2002] investigate the source of brand preference, and find evidence that variation in the consumption environment, and resulting motivations, leads to changes in a unit's preference for a product offering. Motivating conditions are an interesting domain for research as they preexist the marketplace, offering a measure of demand that is independent of marketplace offerings. Other research has documented evidence that the decision process employed by a unit is not necessarily constant throughout a unit's purchase history (Yang and Allenby [2003]). This evidence indicates that the appropriate unit of analysis for marketing is at the level that is less aggregate than a person or respondent, although there is evidence that household sensitivity to marketing variables (Ainslie and Rossi [1998]) and state dependence (Seetharaman et al. [1999]) is constant across categories.

The normal continuous model of heterogeneity appears to do reasonably well in characterizing this heterogeneity but there has not yet been sufficient experimentation with alternative models such as the mixture of normals to draw any definitive conclusions. With the relatively short panels typically found in marketing applications, it may be difficult to identify much more detailed structure beyond that afforded by the normal model. In addition, relatively short panels may produce a confounding of the finding of heterogeneity with various model mis-specifications in the first stage. If only one observation is available for each unit, then the probability model for the unit level is the mixture of the first stage model with the second stage prior:

$$p\left(y|\tau\right) = \int p\left(y|\theta\right) p\left(\theta|\tau\right) d\theta$$

This mixing can provide a more flexible probability model. In the one observation situation, we can never determine whether it is "heterogeneity" or lack of flexibility that causes the Bayesian hierarchical model to fit the data well. Obviously, with more than one observation per unit, this changes and it is possible to separately diagnose first stage model problems and deficiencies in the assumed heterogeneity distribution. However, with short panels there is unlikely to be a clean separation between these problems and it may be the case that some of the heterogeneity detected in marketing data is really due to lack of flexibility in the base model.

There have been some comparisons of the normal continuous model with the discrete approximation approach of a finite mixture model. It is our view that it is conceptually inappropriate to view any population of units of being comprised of only a small number of homogeneous groups and, therefore, the appropriate interpretation of the finite mixture approach is an approximation method. Allenby and Rossi [1999] and Lenk et al. [1996] show some of the shortcomings of the finite mixture model and provide some evidence that the finite mixture model does not recover reasonable unit level parameter estimates.

At the same time that the Bayesian work in the academic literature has shown the ability to produce unit-level estimates, there has been increased interest on the part of practitioners in unit-level analysis. Conjoint researchers have always had an interest in respondent-level part-worths and had various ad hoc schemes for producing these estimates. Recently, the Bayesian hierarchical approach to the logit model has been implemented in the popular Sawtooth conjoint software. Experience with this software and simulation studies have led Rich Johnson, Sawtooth software's founder, to conclude that Bayesian methods are superior to others considered in the conjoint literature.

Retailers are amassing volumes of store-level scanner data. Not normally available to academic researchers, this store-level data is potentially useful for informing the basic retail decisions such as pricing and merchandizing. Attempts to develop reliable models for pricing and promotion have been frustrated by the inability to produce reliable promotion and price response parameters. Thus, the promise of store-level pricing has gone unrealized. Recently, a number of firms have appeared in this space, offering data-based pricing and promotion services to retail customers. At the heart of some of these firms' approach is a Bayesian shrinkage model applied to store-sku-week data obtained directly from the retail client. The Bayesian shrinkage methods produce reasonable and relatively stable store-level parameter estimates. This approach builds directly on the work of Montgomery [1997].

6

Model Choice and Decision Theory

Abstract

This chapter discusses Bayesian Model Choice and decision theory. Bayesian Model Choice involves various approaches to computing the posterior probability of a model. Posterior model probabilities are useful in comparing two or more competing models or in the choice from a class of models. While there are some methods which can use standard MCMC output to approximate these probabilities, most problems require additional computations for accurate evaluation of posterior probabilities. Sections 6.1–6.9 introduce various methods for computation of model probabilities and compare some of the most useful methods in the context of a model comparison motivated by the multinomial probit model. Many marketing problems suggest a natural decision problem (such as profit-maximization) so that there is more interest in non-trivial applications of decision theory. Sections 6.10 and 6.11 introduce a Bayesian decision-theoretic approach to marketing problems and provide an example by considering the valuation of disaggregate sample information.

Most of the recent Bayesian literature in marketing emphasizes the value of the Bayesian approach to inference, particularly in situations with limited information. Bayesian inference is only a special case of the more general Bayesian decision-theoretic approach. Bayesian Decision Theory has two critical and separate components: 1. a loss function and 2. the posterior distribution. The loss function associates a loss with a state of nature and an action, $\ell(a, \theta)$ where a is the action and θ is the state of nature (parameter). The optimal decision maker chooses the action so as to minimize expected loss where the expectation is taken with respect to the posterior distribution.

$$\min_a \bar{\ell}(a) = \int \ell(a, \theta)\, p(\theta|Data)\, d\theta$$

Bayesian Statistics and Marketing, Second Edition. Peter E. Rossi, Greg M. Allenby, and Sanjog Misra
© 2024 John Wiley & Sons Ltd. Published 2024 by John Wiley & Sons Ltd.

As indicated in Chapter 2, inference about θ can be viewed as a special case of decision theory where the "action" is to choose an estimate based on the data. Quadratic loss yields the posterior mean as the optimal (minimum expected loss) estimator.

Model choice can also be thought of as a special case of decision theory where there is a zero-one loss function. If loss is "1" when the correct model is chosen and "0" if not, then the optimal action is to choose the model with highest posterior probability. In Sections 6.1–6.9, we develop methods for computing posterior probabilities for a set of models.

In marketing applications, a natural loss function is the profits of a firm. The firm seeks to determine marketing actions so as to maximize the expected profits which arise from these actions. In Section 6.10, we develop this loss frame work and apply this idea to the valuation of various information sets. An example of targeting couponing is introduced to make these ideas concrete in Section 6.11.

6.1 MODEL SELECTION

In many scientific settings, the action is a choice between competing models. In the Bayesian approach, it is possible to define a set of models $M_1, \ldots, M_k$ and calculate the posterior probability of each of the models. If the loss function is zero when the correct model is chosen and equal to one for all cases in which the incorrect model is chosen, then the optimal Bayesian decision maker chooses the model with the highest posterior probability. The posterior probability of a model can be obtained from the data likelihood and the prior model probability in the usual manner. Throughout this chapter, we will use the notation, "y," to refer to the observed data. This is for ease of reference to the Bayesian model choice literature.

$$p\left(M_i|y\right) = \frac{p\left(y|M_i\right) p\left(M_i\right)}{p\left(y\right)} \tag{6.1.1}$$

If the set of models, $\{M_1, \ldots, M_k\}$ is exhaustive, we can compute the posterior probability of model i as

$$p\left(M_i|y\right) = \frac{p\left(y|M_i\right) p\left(M_i\right)}{\sum_j p\left(y|M_j\right) p\left(M_j\right)} \tag{6.1.2}$$

In many instances, we might wish to compare two models and choose the model with higher posterior probability. For these cases, the ratio of the posterior probabilities (called the posterior odds ratio) is relevant. The posterior odds ratio is the ratio of relative likelihood times the prior odds ratio.

$$\frac{p\left(M_1|y\right)}{p\left(M_2|y\right)} = \frac{p\left(y|M_1\right)}{p\left(y|M_2\right)} \times \frac{p\left(M_1\right)}{p\left(M_2\right)} \tag{6.1.3}$$

$$= \text{Bayes Factor} \times \text{Prior Odds}$$

In a parametric setting, the posterior probability of a model requires that we integrate out the parameters by averaging the density of the data over the prior density of the parameters,

$$p(y|M_i) = \int p(y|\theta, M_i) \, p(\theta|M_i) \, d\theta \tag{6.1.4}$$

Some write this as the expectation of the likelihood with respect to the prior distribution and, thus, call this the "marginal likelihood."

$$\ell^*(y|M_i) = E_{\theta|M_i} \left[\ell(\theta|y, M_i) \right] \tag{6.1.5}$$

However, it should be noted that the likelihood is any function proportional to the data density so that this interpretation is somewhat imprecise. Henceforth in this chapter, when we use the notation $\ell(\bullet)$, we mean the density of the data including all normalizing constants. The intuition is that if the likelihood of the model is high where we think the parameters are (a priori) then the model has high posterior probability.

The Posterior Odds ratio for parametric models can be written:

$$\frac{p(M_1|y)}{p(M_2|y)} = \frac{\int \ell_1(\theta_1) \, p_1(\theta_1) \, d\theta_1}{\int \ell_2(\theta_2) \, p_2(\theta_2) \, d\theta_2} \times \frac{p(M_1)}{p(M_2)} = BF \times \text{Prior Odds} \tag{6.1.6}$$

In the Bayesian approach, the posterior probability only requires specification of the class of models and the priors. There is no distinction between nested and non-nested models as in the classical hypothesis-testing literature. However, we do require specification of the class of models under consideration; there is no omni-bus measure of the plausibility of a given model or group of models vs some unspecified and possibly unknown set of alternative models.

In the classical testing literature, there is an important distinction made between the non-nested and nested cases. The Neyman–Pearson approach to hypothesis testing requires the specification of a specific null hypothesis. In the case of non-nested models, there is no natural "null" model and classical methods for hypothesis testing can lead to contradictory results in which model 1 is rejected in favor of model 2 and vice versa. The Bayesian approach uses the predictive density of the data under each model in the Bayes Factor (6.1.4). The predictive density is the density of the data averaged over the prior distribution of the model parameters. This can be defined for any set of models, nested or non-nested.

Equation (6.1.4) also reveals the sensitivity of the Bayes Factor to the prior. This is not a limitation or weakness of the Bayesian approach but simply a recognition that model comparison depends critically on the assessment of a prior. From a practical point of view, this means that the researcher must carefully select the prior for each model to insure that it does indeed reflect his views regarding the possible set of parameter values. In particular, "standard" diffuse prior settings can be deceptive. As the diffusion of the prior increases for a particular model, the value of the predictive density in (6.1.4) will decline, at least in the limit. This means that relative diffusion of the priors is important in the computation of the Bayes Factor for comparison of two models. If one prior is very diffuse relative to the other, the BF will tend to favor the model with the less diffuse prior. Improper priors are the limiting case of proper, but diffuse priors, and must be avoided in the computation of Bayes Factors.

We also can interpret the marginal density of the data given model i as the normalizing constant of the posterior.

$$p\left(\theta|y, M_i\right) = \frac{\ell\left(\theta|y, M_i\right) p\left(\theta|M_i\right)}{p\left(y|M_i\right)} \tag{6.1.7}$$

If we let $\tilde{p}\left(\theta|y, M_i\right)$ denote the "un-normalized" posterior, then the marginal density of the data can be written:

$$p\left(y|M_i\right) = \int \tilde{p}\left(\theta|y, M_i\right) d\theta = \frac{\tilde{p}\left(\theta|y, M_i\right)}{p\left(\theta|y, M_i\right)} \tag{6.1.8}$$

6.2 BAYES FACTORS IN THE CONJUGATE SETTING

For conjugate models, (6.1.8) can be used to compute Bayes Factors. That is, the full density form for the posterior can be divided by the product of the data density and the prior to obtain the marginal likelihood of the data. This requires, however, fully conjugate priors not just conditionally conjugate priors. Care must be taken to include all appropriate normalizing constants of the posterior, data density, and priors.

$$p\left(y|M_i\right) = \frac{\tilde{p}\left(\theta|y, M_i\right)}{p\left(\theta|y, M_i\right)} = \frac{p\left(y|\theta, M_i\right) p\left(\theta|M_i\right)}{p\left(\theta|y, M_i\right)}$$

For nested hypotheses, a simplification of the Bayes Factor, called the Savage–Dickey density ratio, can be used. Consider the case of comparison of model M_0 and M_1 where M_0 is a restricted version of M_1. Transform[1] θ to ϕ so that the restriction amounts to setting a subvector of ϕ to some specified value (often 0).

$$M_0 : \quad \phi_1 = \phi_1^b$$
$$M_1 : \quad \text{unrestricted}$$

where $\phi' = \left(\phi_1', \phi_2'\right)$. In this case, the Bayes Factor for comparison of M_0 to M_1 is given by

$$\frac{p\left(y|M_0\right)}{p\left(y|M_1\right)} = \frac{\int \ell\left(\phi_2|y\right) p\left(\phi_2\right) d\phi_2}{\iint \ell\left(\phi_1, \phi_2|y\right) p\left(\phi_1, \phi_2\right) d\phi_1 \, d\phi_2} \tag{6.2.1}$$

where $\ell\left(\phi_2|y\right) = \ell\left(\phi_1, \phi_2|y\right)\big|_{\phi_1=\phi_1^b}$.

One "natural" choice for the prior on the unrestricted component, ϕ_2, is the conditional prior derived from the joint prior under M_1.

$$p\left(\phi_2|\phi_1 = \phi_1^b\right) = \frac{p\left(\phi_1, \phi_2\right)}{\int p\left(\phi_1, \phi_2\right) d\phi_2}\bigg|_{\phi_1=\phi_1^b} \tag{6.2.2}$$

---

[1] In most cases, this must be a linear transformation to take full advantage of the Savage–Dickey simplification.

Using the prior in (6.2.2), the BF can be written

$$
BF = \left. \frac{\int \ell\left(\phi_1,\phi_2|y\right) \frac{p(\phi_1,\phi_2)}{p(\phi_1)} d\phi_2}{\iint \ell\left(\phi_1,\phi_2\right) p\left(\phi_1,\phi_2\right) d\phi_1\, d\phi_2} \right|_{\phi_1=\phi_1^b}
$$

$$
= \left. \frac{\int \ell\left(\phi_1,\phi_2|y\right) p\left(\phi_1,\phi_2\right) d\phi_2}{p\left(\phi_1\right) \iint \ell\left(\phi_1,\phi_2|y\right) p\left(\phi_1,\phi_2\right) d\phi_1\, d\phi_2} \right|_{\phi_1=\phi_1^b} \qquad (6.2.3)
$$

$$
= \left. \frac{\int p\left(\phi_1,\phi_2|y\right) d\phi_2}{p\left(\phi_1\right)} \right|_{\phi_1=\phi_1^b}
$$

Thus, the BF can be written as the ratio of the marginal posterior of ϕ_1 to the marginal prior of ϕ_1 evaluated at $\phi_1 = \phi_1^b$. We also note that if the conditional posterior of $\phi_1|\phi_2$ is of known form, then we can write the marginal posterior as the average of the conditional. We can estimate this with MCMC output from the marginal posterior of ϕ_2.

$$
BF = \left. \frac{\int p\left(\phi_1|\phi_2,y\right) p\left(\phi_2|y\right) d\phi_2}{p\left(\phi_1\right)} \right|_{\phi_1=\phi_1^b}
$$

$$
\hat{BF} = \left. \frac{\frac{1}{R}\sum_r p\left(\phi_1|\phi_2^r,y\right)}{p\left(\phi_1\right)} \right|_{\phi_1=\phi_1^b} \qquad (6.2.4)
$$

Outside of the conjugate setting, the computation of Bayes Factors must rely on various numerical methods for computing the requisite integrals unless asymptotic methods are used. For most problems, asymptotic methods which rely on the approximate normality of the posterior are not reliable.[2] We briefly review the asymptotic approach to computation of Bayes Factors.

6.3 ASYMPTOTIC METHODS FOR COMPUTING BAYES FACTORS

Asymptotic methods can be used to approximate model probabilities. The idea is that the posterior converges to a normal distribution and then we can use results for the multivariate normal distribution to approximate the marginal likelihood and, therefore, the posterior model probability.

$$
p\left(y|M_i\right) = \int p\left(y|\theta_i, M_i\right) p\left(\theta_i|M_i\right) d\theta_i \qquad (6.3.1)
$$

We can approximate the integral in (6.3.1) using the normal approximation to the posterior. This is achieved by expanding the log of the un-normalized posterior around its

---

[2] One notable exception is the MNL model whose likelihood, we have already seen, closely resembles a normal density.

mode in a Taylor series.

$$
\begin{aligned}
p\left(y|M_i\right) &= \int \exp\left(\Gamma\left(\theta\right)\right) d\theta \\
&\approx \int \exp\left(\Gamma\left(\tilde\theta\right) - \frac{1}{2}(\theta - \tilde\theta)' H\left(\tilde\theta\right)(\bullet)\right) d\theta = \exp\left(\Gamma\left(\tilde\theta\right)\right)(2\pi)^{p/2}\left|H\left(\tilde\theta\right)\right|^{-1/2} \\
&= p\left(y|\tilde\theta, M_i\right) p\left(\tilde\theta|M_i\right)(2\pi)^{p/2}\left|H\left(\tilde\theta\right)\right|^{-1/2}
\end{aligned}
\tag{6.3.2}
$$

$\tilde\theta$ is the posterior mode and $H\left(\tilde\theta\right) = -\frac{\partial^2 \tilde p(\theta|y)}{\partial\theta\partial\theta'}\Big|_{\theta=\tilde\theta}$, the negative of the Hessian of the un-normalized posterior with $\tilde p\left(\theta|y\right) = \exp\left(\Gamma\left(\theta\right)\right)$. The approximate BF for comparison of two models will depend on the ratio of the likelihoods as well as the ratio of the prior densities evaluated at the posterior mode.

$$
BF = \frac{p\left(y|M_1\right)}{p\left(y|M_2\right)} \approx \frac{p\left(\tilde\theta_1|M_1\right)}{p\left(\tilde\theta_2|M_2\right)} \times \frac{p\left(y|\tilde\theta_1, M_1\right)\left|H_1\left(\tilde\theta_1\right)\right|^{-1/2}}{p\left(y|\tilde\theta_2, M_2\right)\left|H_2\left(\tilde\theta_2\right)\right|^{-1/2}} \times (2\pi)^{(p_1-p_2)/2}
\tag{6.3.3}
$$

If we make one of the priors in (6.3.3) more diffuse, then the prior density evaluated at the posterior mode will decline and the approximate BF will move in favor of the other model.

Of course, it is possible to base the asymptotic approximation by expanding about the MLE rather than the posterior mode.

$$
BF = \frac{p\left(y|M_1\right)}{p\left(y|M_2\right)} \approx \frac{p\left(\hat\theta_{MLE,1}|M_1\right)}{p\left(\hat\theta_{MLE,2}|M_2\right)} \times \frac{p\left(y|\hat\theta_{MLE,1}, M_1\right)\left|Inf_1\left(\hat\theta_{MLE,1}\right)\right|^{-1/2}}{p\left(y|\hat\theta_{MLE,2}, M_2\right)\left|Inf_2\left(\hat\theta_{MLE,2}\right)\right|^{-1/2}} \times (2\pi)^{\frac{p_1-p_2}{2}}
\tag{6.3.4}
$$

$Inf_i\left(\theta\right) = \left[\frac{\partial^2 \log \ell_i}{\partial\theta\partial\theta'}\right]$ is the observed information matrix. Note that this is the observed information in the sample of size N. For vague priors, there will be little difference. However, for informative priors, maximum accuracy can be obtained by expanding about the posterior mode. Computation of the posterior mode is no more difficult or time intensive than computation of the MLE. Moreover, the posterior may be a more regular surface to maximize over than the likelihood. An extreme example of this occurs when the maximum of the likelihood fails to exist or in cases of non-identified parameters.

The approximate BFs in (6.3.3) and (6.3.4) differ only by constants which do not depend on n. While these two approximations are asymptotically equivalent, we have reason to believe that the BF based on the posterior mode may be more accurate. Both expressions are dependent on the ordinate of the prior, while the expression in (6.3.3) also depends on the curvature of the prior (the curvature of the log posterior is the sum of the curvature of the prior and the likelihood). It is possible to define an asymptotic approximation to the marginal density of the data which depends only on the dimension of the model (see Schwarz [1978]). If we expand about the MLE and rewrite the information matrix in terms of the average information, the posterior probability of model i can be written as

$$
\begin{aligned}
p\left(M_i|y\right) &\propto p\left(y|M_i\right) \\
&\approx k_i p\left(\hat\theta_{MLE,i}|M_i\right) p\left(y|\hat\theta_{MLE,i}, M_i\right)(2\pi)^{\frac{p_i}{2}} n^{-\frac{p_i}{2}}\left|Inf_i\left(\hat\theta_{MLE,i}\right)/n\right|^{-\frac{1}{2}}
\end{aligned}
\tag{6.3.5}
$$

p_i is the dimension of the parameter space for model i. Asymptotically, the average information converges to the expected information in one observation. If we drop everything that is not of order n, then (6.3.5) simplifies to

$$p\left(M_i|y\right) \approx p\left(y|\hat{\theta}_{MLE,i}, M_i\right) n^{-p_i/2} \qquad (6.3.6)$$

Equation (6.3.6) is often computed in log form to select a model from a group of models by picking the model with highest approximate log posterior probability.

$$\log\left(p\left(M_i|y\right)\right) \approx \log\left(\ell_i\left(\hat{\theta}_{MLE,i}\right)\right) - \frac{p_i}{2}\log\left(n\right) \qquad (6.3.7)$$

Equation (6.3.7) is often called the Bayesian Information Criterion (BIC) or Schwarz criterion. The prior has no influence on the BIC expression. In some situations, the BIC is used to compute an approximation to the Bayes Factor.

$$BF \approx \log\left(\ell_1\left(\hat{\theta}_1\right)\right) - \log\left(\ell_2\left(\hat{\theta}_2\right)\right) - \left(\frac{p_1 - p_2}{2}\right)\log\left(n\right)$$
$$= \log\left(LR_{1,2}\right) - \frac{\Delta p}{2}\log\left(n\right) \qquad (6.3.8)$$

The BIC can be extremely inaccurate and should be avoided whenever possible. However, the expression in (6.3.8) is useful to illustrate a fundamental intuition for Bayes Factors. The posterior model probability includes an "automatic" or implicit penalty for models which have higher dimensional parameters. The BF recognizes that adding parameters can simply "overfit" the data and this is automatically accounted for without resort to ad hoc procedures such as out-of-sample validation.

6.4 COMPUTING BAYES FACTORS USING IMPORTANCE SAMPLING

The marginal density of the data for each model can be written as the integral of the unnormalized posterior over the parameter space as in (6.1.8). We can apply importance sampling techniques to this problem (as in Gelfand and Dey [1994], eqn (23)).

$$p\left(y|M_i\right) = \int \tilde{p}\left(\theta|y\right) d\theta = \int \frac{\tilde{p}\left(\theta|y\right)}{q\left(\theta\right)} q\left(\theta\right) d\theta \qquad (6.4.1)$$

Here $q\left(\bullet\right)$ the importance sampling density. We note that, unlike applications of importance sampling to computing posterior moments, we require the full importance sampling density form, including normalizing constants. Using draws from the importance sampling density, we can estimate the BF as a ratio of integral estimates. We note that a separate importance density will be required for both models.

$$\hat{BF} = \frac{\frac{1}{R}\sum_r w_r\left(M_1\right)}{\frac{1}{R}\sum_r w_r\left(M_2\right)}$$
$$w_r\left(M_i\right) = \frac{\tilde{p}\left(\theta_i^r|y, M_i\right)}{q_i\left(\theta_i^r\right)}; \quad \theta_i^r \sim q_i \qquad (6.4.2)$$

Choice and calibration of the importance sampling density is critical to the accuracy of the importance sampling approach. Two suggestions can be helpful. First, we should transform to a parameterization that is unrestricted in order to use an elliptically symmetric importance density. For example, we should transform variance/scale parameters to an unrestricted parameterization. For the scalar case, we can write $\theta = \exp(\gamma)$ and use the Jacobian, $\exp(\gamma)$. For covariance matrices, we can transform to the nondiagonal elements of the Cholesky root and exp of the diagonal elements, denoted by the matrix, Γ. This transformation and associated Jacobian is given by:

$$\Sigma = U'U$$

$$U = \begin{bmatrix} e^{\gamma_{1,1}} & \gamma_{1,2} & \cdots & \gamma_{1,p} \\ 0 & e^{\gamma_{2,2}} & \ddots & \vdots \\ \vdots & \ddots & \ddots & \gamma_{p-1,p} \\ 0 & \cdots & 0 & e^{\gamma_{p,p}} \end{bmatrix} \tag{6.4.3}$$

$$J(\Gamma) = 2^p \prod_{i=1}^{p} e^{\gamma_{ii}(p-i+1)} \prod_{i=1}^{p} e^{\gamma_{ii}} = 2^p \prod_{i=1}^{p} e^{\gamma_{ii}(p+2-i)} \tag{6.4.4}$$

Typically, we would use a multivariate Student t importance density with a moderate degrees of freedom (to insure fatter tails than the unnormalized posterior) and use the MCMC draws from the posterior to assess a mean and covariance matrix in the transformed parameters.

$$\gamma \sim MSt\left(\upsilon, \bar{\gamma}, s^2 \hat{V}\right)$$

$$\bar{\gamma} = \frac{1}{R}\sum_r \gamma_{mcmc}^r \tag{6.4.5}$$

$$\hat{V} = \frac{1}{R}\sum_r \left(\gamma_{mcmc}^r - \bar{\gamma}\right)\left(\gamma_{mcmc}^r - \bar{\gamma}\right)'$$

The importance density would be tuned by selecting the constant s to insure that the distribution of the importance weights is reasonable (not driven by a small number of outliers). The importance sampling estimate of model probability would now be expressed as

$$\hat{p}(y) = \frac{1}{R}\sum_r \frac{\ell(\theta(\gamma^r))\, p(\theta(\gamma^r))\, J(\theta(\gamma^r))}{q(\gamma^r)} \tag{6.4.6}$$

6.5 BAYES FACTORS USING MCMC DRAWS FROM THE POSTERIOR

Typically, we have an MCMC method implemented for each model under consideration. This gives us the ability to simulate from the posterior distribution of each model's parameters. Therefore, there is a natural interest in methods that can express Bayes Factors as the expectation of quantities with respect to the posterior distribution. These identities allow for the "re-use" of already existing posterior draws for the purpose of estimating the posterior expectation.

Gelfand and Dey [1994] provide one such basic identity.

$$\int \frac{q(\theta)}{\tilde{p}(\theta|y, M_i)} p(\theta|y, M_i)\, d\theta = \frac{1}{p(y|M_i)} \qquad (6.5.1)$$

Equation (6.5.1) can be verified as follows:

$$\int \frac{q(\theta)}{\tilde{p}(\theta|y, M_i)} p(\theta|y, M_i)\, d\theta = \int \frac{q(\theta)}{p(\theta|M_i)\, p(y|\theta, M_i)} p(\theta|y, M_i)\, d\theta$$

$$= \frac{1}{p(y|M_i)} \int q(\theta)\, d\theta = \frac{1}{p(y|M_i)}$$

This derivation makes it clear that q must be a proper density. Equation (6.5.1) can be used to express the Bayes Factor as a ratio of posterior expectations.

$$BF\left(M_1 \; vs \; M_2\right) = \frac{p(y|M_1)}{p(y|M_2)} = \frac{E_{\theta|y, M_2}\left[\dfrac{q_2(\theta)}{\tilde{p}(\theta|y, M_2)}\right]}{E_{\theta|y, M_1}\left[\dfrac{q_1(\theta)}{\tilde{p}(\theta|y, M_1)}\right]} \qquad (6.5.2)$$

We can estimate each of the marginal densities of the data by

$$\hat{p}(y|M_i) = \frac{1}{\dfrac{1}{R}\sum_{r=1}^{R} \dfrac{q_i(\theta^r)}{\ell(\theta^r|M_i)\, p(\theta^r|M_i)}} \qquad (6.5.3)$$

It should be noted that for some models evaluation of the likelihood can be computationally demanding and (6.5.3) will require many thousands of likelihood evaluations.

The $q(\bullet)$ function above plays a role analogous to the reciprocal of an importance function. As with an importance function, it is important that the q function "match" or mimic the posterior as closely as possible. This will minimize the variance of the "weights"

$$\hat{p}(y|M_i) = \frac{1}{\dfrac{1}{R}\sum_{r=1}^{R} w^r(M_i)}$$

$$w^r(M_i) = \frac{q_i(\theta^r)}{\ell(\theta^r|M_i)\, p(\theta^r|M_i)}$$

However, the desirable tail behavior for the q function is exactly the opposite of that of an importance density. The tails of the q function serve to attenuate the influence of small values of the posterior density on the estimator. Because of the reciprocal formula, small values of the posterior density can create an estimator with infinite variance. On a practical level, a few "outliers" in the $\{\theta^r\}$ can dominate all other draws in the estimate of the marginal data density in (6.5.3). For this reason, it is important to choose a q function with thin tails relative to the posterior. For problems in high dimensions, it may be difficult to select a q function that works well in the sense of matching the posterior while still having thin tails.

A special case of (6.5.2)/(6.5.3) is the estimator of Newton and Raftery [1994] where $q(\theta){=}p(\theta|M_i)$.

$$\hat{p}(y|M_i) = \frac{1}{\frac{1}{R}\sum_{r=1}^{R}\frac{1}{\ell(\theta^r|M_i)}} \tag{6.5.4}$$

Equation (6.5.4) is the harmonic mean of the likelihood values evaluated at the posterior draws. The function, `logMargDenNR`, in `bayesm` computes this estimator. Thus, only the likelihood must be evaluated to compute the N–R estimate. Many researchers examine the sequence plot of the log-likelihood over the MCMC draws as an informal check on the model fit and convergence. The N–R estimate uses these same likelihood evaluations. In our experience, the sequence plots of log-likelihood values for two or more competing models can be more informative than the computation of Bayes Factors via the N–R method.

The N–R estimate has been criticized as having undesirable sampling properties. In many applications, only handful of draws determine the value of the N–R estimate. If the data is not very informative about the parameters and vague or relative diffuse priors are used, then some of the posterior draws can give rise to very small values of the likelihood and make the N–R estimate unstable. More carefully assessed informative priors can improve the performance of the N–R and Gelfand–Dey estimates.

Another useful identity for the computation of Bayes Factors from MCMC output can be found in the case of nested models:

$$\theta' = (\theta'_1, \theta'_2)$$
$$M_0: \ \theta_1 = \theta_1^b$$
$$M_1: \ \text{unrestricted}$$

We can write the BF as

$$BF = \frac{p(y|M_0)}{p(y|M_1)} = \frac{\int p(y|\theta_2, M_0)\, p(\theta_2|M_0)\, d\theta_2}{p(y|M_1)} \tag{6.5.5}$$

If the BF expression in (6.5.5) is integrated over the marginal prior distribution of θ_1 under M_1, we will still obtain the BF as it is a constant not dependent on model parameters.

$$BF = \int \left[\frac{\int p(y|\theta_2, M_0)\, p(\theta_2|M_0)\, d\theta_2}{p(y|M_1)}\right] p(\theta_1|M_1)\, d\theta_1 \tag{6.5.6}$$

Using the relationship, $p(y|M_1) = \frac{p(y|\theta_1,\theta_2,M_1)p(\theta_1,\theta_2|M_1)}{p(\theta_1,\theta_2|M_1)}$, we can write (6.5.6) as

$$BF = \iint p(y|\theta_2, M_0)\, p(\theta_2|M_0)\, p(\theta_1|M_1)$$
$$\times \frac{p(\theta_1,\theta_2|y, M_1)}{p(y|\theta_1,\theta_2, M_1)\, p(\theta_1,\theta_2|M_1)}\, d\theta_1\, d\theta_2 \tag{6.5.7}$$

Recognizing that $p(\theta_1|M_1)\big/p(\theta_1,\theta_2|M_1) = 1\big/p(\theta_2|\theta_1,M_1)$, (6.5.7) becomes

$$BF = \iint \frac{p\left(y|\theta_2, M_0\right) p\left(\theta_2|M_0\right)}{p\left(y|\theta_1, \theta_2, M_1\right) p\left(\theta_2|\theta_1, M_1\right)} p\left(\theta_1, \theta_2|y, M_1\right) d\theta_1 \, d\theta_2 \qquad (6.5.8)$$

We note that $p\left(y|\theta_2, M_0\right) = p\left(y|\theta_1, \theta_2, M_1\right)\big|_{\theta_1=\theta_1^b}$.

Equation (6.5.8) suggests that we can use the posterior draws to form an estimated BF of the form.

$$\hat{B}F = \frac{1}{R}\sum_r \frac{p\left(y|\theta_1, \theta_2^r\right)\big|_{\theta_1=\theta_1^b} p\left(\theta_2^r|M_0\right)}{p\left(y|\theta_1^r, \theta_2^r\right) p\left(\theta_2^r|\theta_1^r, M_1\right)} \qquad (6.5.9)$$

where $\{\theta^r\}$ are MCMC draws from the posterior under the unrestricted model. We note that, unlike the Savage–Dickey set-up, the priors under and need not be linked via conditioning. What is required, however, is that the conditional prior for $\theta_2|\theta_1$ under must be available as a normalized density.

6.6 BRIDGE SAMPLING METHODS

Meng and Wong [1996] provide an identity which links together methods that rely on expectations with respect to the posterior with importance sampling methods. A hybrid procedure which relies on both is termed "bridge sampling." The bridge sampling identity starts with a pair of functions $\alpha(\theta)$ and $q(\theta)$ such that $\int \alpha(\theta)\,p(\theta|y)\,q(\theta)\,d\theta > 0$.

$$1 = \frac{\int \alpha(\theta)\,p(\theta|y)\,q(\theta)\,d\theta}{\int \alpha(\theta)\,q(\theta)\,p(\theta|y)\,d\theta} = \frac{E_q\left[\alpha(\theta)\,p(\theta|y)\right]}{E_p\left[\alpha(\theta)\,q(\theta)\right]} \qquad (6.6.1)$$

Using the relationship between the marginal density of the data and the unnormalized posterior, we can establish the following identity.

$$p(y) = \frac{E_q\left[\alpha(\theta)\,\tilde{p}(\theta|y)\right]}{E_p\left[\alpha(\theta)\,q(\theta)\right]} \qquad (6.6.2)$$

We can estimate (6.6.2) by approximating both expectations in the numerator and denominator by iid draws from q and MCMC draws from p.

$$\hat{p}(y) = \frac{\frac{1}{R_q}\sum \alpha\left(\theta_q^r\right)\tilde{p}\left(\theta_q^r|y\right)}{\frac{1}{R_p}\sum \alpha\left(\theta_p^r\right)q\left(\theta_p^r\right)} \qquad (6.6.3)$$

Meng and Wong point out that the Gelfand–Dey estimator and the importance sampling estimator are special cases of bridge sampling with a choice of $\alpha(\bullet)$ to be either the importance density or the un-normalized posterior. They consider the question of an "optimal" choice (in the sense of smallest MSE in estimation) and provide an iterative scheme for constructing $\alpha(\bullet)$ as an weighted combination of Gelfand–Dey and importance sampling. Fruhwirth-Schnatter [2004] applies this iterative scheme to the construction of BF for mixtures of normals and finds the Meng and Wong estimator to be an improvement over standard procedures.

6.7 POSTERIOR MODEL PROBABILITIES WITH UNIDENTIFIED PARAMETERS

We have considered a number of models with unidentified parameters. We have also seen that, in some cases, it is desirable to navigate in the unidentified parameter space and then margin down or "post-process" the MCMC draws to make inferences regarding the identified parameters. A reasonable question to ask is whether draws of the unidentified parameters can be used to estimate posterior model probabilities. From a purely theoretical point of view, this can be justified. Let us assume that θ is not identified but, $\tau(\theta)$, is.

We can then define a transformation, $g(\theta)$, which partitions the transformed parameters into those which are identified and those which are not.

$$\delta = g(\theta) = \begin{bmatrix} \upsilon(\theta) \\ \tau(\theta) \end{bmatrix} = \begin{bmatrix} \delta_1 \\ \delta_2 \end{bmatrix} \tag{6.7.1}$$

δ_1 is a subvector of length k – containing the unidentified parameters. We start with a prior over the full vector of parameters, $p_\theta(\bullet)$, and then compute the induced prior over the identified parameters.

$$p_{\delta_2}(\delta_2) = \int p_\delta(\delta_1, \delta_2)\, d\delta_1 \tag{6.7.2}$$

$$p_\delta(\delta_1, \delta_2) = p_\theta(g^{-1}(\delta))\, J_{\theta \to \delta}$$

We can compute the marginal density of the data in two ways. We can compute the density in the space of unidentified parameters or directly on the identified parameters using the induced prior in (6.7.2).

Working in the identified parameter space,

$$p'(y) = \int p(y|\delta_2)\, p_{\delta_2}(\delta_2)\, d\delta_2 \tag{6.7.3}$$

Now consider the marginal density of the data computed in the full, unidentified parameter space.

$$p(y) = \iint p(y|\delta_1, \delta_2)\, p_\delta(\delta_1, \delta_2)\, d\delta_1\, d\delta_2$$

$$= \int \left[\int p(y|\delta_2)\, p_{\delta_2}(\delta_2)\, d\delta_2 \right] p(\delta_1|\delta_2)\, d\delta_1$$

$$= \int p'(y)\, p(\delta_1|\delta_2)\, d\delta_1 = p'(y)$$

Here we are using the fact that, $p(y|\delta_1, \delta_2) = p(y|\delta_2)$ since δ_1 represents the unidentified parameters.

Thus, we are theoretically justified in using MCMC draws from the unidentified parameter space in the methods considered in Section 6.6. However, in practice we recognize that the draws of the un-identified parameters may exhibit a great deal of variation, especially in situations with vague or very diffuse priors. Again, there is a pay-off to assessing realistic priors.

6.8 CHIB'S METHOD

Chib [1995] proposes a method which uses MCMC output to estimate the marginal density of the data. This method is particularly appropriate for models that have a conjugate structure conditional on the value of augmented latent variables. Chib starts with the basic identity relating the normalized and unnormalized posteriors. This identity holds for any value of θ, indicated by θ^* below

$$p(y) = \frac{\tilde{p}(\theta^*|y)}{p(\theta^*|y)} = \frac{p(y|\theta^*) p(\theta^*)}{p(\theta^*|y)} \tag{6.8.1}$$

The key insight of Chib [1995] is that, for certain conditionally conjugate models, the denominator of (6.8.1) can be expressed as an average of densities which are known up to and including normalizing constants.

For example, consider the archetypal data augmentation model.

$$\begin{aligned} y|z \\ z|\theta \\ \theta \end{aligned} \tag{6.8.2}$$

For the model in (6.8.2), we can write the ordinate of the posterior at θ^* as the average of the posterior conditional on the latent z over the marginal posterior distribution of z.

$$p(\theta|y) = \int \frac{p(y, z, \theta)}{p(y)} dz = \int \frac{p(\theta|y, z) p(y, z)}{p(y)} dz$$
$$= \int \frac{p(\theta|y, z) p(z|y) p(y)}{p(y)} dz = \int p(\theta|y, z) p(z|y) dz \tag{6.8.3}$$

Equation (6.8.3) suggests that we can estimate the marginal density of the data as follows:

$$\hat{p}(y) = \frac{p(y|\theta^*) p(\theta^*)}{\hat{p}(\theta^*|y)} = \frac{p(y|\theta^*) p(\theta^*)}{\frac{1}{R} \sum_r p(\theta^*|y, z^r)} \tag{6.8.4}$$

$\{z^r\}$ are draws from the marginal posterior of the latent variables. θ^* is usually taken to be the posterior mean or mode, computed from the MCMC draws. Equation (6.8.4) requires that we simply save the latent draws from our MCMC run and that we be able to evaluate the data density (likelihood) and the prior densities with all normalizing constants. We should note that (6.8.4) requires evaluation of the marginal likelihood (without the latents). For some models, this can be computationally challenging. However, we should point out that the Chib method only requires one likelihood evaluation whereas Gelfand/Dey style methods would require R evaluations.

The Chib method in (6.8.4) requires that we be able to evaluate, $p\left(\theta^{*}|y,z\right)$ including all normalizing constants. For some models, we can use the fact that $p\left(\theta|y,z\right)=p(\theta|z)$ and that we have a conjugate set-up conditional on z. However, in other applications, this will not be possible. In some cases, $\theta|z$ is not fully conjugate but can be broken into two conjugate blocks. Consider the case where θ is partitioned into $\left(\theta_{1},\theta_{2}\right)$. We can then estimate $p\left(\theta^{*}|y,z\right)$ using the identity, $p\left(\theta_{1},\theta_{2}|y\right)=p\left(\theta_{1}|y\right)p\left(\theta_{2}|\theta_{1},y\right)$. We can compute the marginal posterior density of θ_{1} by averaging the conditional density.

$$p\left(\theta_{1}|y\right)=\int p\left(\theta_{1}|\theta_{2},z,y\right)p\left(\theta_{2},z|y\right)dz\,d\theta_{2} \tag{6.8.5}$$

Equation (6.8.5) can be estimated by averaging the conditional density over the MCMC draws of θ_{2} and z.

$$\hat{p}\left(\theta_{1}^{*}|y\right)=\frac{1}{R}\sum_{r}p\left(\theta_{1}^{*}|\theta_{2}^{r},z^{r}\right) \tag{6.8.6}$$

However, the conditional posterior density of $\theta_{2}|\theta_{1}$ cannot be estimated by averaging the conditional with respect to the marginal posterior as in (6.8.5).

$$p\left(\theta_{2}|\theta_{1},y\right)=\int p\left(\theta_{2}|\theta_{1},z,y\right)p\left(z|\theta_{1},y\right)dz \tag{6.8.7}$$

To estimate this density at the point $\left(\theta_{1}^{*},\theta_{2}^{*}\right)$ requires a modified MCMC sampler for $\left(z,\theta_{2}\right)$ given $\theta_{1}=\theta_{1}^{*}$. As Chib points out, this is simple to achieve by shutting down the draws for θ_{1}. If $z_{\theta_{1}^{*}}^{r}$ are draws from the marginal posterior of z given $\theta_{1}=\theta_{1}^{*}$, then we can estimate (6.8.7) by

$$\hat{p}\left(\theta_{2}^{*}|\theta_{1}^{*},y\right)=\frac{1}{R}\sum_{r}p\left(\theta_{2}^{*}|\theta_{1}^{*},z_{\theta_{1}^{*}}^{r},y\right) \tag{6.8.8}$$

To estimate the marginal density of the data, we put together (6.8.6) and (6.8.8).

$$\begin{aligned}\hat{p}\left(y\right)&=\frac{p\left(y|\theta^{*}\right)p\left(\theta^{*}\right)}{\hat{p}\left(\theta_{1}^{*}|y\right)\hat{p}\left(\theta_{2}^{*}|\theta_{1}^{*},y\right)}\\[2mm]&=\frac{p\left(y|\theta^{*}\right)p\left(\theta^{*}\right)}{\left(\frac{1}{R}\sum_{r}p\left(\theta_{1}^{*}|\theta_{2}^{r},z^{r}\right)\right)\left(\frac{1}{R}\sum_{r}p\left(\theta_{2}^{*}|\theta_{1}^{*},z_{\theta_{1}^{*}}^{r},y\right)\right)}\end{aligned} \tag{6.8.9}$$

6.9 AN EXAMPLE OF BAYES FACTOR COMPUTATION: DIAGONAL MNP MODELS

We will illustrate the methods of importance sampling, Newton–Raftery, and Chib using an example of comparison of different MNP models. The estimation of the off-diagonal covariance (correlation) elements in the MNP is often difficult. We typically find that the diagonal elements (relative variances) are much more precisely estimated. In addition, we often find large differences in relative variance between choice alternatives. This is consistent with a view that X variables explain different portions of the utility of each choice. For example, some choice alternatives may have utility explained well by price while others may have utility that depends on attributes not measured in our data. For this reason, the MNP model with a non-scalar but diagonal covariance matrix could be considered a central model. We will develop Bayes Factors to compare the diagonal MNP with an "identity" MNP or a model with an Identity covariance matrix. This would closely approximate the IIA properties of the MNL model.

Diagonal MNP:

$$y_i = \sum_{j=1}^{p} I\left(z_{i,j} = \max\left(z_i\right)\right)$$

$$z_i = X_i\beta + \varepsilon_i$$

$$\varepsilon_i \sim N\left(0, \Lambda\right)$$

$$\Lambda = \begin{bmatrix} 1 & & & \\ & \sigma_{22} & & \\ & & \ddots & \\ & & & \sigma_{pp} \end{bmatrix}$$

Identity MNP:

$$y_i = \sum_{j=1}^{p} I\left(z_{i,j} = \max\left(z_i\right)\right)$$

$$z_i = X_i\beta + \varepsilon_i$$

$$\varepsilon_i \sim N\left(0, I_p\right)$$

We use normal priors on β and independent scaled inverted chi-squared priors on the diagonal elements $2, \ldots, p$. The first diagonal element is fixed at 1.

$$\beta \sim N\left(\bar{\beta}, A^{-1}\right)$$

$$\sigma_{j,j} \sim ind\ v_0 s_0^2 / \chi_{v_0}^2$$

$$(6.9.1)$$

To apply the Newton–Raftery method, we must implement MCMC samplers for each of the two MNP models and compute the likelihood at each draw as in (6.5.4). The diagonal MNP sampler and Identity MNP samplers are special cases of the algorithms given in Section 4.2. To compute the likelihood, we must compute the choice probabilities for each observation. The general MNP likelihood requires evaluation of the integral of a correlated normal random variable over a rectangular region as discussed in Section 5.2.1. However, for the diagonal MNP model, the choice probabilities can be simplified to the average of normal cdfs with respect to a univariate normal distribution. Both importance sampling and N–R methods will be computationally intensive due to the evaluation of the likelihood over the set of MCMC draws or over draws from the importance density. This argues in favor of the Chib method which only requires one evaluation of the likelihood.

To implement an importance function approach, we first transform the variance parameters to an unrestricted space. This only applies to the Diagonal MNP model. If θ

is the stacked vector of the parameters of the Diagonal MNP model, $\theta' = (\beta, \text{diag}(\Lambda)')$, then we define the transformed vector η by

$$\eta = \begin{bmatrix} \eta_1 \\ \eta_2 \end{bmatrix} = \begin{bmatrix} \beta \\ \ln(\text{diag}(\Lambda)) \end{bmatrix}; \quad \theta = \begin{bmatrix} \eta_1 \\ e^{\eta_2} \end{bmatrix} \tag{6.9.2}$$

with Jacobian:

$$J_{\theta \to \eta} = \left\| \begin{matrix} I_{\text{dim}(\beta)} & & & 0 \\ & e^{\eta_{2,2}} & & \\ 0 & & \ddots & \\ & & & e^{\eta_{2,p}} \end{matrix} \right\| = \prod_{i=2}^{p} e^{\eta_{2,i}} \tag{6.9.3}$$

To implement the importance sampling method, we use a normal importance function with location and scale chosen using the MCMC draws in the transformed parameter η.

$$q(\eta) = \phi\left(\bar{\eta}, s^2 \hat{V} \right)$$
$$\bar{\eta} = \frac{1}{R} \sum_r \eta_{mcmc}^r \tag{6.9.4}$$
$$\hat{V} = \frac{1}{R} \sum_r \left(\eta_{mcmc}^r - \bar{\eta} \right) \left(\eta_{mcmc}^r - \bar{\eta} \right)'$$

where ϕ denotes a normal density. We approximate the marginal density of the data using draws from q as follows:

$$\hat{p}(y) = \frac{1}{N} \sum_i \frac{p\left(y | \theta\left(\eta^i \right) \right) p\left(\theta\left(\eta^i \right) \right) J_{\theta \to \eta}\left(\theta\left(\eta^i \right) \right)}{q\left(\eta^i \right)} = \frac{1}{N} \sum_i w_i \tag{6.9.5}$$

We note that the q density in the denominator of (6.9.5) is the full normalized density. The importance sampling method can be tuned by choice of the scaling parameter, s.

The Chib method can be implemented directly using (6.8.4) for the Identity MNP model. However, for the Diagonal MNP model, we use the variant of the Chib method which breaks the parameter vector into two parts, $\theta_1 = \beta$, $\theta_2 = \text{diag}(\Lambda)$ in equation (6.8.9).

As an illustration, we simulate data from a Diagonal MNP model with $p = 3$ and $\text{diag}(\Lambda) = (1, 2, 3)$, $N = 500$. X contains $p - 1$ intercepts and two regressors which are simulated unif$(-1,1)$. We compute the BF factors for the Diagonal MNP vs the Identity MNP. We use 150,000 draws from the MCMC samplers and the importance sampling density. We assess modestly informative priors with $\bar{\beta} = 0$, $A = 0.25I$, $v_0 = 10$, and $s_0^2 = 1$.

Figure 6.1 shows the log-likelihood values for the Diagonal MNP (dark) and the Identity MNP (light) plotted for every 75th draw. The figure clearly shows that the Diagonal MNP fits the data better than the Identity MNP. Even though inspection of this plot is somewhat "informal," we highly recommend this as a rough assessment of the models as well as the convergence of the MCMC algorithm. In fact, in our experience this sort of plot can be more informative than formal Bayes Factor computations.

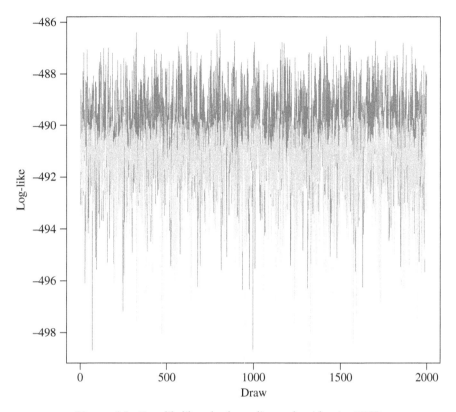

Figure 6.1   Log-likelihood values: diagonal vs identity MNP

Below we show the results of Bayes Factor computations using all three methods.

Method	BF
NR	3.4
IS ($s = 1.5$)	2.9
IS ($s = 2.0$)	2.9
Chib	3.4

These numbers do not convey the sampling error in each of the estimates. As an informal assessment, we made more than 10 different runs and observed very little variation in the IS sampling or Chib numbers and a wide range of NR BFs from 2.0 to 3.5. We note that our prior is more informative than most and we expect that with more traditional "vague" prior settings we would see even more variation in the NR numbers.

The NR approach is frequently criticized in the literature due to the fact that the estimator has an infinite variance under some conditions. However, the convenience of the NR approach accounts for its widespread popularity. Figure 6.2 illustrates the problems with the NR approach more dramatically. The NR estimate is driven by a small

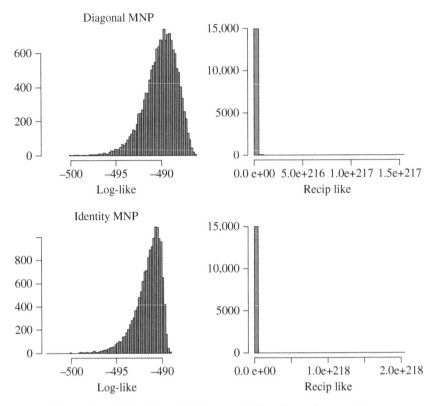

Figure 6.2   Difficulties with Newton–Raftery Bayes factor estimator

portion of the values of the reciprocal of the likelihood as shown by the histogram of the reciprocal of the likelihood for every 10th draw (right hand side panel). While the distribution of log-likelihood values is not particularly abnormal, the harmonic mean estimator is particularly vulnerable to outlying observations.

The Importance Sampling approach does not have as severe problem with outlying observations as the NR approach. However, even with careful choice of scaling constant,[3] the weights have outliers as shown in Figure 6.3. Figure 6.3 shows the distribution of the importance sampling weights normalized by the median of the weight. This figure does not instill confidence in the importance sampling approach either. The parameter space is only of dimension seven in this example and we might expect the problems with the importance sampling approach to magnify in larger dimension parameter spaces.

The Chib approach to calculating BF relies on various estimates of the ordinate of posterior or conditional posterior densities. For the Diagonal MNP, the Chib approach averages $p\left(\beta^*|\Lambda, z, y\right)$ over MCMC draws of Λ and z. $p\left(\Lambda^*|\beta^*, z, y\right)$ is averaged over draws of $z|\beta^*, y$. For the Identity MNP, only the density of β is averaged. Figure 6.4 shows the distribution of these densities (right hand side) and the log of these densities.

---

[3] In this application, choice of the scaling constant is critical. We experimented with student t importance densities and found little benefit.

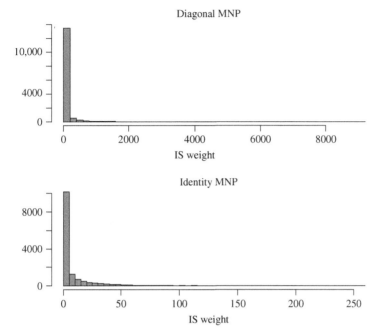

Figure 6.3 Importance sampling weight distribution

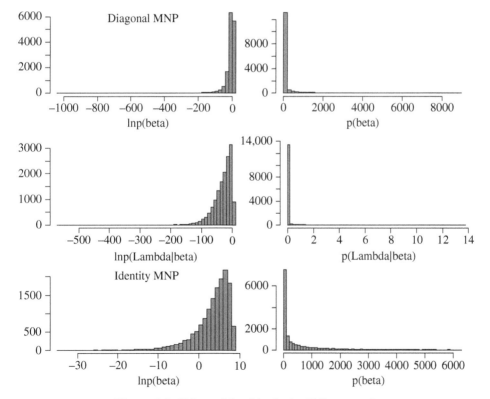

Figure 6.4 Values of densities in the Chib approach

Again, there are very large outliers which influence the averages. These outliers contribute to instability of this approach.

In this example, the Importance Sampling approach provides the most reliable results. However, it is important to note that all three approaches are sensitive to outliers. The "numerical standard error" formulas used to measure the sampling errors in the IS and Chib approach are unlikely to be reliable in the presence of such large outliers.

6.10 MARKETING DECISIONS AND BAYESIAN DECISION THEORY

Computation of posterior model probabilities is motivated as a special case of Bayesian decision theory with a zero-one loss function. In marketing problems, the profit function of the firm provides a more natural choice of loss. In addition, there is often considerable parameter or modeling uncertainty. Bayesian decision theory is ideally suited for application to many marketing problems in which a decision must be made given substantial parameter or modeling uncertainty. In these situations, the uncertainty must factor into the decision itself. The marketing decision maker takes an action by setting the value of various variables designed to quantify the marketing environment facing the consumer (such as price or advertising levels). These decisions should be affected by the level of uncertainty facing the marketer. To make this concrete, begin with a probability model that specifies how the outcome variable (y) is driven by the explanatory variables (x) and parameters θ.

$$p\left(y|x,\theta\right)$$

The decision maker has control over a subset of the x vector, $x' = \left[x'_d, x'_{cov}\right]$. x_d represents the variables under the decision maker's control and x_{cov} are the covariates. The decision maker chooses x_d so as to maximize the expected value of profits where the expectation is take over the distribution of the outcome variable. In a fully Bayesian Decision Theoretic treatment, this expectation is taken with respect to the posterior distribution of θ as well as the predictive conditional distribution $p(y|x_d, x_{cov})$.

$$\pi^*\left(x_d|x_{cov}\right) = E_\theta\left[E_{y|\theta}\left[\pi\left(y|x_d\right)\right]\right]$$

$$= E_\theta\left[\int \pi\left(y|x_d\right) p\left(y|x_d, x_{cov}, \theta\right) dy\right] \qquad (6.10.1)$$

$$= E_\theta\left[\bar{\pi}\left(x_d|x_{cov}, \theta\right)\right]$$

The decision maker chooses x_d to maximize profits π^*. In general, the decision maker can be viewed as minimizing expected loss which is frequently taken as – profits but need not be in all cases (see, for example, Steenburgh et al. [2003]).

6.10.1 Plug-In vs Full Bayes Approaches

The use of the posterior distribution of the model parameters to compute expected profits is an important aspect of the Bayesian approach. In an approximate or conditional Bayes approach, the integration of the profit function with respect to the posterior distribution of θ is replaced by an evaluation of the function at the posterior mean or mode of the parameters. This approximate approach is often called the "plug-in" approach, or according to Morris [1983], "Bayes Empirical Bayes."

$$\pi^* \left(x_d \right) = E_{\theta|y} \left[\bar{\pi} (x_d | \theta) \right] \neq \bar{\pi} (x_d | \hat{\theta} = E_{\theta|y} [\theta]) \tag{6.10.2}$$

When the uncertainty in θ is large and the profit function is non-linear, errors from the use of the plug-in method can be large. In general, failure to account for parameter uncertainty will overstate the potential profit opportunities and lead to "over-confidence" that results in an overstatement of the value of information (c.f. Montgomery and Bradlow [1999]).

6.10.2 Use of Alternative Information Sets

One of the most appealing aspects of the Bayesian approach is the ability to incorporate a variety of different sources of information. All adaptive shrinkage methods utilize the similarity between cross-sectional units to improve inference at the unit level. A high level of similarity among units leads to a high level of information shared. Since the level of similarity is determined by the data via the first-stage prior, the shrinkage aspects of the Bayesian approach adapt to the data. For example, Neelameghan and Chintagunta [1999] show that similarities between countries can be used to predict the sales patterns following the introduction of new products.

The value of a given information set can be assessed using a profit metric and the posteriors of θ corresponding to the two information sets. For example, consider two information sets A and B along with corresponding posteriors, $p_A(\theta), p_B(\theta)$. We solve the decision problem using these two posterior distributions.

$$\Pi_l = \max_{x_d} \pi_l^* \left(x_d | x_{cov} \right) = \max_{x_d} \int \bar{\pi} \left(x_d | x_{cov}, \theta \right) p_l(\theta) \, d\theta \tag{6.10.3}$$

$$l = A, B$$

We now turn to the problem of valuing disaggregate information.

6.10.3 Valuation of Disaggregate Information

Once a fully decision-theoretic approach has been specified, we can use the profit metric to value the information in disaggregate data. We compare profits that can be obtained via our disaggregate inferences about $\{\theta_i\}$ with profits that could be obtained using only aggregate information. The profit opportunities afforded by disaggregate data will depend both on the amount of heterogeneity across the units in the panel data as well as the level of information at the disaggregate level.

To make these notions explicit, we will lay out the disaggregate and aggregate decision problems. As emphasized in Chapter 5, Bayesian methods are ideally suited for inference about the individual or disaggregate parameters as well as the common parameters.

Recall the profit function for the disaggregate decision problem.

$$\pi_i^* \left(x_{d,i} | x_{cov,i} \right) = \int \overline{\pi} \left(x_{d,i} | x_{cov,i}, \theta_i \right) p \left(\theta_i | Data \right) d\theta_i \qquad (6.10.4)$$

Here we take the expectation with respect to the posterior distribution of the parameters for unit "i." Total profits from the disaggregate data are simply the sum of the maximized values of the profit function above. $\Pi_{disagg} = \sum_i \pi_i^* \left(\tilde{x}_{d,i} | x_{cov,i} \right)$ where $\tilde{x}_{d,i}$ is the optimal choice of $x_{d,i}$.

Aggregate profits can be computed by maximizing the expectation of the sum of the disaggregate profit functions with respect to the predictive distribution of θ_i

$$\pi_{agg} \left(x_d \right) = E_\theta \left[\sum \overline{\pi} \left(x_d | x_{cov,i}, \theta \right) \right] = \int \sum \overline{\pi} \left(x_d | x_{cov,i}, \theta \right) \overline{p} \left(\theta \right) d\theta$$

$$\Pi_{agg} = \pi_{agg} (\tilde{x}_d) \qquad (6.10.5)$$

The appropriate predictive distribution of θ, $\overline{p}(\theta)$, is formed from the marginal of the first stage prior with respect to the posterior distribution of the model parameters.

$$\overline{p}(\theta) = \int p(\theta | \tau) p(\tau | Data) d\tau$$

Comparison of Π_{agg} with Π_{disagg} provides a metric for the achievable value of the disaggregate information.

6.11 AN EXAMPLE OF BAYESIAN DECISION THEORY: VALUING HOUSEHOLD PURCHASE INFORMATION

As emphasized in Section 6.10, valuation of information must be made within a decision context. Rossi et al. [1996] consider the problem of valuing household purchase information using a targeted couponing problem. In traditional couponing exercises, neither retailers nor manufacturers have access to information about individual consumers. Instead, a "blanket" coupon is distributed (via mass mailings or inserts in newspapers). This blanket coupon is, in principle, available to all consumers (note: the large denomination "rebates" available on consumer durable goods are the same idea with a different label). In order to be profitable, the issuer of the coupon relies on a indirect method of price discrimination where the consumers with lower willingness to pay use the coupon with higher probability that those with higher willingness to pay. In the late 1980s, the technology for issuing customized coupons became available. For example, Catalina Marketing Incorporated started a highly successful business by installing coupon printers in grocery stores that were connected to the point of sale terminal. This technology opened the possibility of issuing coupons based on purchase history information. At the most elementary level, it was now possible to issue coupons to consumers who exhibited interest in a product category by purchasing some product in this category. For example, an ice-cream manufacturer could pay Catalina to issue coupons for its ice cream products to anyone observed to be purchasing ice cream.

Frequent shopper programs adopted by many retailers also provide a source of purchase and demographic information. By linking purchase records via the frequent shopper id, it is possible to assemble a panel of information about a large fraction of the retailers' customers. This allows for much more elaborate coupon trigger strategies which involve estimated willingness to pay. That is, we might be able to directly estimate a consumers' willingness to pay for a product by estimating a demand model using purchase history data. In addition, frequent shopper programs sometimes collect limited demographic information either directly on enrollment applications or via indirect inference from the members address (so-called geo-demographic information).

These developments mean that the issuer of the coupon now has access to a rich information set about individual consumers. How much should the issuer be willing to pay for this information? Or, what is the value of targeted couponing relative to the traditional blanket approach? To answer these questions, a model of purchase behavior is required as well as a loss function. The natural loss function for this problem is expected incremental profits from the issue of a coupon. Rossi et al. postulate a hierarchical probit model similar to the hierarchical logit model of Chapter 5 in which demographic information enters the hierarchy. The hierarchical probit model allows the issue of the coupon to make inferences about consumer level purchase probabilities given various information sets. Clearly, only a very limited set of information is available for each consumer. This means that these inferences will imprecise and the degree of "targeting" or customization of the face value of the coupon will depend not only on how different consumers are in willingness to pay but also on how precisely we can measure these differences. This full accounting for parameter uncertainty is a key feature of a Bayesian decision-theoretic approach. This will guard against the overconfidence that can arise from "plug-in" approaches that simply estimate parameters without accounting for uncertainty.

Rossi et al. start with a multinomial probit model at the household or consumer level. Households are confronted with a choice from among p brands with covariates taken to be measures of the marketing mix facing the consumer (price and advertising variables).

$$y_{h,t} = j \ if \ \max \left(z_{h,t} = X_{h,t}\beta_h + \varepsilon_{h,t} \right) = z_{j,h,t} \qquad (6.11.1)$$

$$\varepsilon_{h,t} \sim N\left(0, \Lambda\right) \qquad (6.11.2)$$

In Rossi et al. (in contrast to the standard MNP in Section 4.2), the covariance matrix of the latent errors (Λ) is taken to be diagonal. This means that we do not have to difference the system and that identification is achieved by setting $\lambda_{11} = 1$. Rossi et al. make this simplification for practical and data-based reasons. Given β_h, Λ, choice probabilities for the diagonal MNP model are simple to compute, requiring only univariate integration. Since these choice probabilities figure in the loss/profit function and will be evaluated for many 1000s of draws, it is important that these probabilities can be computed at low cost. This is not as much of a consideration now as it was in the early 1990s when this research was conducted. However, this is not the only reason for using a diagonal covariance. Much of the observed correlation in non-hierarchical probit models can be ascribed to heterogeneity. Once heterogeneity is taken into account via the hierarchy, the errors in the unit-level probit are much less correlated. However, they can often be very heteroskedastic as pointed out in Section 6.9.

This "unit-level" model is coupled with the, by now, standard normal model of heterogeneity.

$$\beta_h = \Delta' z_h + v_h$$
$$v_h \sim N\left(0, V_\beta\right)$$

(6.11.3)

z_h is a vector of demographic information. The key task is to compute the predictive distribution of the household parameters and choice probabilities for various information sets. That is we must compute, $p\left(\beta_h, \Lambda|\Omega^*\right)$, where Ω^* denotes a particular information set. This predictive distribution will be used in the decision problem to determine what is the optimal face value coupon to issue to household h.

Given the proper choice of priors on the common parameters, Λ, Δ, V_β, we can define an MCMC method to draw from the joint posterior of $\{\beta_1, \dots, \beta_h, \dots, \beta_H\}, \Lambda, \Delta, V_\beta$. Rossi et al. observed that one can develop a Gibbs Sampler for this problem by using the McCulloch and Rossi [1994] sampler coupled with the standard normal hierarchical set-up. That is, given $\{\beta_h\}$, we have standard normal, IW draws for Δ, V_β as in Section 3.7 or 5.3.

A DAG for the model is given below:

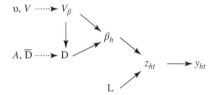

Thus, we can obtain draws from the marginal posterior distributions of all household and common parameters.

Consider three information sets:

1. "Full": Purchase history information on household h and demographics

2. "Demographics Only": Just knowledge of the demographic vector for household h

3. "Blanket": No information about household h but only information about the "population" distribution of households obtained from a sample of households.

Rossi et al. consider a "choices-only" information set which we will not discuss here. We must compute the "predictive" or posterior distribution of β_h for each of the three information sets. The first and richest information set is a natural byproduct of the Gibbs Sampler defined above. That is, when we compute the posterior distribution of each β_h, we will take into account each households demographics and purchase history. In addition, information from other households will influence our views about a particular β_h via inferences about the common parameters. We note that by looking at the marginal distribution of the posterior draws, we are integrating out or averaging over draws of the common parameters, accounting for uncertainty in these quantities. That is,

$$p\left(\beta_h| \{y_1, \dots, y_h, \dots, y_H, X_1, \dots, X_h, \dots, X_H\}, Z\right)$$

$$= \int p\left(\beta_h | y_h, X_h, \Lambda, \Delta, V_\beta\right)$$

$$\times \ p\left(\Lambda, \Delta, V_\beta | \left\{y_1, \ \dots \ , y_h, \ \dots \ , y_H, X_1, \ \dots \ , X_h, \ \dots \ , X_H\right\}, Z\right) d\Lambda d\Delta dV_\beta$$

Z is the matrix of all household demographics. We cannot simply "plug-in" estimates of the common parameters.

$$p\left(\beta_h | \left\{y_1, \ \dots \ , y_h, \ \dots \ , y_H, X_1, \ \dots \ , X_h, \ \dots \ , X_H\right\}, Z\right) \neq p\left(\beta_h | y_h, X_h, \hat{\Lambda}, \hat{\Delta}, \hat{V}_\beta\right)$$

The second and third information sets require some thought. If we only have demographic information but no purchase information on household h, we then must compute an appropriate predictive distribution from the model in (6.11.3).

$$\int p\left(\beta_h | z_h, \Delta, V_\beta\right) p\left(\Delta, V_\beta | Info\right) d\Delta dV_\beta \tag{6.11.4}$$

In order to undertake the computation in (6.11.4), we must specify an information set on which we base our inferences on the parameters of the heterogeneity distribution. Our idea is that we might have a "pet" panel of households, not including household h, on which we observed purchases and demographics so that we can gauge the distribution of probit parameters. That is, we have a sample of data which enables us to gauge the extent of differences among households in this population. This distribution could also simply reflect prior beliefs on the part of managers regarding the distribution. To implement this, we use the posterior distribution from our sample of households. We can define a simulator for (6.11.4) by using all R draws of Δ, V_β and drawing from the appropriate normal draw.

$$\beta_h^r | \Delta^r, V_\beta^r \sim N\left(\Delta^{r\prime} z_h, V_\beta^r\right) \tag{6.11.5}$$

The third and coarsest information set is the same information set used in setting a blanket coupon face value. No information is available about household h. The only information available is information regarding the distribution of the β parameters and demographics in the population. In this situation, we simply integrate over the distribution of demographics as well as the posterior distribution of Δ, V_β.

$$\int p\left(\beta_h | z_h, \Delta, V_\beta\right) p\left(z_h\right) p\left(\Delta, V_\beta | Info\right) dz_h d\Delta dV_\beta \tag{6.11.6}$$

Rossi et al. use the empirical distribution of z to perform this integral.

Thus, we now have the ability to compute the predictive distribution of each household parameter vector for each of the three information sets. We must now pose the decision problem. The problem is to choose the face value of a coupon to as to maximize total profits. We model the effect of a coupon as simply a reduction in price for the brand for which the coupon is issued. If the coupon is issued for brand i, the decision problem as be written as follows:

$$\max_F \pi\left(F\right) = \int \Pr\left[i | \beta_h, \Lambda, X\left(F\right)\right] \left(M - F\right) p\left(\beta_h, \Lambda | \Omega^*\right) d\beta_h d\Lambda \tag{6.11.7}$$

$p\left(\beta_h, \Lambda | \Omega^*\right)$ is the predictive or posterior distribution for information set Ω^*. M is the margin on the sale of brand i without consideration of the coupon cost. $X(F)$ denotes the value of the marketing mix variables with a coupon of face value, F. We assume that the effect of the coupon is to reduce the price of alternative i by F. In many cases, the amount of information available about β_h is very small so that the predictive distribution will be very diffuse. It is therefore, extremely important that the integration in (6.11.7) be performed. If "plug-in" estimates of β_h are used, profits can be dramatically overstated due to overconfidence. At the plug-in estimate, the probabilities can be more extreme, suggesting that the coupon will have a greater effect on purchase behavior. In the case of the most popular Catalina product, coupons are issued on the basis of only *one* purchase observation so the extent of the effect can be huge.

Figure 6.5 shows the distribution of expected revenue for a specific household and various coupons face values in an example drawn from Rossi et al. The product is canned tuna fish and the coupon face values are given in cents and restricted to be multiples of 5 cents. The boxplots show the considerable uncertainty regarding expected revenue based on very imprecise inferences about β_h. The solid dots in the figure correspond to predicted revenue as the "plug-in" estimate of β_h equal to the posterior mean. The figure shows the "overconfidence" aspect of plug estimates.

Rossi et al. demonstrate that, relative to no household-specific information, various information sets regarding households have the potential for large value. That is, even with a small amount of information, the ability to customize the face value of the coupon is high. Revenues from even one observation are over 50% higher than in the blanket coupon condition. Revenues from longer purchase histories can be even greater, exceeding 100% larger.

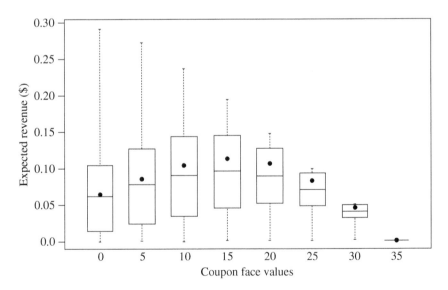

Figure 6.5 Posterior distribution of expected revenues for various coupon values. (Reprinted by permission, Rossi et al. [1996]. Copyright 1996, the Institute for Operations Research and the Management Sciences, 7240 Parkway Drive, Suite 310, Hanover, MD 21076 USA.)

7

Simultaneity

Abstract

This chapter discusses the problem of Bayesian inference for models in which both the response variable and some of marketing mix variables are jointly determined. We can no longer focus only on the model of the distribution of the response variable conditional on the marketing mix variables. We must build a model which (at least implicitly) specifies the joint distribution of these variables conditional on a set of driving or "exogenous" variables. Section 7.1 provides a Bayesian treatment of the linear "instrumental" variables problem, a problem for which standard classical asymptotic methods have proved inadequate. Section 7.2 considers a system of supply and demand where the demand system is built up by aggregating consumer level choice models. Section 7.3 considers the situation in which simultaneity is present in a hierarchical model.

At the base of all the models considered so far is the distribution of a dependent variable (or vector) conditional on a set of independent factors. The classic marketing example is a sales response model in which quantity demanded is modeled conditional on marketing mix variables which typically include price and advertising measures. However, we should recognize that firms may set these marketing mix variables in a strategic fashion. For example, firms may consider strategic considerations or the response of other competitors in setting price. Firms may also set the levels of marketing mix variables by optimizing a profit function that involves the sales response parameters. These considerations lead toward a joint or "simultaneous" model of the entire vector of sales responses and marketing mix variables. In this chapter, we will consider three approaches to this problem. We will first consider a Bayesian version of the instrumental variables or "limited information" approach. We will then consider a joint or simultaneous approach. Finally, we will consider the implications of optimization in the selection of the marketing mix on the estimation of conditional sales response models.

Bayesian Statistics and Marketing, Second Edition. Peter E. Rossi, Greg M. Allenby, and Sanjog Misra
© 2024 John Wiley & Sons Ltd. Published 2024 by John Wiley & Sons Ltd.

7.1 A BAYESIAN APPROACH TO INSTRUMENTAL VARIABLES

Instrumental variables techniques are widely used in economics to avoid the "endogeneity" bias of including an independent variable that is correlated with the error term. It is probably best to start with a simple example.[1]

$$x = \delta z + \varepsilon_1 \qquad (7.1.1)$$

$$y = \beta x + \varepsilon_2 \qquad (7.1.2)$$

If $\varepsilon_1, \varepsilon_2$ are independent, then both (7.1.2) and (7.1.1) are valid regression equations and form what is usually termed a recursive system. From our perspective, we can analyze each equation separately using the standard Bayesian treatment of regression. However, if $\varepsilon_1, \varepsilon_2$ are dependent, then (7.1.2) is no longer a regression equation in the sense that the conditional distribution of ε_2 given x depends on x. In this case, x is often referred to as an "endogenous" variable. If (7.1.1) is still a valid regression equation, econometricians call z an "instrument." z is a variable related to x but independent of ε_2.

The system in (7.1.1)–(7.1.2) can be motivated by the example of a sales response model. If y is sales volume in units and x is price, then there are situations in which x can depend on the value of ε_2. One situation (c.f. Villas-Boas and Winer [1999]) has a shock that affects the demand of all consumers and which is known to the firm setting prices. For example, a manufacturer coupon will be dropped in a market. The drop of a blanket coupon will presumably increase retail demand for the product for all consumers. If retailers know about the coupon drop, they may adjust retail price (x) in order to take advantage of this knowledge. This would create a dependence or correlation between x and ε_2. Another classic example, in this same vein, is the "unobserved" characteristics argument of Berry et al. [1995]. In this example, there are characteristics of a product which drive demand but are not observed to the econometrician. These unobserved characteristics also influence price. If we are looking across markets, then these characteristics could be market-specific demand differences. If the data are a time series, then these characteristics would have to vary across time. In both cases, this motivates a dependence between x and ε_2 in the same manner as the common demand shock argument.

These examples provide an "omitted" variables interpretation of the "endogeneity" problem. If we were able to observe the omitted variable, then we would have a standard multivariate regression system.

$$x = \delta z + \alpha_x w + u_1 \qquad (7.1.3)$$

$$y = \beta x + \alpha_y w + u_2 \qquad (7.1.4)$$

If w is unobserved, this induces a dependence or correlation between x and $\varepsilon_2 = \alpha_y w + u_2$. w would be interpreted as the common demand shock or unobserved characteristic (s) in the examples mentioned above. The consequence of ignoring the dependence between x and ε_2 and analyzing (7.1.2) using standard conditional models is a so-called "endogeneity" bias. Only part of the movement in x helps us to identify β. The part of the movement in x that is correlated with ε_2 should not be used

---

[1] See also Lancaster [2004], chapter 8 for an excellent introduction to instrumental variables as well as a lucid discussion of the classic returns to education endogeneity problem.

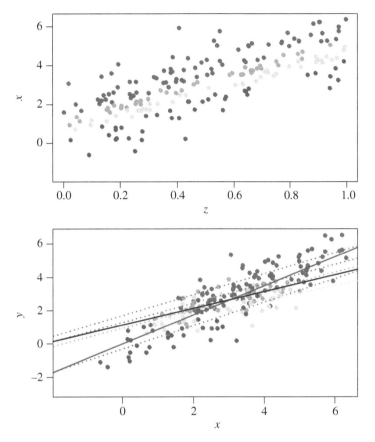

Figure 7.1 Illustration of instrumental variables method

in inferring about β. This means that if the correlation between ε_1 and ε_2 is positive, there will be a positive "endogeneity" bias.

To illustrate this situation, Figure 7.1 plots x vs z and y vs x using a "brushing" technique. The data is simulated from a model in which errors are bivariate normal with correlation $=.8$. Points of the same color are linked between the two scatterplots. In the graph of x vs z, we hold the variation in the residuals of x on z relatively constant. Thus, within each shade group we see variation in x that has been created by movements in the instrument z. The dotted lines in the y vs x graph show a least squares estimate of β for each brushing group. The solid grey line shows the biased least squares estimate of β from a simple regression of y on x while the solid black line shows the true value of β. The role of the instrument is to inject variation into x that is independent of ε_2.

It is easy to see that if the instrument is "weak" in the sense that it induces little variation in x, then there will be little information in the sample regarding β. What is less obvious is that, as instruments become weaker and weaker, we approach an unidentified case. To see this, we need to write down the likelihood for this model. The likelihood is the joint distribution of (x, y) given z. This is often called the "reduced" form. In order to proceed further, we need to make assumptions regarding the joint distribution of $\varepsilon_1, \varepsilon_2$. Given the regression specification, a natural starting point would be to assume that $\varepsilon_1, \varepsilon_2$ are bivariate normal.

The joint distribution of (x, y) given z can be written down by substituting (7.1.1) into (7.1.2).

$$x = \delta z + \varepsilon_1$$
$$y = \beta \delta z + (\beta \varepsilon_1 + \varepsilon_2) \tag{7.1.5}$$

or

$$x = \pi_x z + v_1$$
$$y = \pi_y z + v_2 \tag{7.1.6}$$

Here $\beta = \pi_y / \pi_x$ and we can think of the model as a multivariate regression with a restriction on the coefficient matrix. This is the approach taken by Lancaster [2004]. However, we must recognize that the covariance of v_1, v_2 depends on β and elements of the covariance matrix of the "structural" equation errors, ε_1 and ε_2. This is also the source of potential identification problems in this model. Consider the case with $\delta = 0$. Here

$$x = \varepsilon_1$$
$$y = \beta \varepsilon_1 + \varepsilon_2 \tag{7.1.7}$$

Variances and covariances are identified via observation of x and y. Equation (7.1.7) implies that

$$\frac{\text{cov}(x, y)}{\text{var}(x)} = \beta + \frac{\sigma_{12}}{\sigma_{11}} \quad \text{with } \sigma_{12} = \text{cov}(\varepsilon_1, \varepsilon_2) \tag{7.1.8}$$

The identified quantity on the left hand side of (7.1.8) can be achieved by many different combinations of β and $\frac{\sigma_{12}}{\sigma_{11}}$. Thus, the model, in the limiting case of weak instruments, is not identified. In the case where variation in is a small fraction of the total variation in δx, there will be a "ridge" in the likelihood, reflecting the trade-off between β and $\frac{\sigma_{12}}{\sigma_{11}}$.[2]

Consider a more general version of (7.1.1)–(7.1.2),

$$x = z' \delta + \varepsilon_1 \tag{7.1.9}$$

$$y = \beta x + w' \gamma + \varepsilon_2 \tag{7.1.10}$$

$$\begin{pmatrix} \varepsilon_1 \\ \varepsilon_2 \end{pmatrix} \sim N(0, \Sigma) \tag{7.1.11}$$

This system shows the case a structural equation (7.1.10) with one "endogenous" variable, multiple instruments, and an arbitrary number of other regressors. We will put standard conditionally conjugate priors on these parameters.

$$\Sigma \sim IW(v, V)$$
$$\delta \sim N\left(\bar{\delta}, A_\delta^{-1}\right)$$
$$\begin{pmatrix} \beta \\ \gamma \end{pmatrix} \sim N\left(\begin{pmatrix} \bar{\beta} \\ \gamma \end{pmatrix}, A_{\beta\gamma}^{-1}\right) \tag{7.1.12}$$

---

[2] Note that this is different from the intuition in Lancaster [2004]. Lancaster parameterizes the covariance of the reduced form errors, ignoring the relationship between the structural errors and the reduced form errors. The dependence between δ and β given the correlation in reduced form errors shown in Lancaster is, in fact, due to the correlation between β and $\frac{\sigma_{12}}{\sigma_{11}}$ in our parameterization.

We believe that it is appropriate to put a prior on the covariance matrix of the structural errors. Lancaster [2004] puts an IW prior on the covariance matrix of the reduced form errors and makes this prior independent of the prior on β. The reduced form errors are related to the structural errors by the transformation:

$$\begin{pmatrix} v_1 \\ v_2 \end{pmatrix} = \begin{bmatrix} 1 & 0 \\ \beta & 1 \end{bmatrix} \begin{pmatrix} \varepsilon_1 \\ \varepsilon_2 \end{pmatrix} \qquad (7.1.13)$$

Given the relationship in (7.1.13), we do not feel that a prior on the covariance of the reduced form errors should be independent of the prior on β. In our prior, we induce a dependence via (7.1.13).

We can easily develop a Gibbs Sampler for the model in (7.1.9)–(7.1.12).[3] The basic three sets of conditionals are given by

$$\beta, \gamma \mid \delta, \Sigma, x, y, w \qquad (7.1.14)$$

$$\delta \mid \beta, \gamma, \Sigma, x, y, w \qquad (7.1.15)$$

$$\Sigma \mid \beta, \gamma, \delta, x, y, w \qquad (7.1.16)$$

The first conditional in (7.1.14) can easily be accomplished by a standard Bayesian regression analysis. The key insight is to recognize that given δ, we can "observe" ε_1. We can then condition our analysis of (7.1.10) on ε_1.

$$\begin{aligned} y &= \beta \left(z'\delta + \varepsilon_1 \right) + w'\gamma + \varepsilon_2 \mid \varepsilon_1 \\ &= \beta \left(z'\delta + \varepsilon_1 \right) + w'\gamma + \frac{\sigma_{12}}{\sigma_{11}} \varepsilon_1 + v_{2\mid 1} \end{aligned} \qquad (7.1.17)$$

Since $var\left(v_{2\mid 1} \right) \equiv \sigma_{2\mid 1}^2 = \sigma_{22} - \frac{\sigma_{12}^2}{\sigma_{11}}$, we can rewrite (7.1.17) so that we can use standard Bayes regression with a unit variance error term.

$$\frac{y - (\sigma_{12}/\sigma_{11})\varepsilon_1}{\sigma_{2\mid 1}} = \beta x/\sigma_{2\mid 1} + (w/\sigma_{2\mid 1})'\gamma + \zeta \qquad \zeta \sim N(0, 1) \qquad (7.1.18)$$

The second conditional in (7.1.15) can be handled by transforming to the reduced form which can be written as a regression model with "double" the number of observations.

$$x = z'\delta + \varepsilon_1$$

$$\tilde{y} = \left(\frac{y - w'\gamma}{\beta} \right) = z'\delta + \left(\varepsilon_1 + \frac{\varepsilon_2}{\beta} \right) \qquad (7.1.19)$$

We can transform the system above to an uncorrelated set of regressions by computing the covariance matrix of the vector of errors in (7.1.19).

---

[3] Geweke [1996] considers a model similar to ours but with a "shrinkage" prior that specifies that each of the regression coefficients are independent with zero mean and the same variance.

$$Var \left(\begin{array}{c} \varepsilon_1 \\ \varepsilon_1 + \frac{1}{\beta}\varepsilon_2 \end{array} \right) = A\Sigma A' = \Omega = U'U$$

$$A = \left[\begin{array}{cc} 1 & 0 \\ 1 & \frac{1}{\beta} \end{array} \right]$$

(7.1.20)

Pre-multiplying (7.1.19) with (U^{-1}), reduces the system to a bivariate system with unit covariance matrix and we can simply stack it up and perform a Bayes regression analysis with unit variance. That is,

$$(U^{-1}), \left(\begin{array}{c} x \\ \tilde{y} \end{array} \right) = (U^{-1}), \left[\begin{array}{c} z' \\ z' \end{array} \right] \delta + u \qquad Var(u) = I_2$$

(7.1.21)

The draw of Σ given the other parameters can be accomplished by computing the matrix of residuals and doing a standard IW draw.

$$\Sigma \,|\, \delta, \beta, \gamma, x, y, w \sim IW(v + n, S + V)$$
$$S = \sum_{i=1}^{n} \varepsilon_i \varepsilon_i'$$
$$\varepsilon_i = \left(\begin{array}{c} \varepsilon_{1,i} \\ \varepsilon_{2,i} \end{array} \right)$$

(7.1.22)

The Gibbs sampler defined by (7.1.14)–(7.1.16) is implemented in the function

R rivGibbs in our R package, bayesm. In the case of weak instruments and high correlation between the two structural errors (high "endogeneity"), we might expect that the Gibbs Sampler will exhibit the highest autocorrelation due to the "ridge" in the likelihood between β and $\frac{\sigma_{12}}{\sigma_{11}}$. The "hem-stitching" behavior of the Gibbs sampler might induce slower navigation and, hence, autocorrelation in this case.

To illustrate the functioning of the sampler and to gain insight into this model, we consider a simulated example.

$$x = \lambda_x + \delta z + \varepsilon_1$$
$$y = \lambda_y + \beta x + \varepsilon_2$$
$$\left(\begin{array}{c} \varepsilon_1 \\ \varepsilon_2 \end{array} \right) \sim N \left(0, \left[\begin{array}{cc} 1 & \rho \\ \rho & 1 \end{array} \right] \right)$$

(7.1.23)

Z is a vector of 200 unif(0,1) draws and $\lambda_x = \lambda_y = 1$. We use relatively diffuse priors for coefficients, $A = 0.04I$, and for the covariance matrix, $\Sigma \sim IW(3, 3I_3)$. We first consider the case of "strong instruments" and a high degree of "endogeneity." This is achieved by simulating data with $\delta{=}4$ and $\rho = 0.8$. In this situation, much of the variation in x is "exogenous" due to variation in δz and we would expect to see good performance of the sampler. Figure 7.2 shows plots of the posterior draws for each pair of the three key parameters, $\left(\beta, \delta, \frac{\sigma_{12}}{\sigma_{11}} \right)$. In addition, we display the sequence of the beta draws on the bottom right. 5000 draws were used with every 5th draw plotted. The dotted lines in each of the scatterplots represent the true values used to simulate the data. The sampler performs very well with a very short "burn-in" period and high numerical efficiency

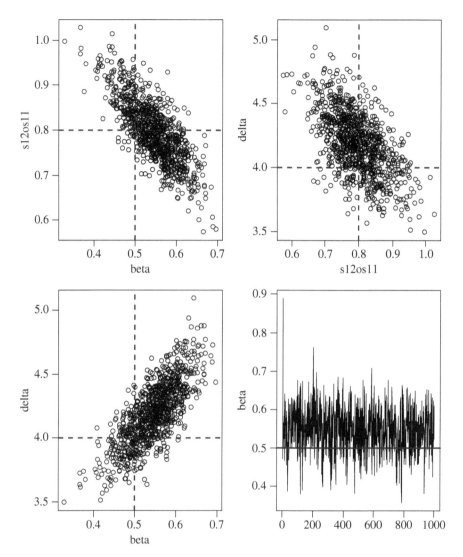

Figure 7.2 MCMC draws for the strong instrument/high endogeneity case

relative to an iid sampler ($\sqrt{f} = 1.6$). As expected, there is a negative correlation between β and $\frac{\sigma_{12}}{\sigma_{11}}$ but it is not too high.

It is well known in the classical econometrics literature that "weak" instruments present problems for the standard asymptotic approximations to the distribution of instrumental variables estimators. This problem is caused by the fact that the IV estimator is a ratio of sample moments. In a likelihood-based procedure such as ours, the weak instruments case creates a situation of near non-identification as indicated in equation (7.1.8). This means there will be a ridge in the likelihood function that will create problems for asymptotic approximations. Since our methods do not rely on asymptotic approximations of any kind, we should obtain more accurate inferences.

However, we should point out that since there is a near nonidentification problem, the role of the prior will be critical in determining the posterior. In addition, our intuition suggests that the MCMC draws may become more autocorrelated due to the ridge in the likelihood function.

In the simulated example, the weak instruments case corresponds to situations in which the variation in δz is small relative to the total variation in x. If we set $\delta = .1$, we create a situation with extremely weak instruments (the population "R-squared" of the instrument regression is .01). Figure 7.3 shows the distribution of the posterior draws for $\delta = 0.1$ and $\rho = 0.8$ that we dub the case of weak instruments and high endogeneity. As expected, there is a very high negative correlation between the draws of β and $\frac{\sigma_{12}}{\sigma_{11}}$.

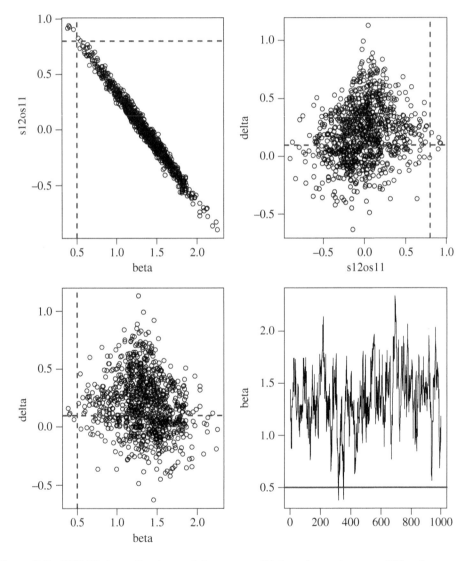

Figure 7.3 MCMC draws for the weak instrument/high endogeneity case: diffuse but proper prior

This creates a sampler with higher autocorrelation ($\sqrt{f} = 10$), or a relative numerical efficiency of $1/10$th of an iid sampler. However, more than 1 million draws of the sampler can be achieved in less than one hour of computing time so that we do not think this is an important issue (Figure 7.2 is based on 10,000 draws with every 10 draw plotted).

It is important to note the influence of the prior for the weak instrument case. We note that the posterior distributions of β and $\frac{\sigma_{12}}{\sigma_{11}}$ is centered away from the true values. This is due to the prior on Σ. Even though it is set to barely proper values, the prior is centered on the identity matrix. This means that $\frac{\sigma_{12}}{\sigma_{11}}$ will be "shrunk" toward zero. There is so little information in the likelihood that this prior "shows" through. However, it is important to realize that the posterior of both quantities is very diffuse, revealing the near lack of identification. Figure 7.4 shows the same situation with an improper prior on Σ. Now the posterior is centered closer to the true values but with huge variability.

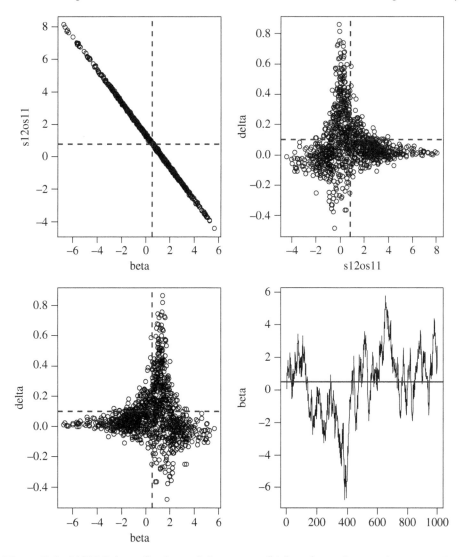

Figure 7.4 MCMC draws for the weak instrument/high endogeneity case: improper prior

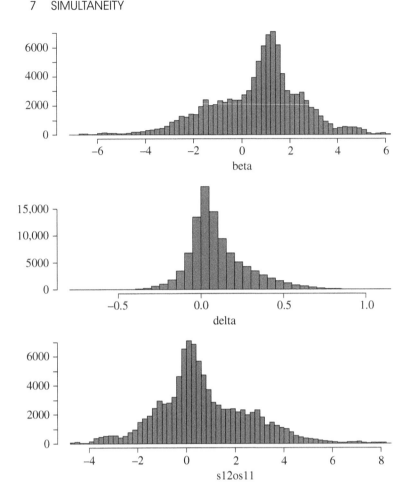

Figure 7.5 Marginal posteriors for weak instruments/high endogeneity case: improper prior, 100,000 draws

Figure 7.5 shows the marginal posterior distributions for a longer run of 100,000 draws. We can see that the marginal posteriors for β and $\frac{\sigma_{12}}{\sigma_{11}}$ have a mode at the true value and a huge "shoulder" representing high uncertainty.

One might argue that the case of weak instruments and a modest or low degree of endogeneity is more representative of the applications of instrumental variable models. For this reason, we considered the case $\delta = 0.1$ and $\rho = 0.1$. Figure 7.6 shows a plot of the posterior draws for these parameter settings, using the proper but diffuse prior on Σ. There is still a high degree of dependence between β and $\frac{\sigma_{12}}{\sigma_{11}}$, reflecting the low amount of information in the weak instrument case. The sampler is slightly less auto-correlated than the weak instruments/high endogeneity case with a relative numerical efficiency of 1/8th of an iid sample. The posterior distributions are very spread out, properly reflecting the small amount of sample information. In addition, the bivariate posteriors have a decidedly nonelliptically symmetric shape, reminiscent of the shape of a double exponential distribution.

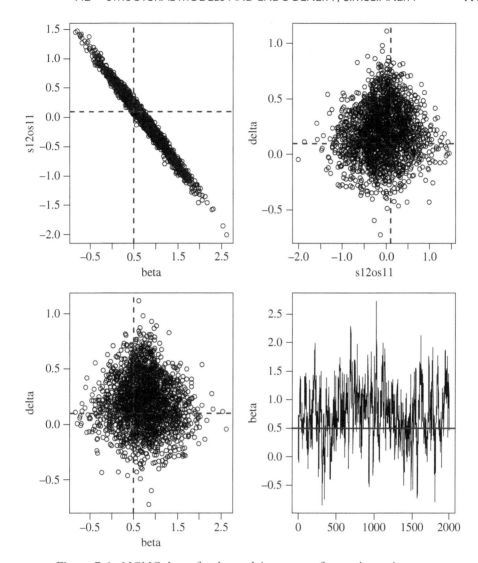

Figure 7.6   MCMC draws for the weak instrument/low endogeneity case

7.2 STRUCTURAL MODELS AND ENDOGENEITY/SIMULTANEITY

The basic "instrumental variable" model can also be interpreted as resulting from a structural model with attention "limited" to one structural equation. The classic example is the example of supply and demand. If we observed marketing clearing quantities and prices along with a "demand" shifter or "exogeneous" variable, then we can write down a model that is identical to our simple example which we started Section 7.1 with.

$$
\begin{aligned}
q_d &= \alpha_0 + \alpha_1 p + \alpha_2 z + \varepsilon_d \\
q_s &= \beta_0 + \beta_1 p + \varepsilon_s
\end{aligned}
\tag{7.2.1}
$$

Here z is a variable such as advertising or promotion that shifts demand but does not alter supply conditions. If we impose the market clearing condition that $q_d = q_s$, then we can rewrite the first equation as a regression and obtain the "limited" information or instrumental variables model.

$$p = \frac{\alpha_0 - \beta_0}{\beta_1 - \alpha_1} + \frac{\alpha_2}{\beta_1 - \alpha_1} z + \frac{\varepsilon_d - \varepsilon_s}{\beta_1 - \alpha_1}$$

$$q = \beta_0 + \beta_1 p + \varepsilon_s$$

$$(7.2.2)$$

Thus, one possible justification of "instrumental" variables approach is that the model arises from some joint or simultaneous model. The joint model imposes the condition of market equilibrium and makes an assumption about the joint distribution of $(\varepsilon_s, \varepsilon_d)$. This allows us to formulate the likelihood that is the joint distribution of (q, p) given z via the Jacobian of the transformation from $(\varepsilon_s, \varepsilon_d)$ to (q, p). This idea can be extended to more "realistic" models of demand in which the demand equation at the aggregate level is the sum of the demands over heterogeneous consumers. Yang et al. [2003b] tackle this problem with a demand model aggregated from heterogenous logits.

The estimation of simultaneous demand and supply with household heterogeneity is challenging. In a full information analysis, where specific assumptions are made about the nature of competition among firms, the supply-side equation is a complex function of household-level parameters and common error terms, or shocks. To date, likelihood-based approaches have not been developed for models with unobserved heterogeneity specified as random-effects. The reason is that a frequentist (i.e., non-Bayesian) approach to analysis does not view the shocks and random-effects as parameters, and analysis proceeds by first integrating them out of the likelihood function. This integration is computationally demanding when the shocks and random-effects are jointly present, and researchers have instead controlled for consumer heterogeneity by incorporating past purchases, or other exogenous (e.g., demographic) variables, into the model specification. In the context of demand analysis using aggregate level data, it has been shown that it is important to control for both price endogeneity and heterogeneity to avoid potential biases in demand side parameter estimates.

We first set notation by developing the demand and supply models used in our study of consumer brand choice and retailer pricing. Consumers are assumed to make brand choice decisions according to a standard discrete choice model. On the supply side, we develop specifications derived from profit-maximizing assumptions made about manufacturer and retailer behavior.

7.2.1 Demand Model

The disaggregate demand function is specified as a logistic normal regression model. Suppose we observe purchase incidences and choices (y) for a group of individuals ($i = 1, \dots, I$) for J brands ($j = 0, \dots, J$) in a product category over T time periods ($t = 1, \dots, T$). The utility of consumer i for brand j at time t is specified as:

$$u_{ijt} = \beta_i' x_{jt} + \alpha_i p_{jt} + \xi_{jt} + \varepsilon_{ijt}$$

$$(7.2.3)$$

and in the case of no purchase from the J available brands, we denote $j = 0$ and the associated utility function as:

$$u_{i0t} = \varepsilon_{i0t} \tag{7.2.4}$$

where x_{jt} is a vector with observed product characteristics including brand intercepts, feature and display variables, p_{jt} is the unit price for brand j at time t, β_i and α_i are individual-level response coefficients, ξ_{jt} is an unobserved demand shock for brand j at time period t, and ε_{ijt} is the unobserved error term that is assumed to be uncorrelated with price. We make assumptions on the error terms and response coefficients as the following:

$$\varepsilon_{ijt} \sim \text{Extreme Value}(0, 1)$$
$$\xi_t \sim MVN(0, \Sigma_d) \tag{7.2.5}$$
$$\theta_i = (\alpha_i, \beta_i')' \sim MVN(\bar{\theta}, \Sigma)$$

The type I extreme value specification of ε_{ijt} leads to a standard logit choice probability for person i choosing brand j at time t,

$$Pr(y_{ijt} = 1) = s_{ijt} = \frac{\exp(V_{ijt})}{1 + \sum_k \exp(V_{ikt})} \tag{7.2.6}$$

where $V_{ijt} = \beta_i' x_{jt} + \alpha_i p_{jt} + \xi_{jt}$
Assuming the sample is representative of the market and households do not make multiple purchases, we obtain market share for brand j at time t as,

$$s_{jt} = \sum_i s_{ijt} / I \tag{7.2.7}$$

If firms use expected demand (s_{jt}) to set price, then price is not exogenously determined. Price and the demand-side error, ξ_t in equation (7.2.3), will be correlated because expected demand, used to set prices, is a function of both. That is, it is not possible to write the joint distribution of prices and demand as a conditional demand distribution and a marginal price distribution. Demand and prices are both functions of consumer price sensitivity and other model parameters. Not accounting for the endogenous nature of price will produce biased estimates of model parameters, including household price sensitivity.

7.2.2 Supply Model – Profit Maximizing Prices

We illustrate the use of Bayesian methods to estimate simultaneous demand and supply models using a simple supply-side model. More complicated models of manufacturer and retailer behavior can be found in Yang et al. [2003b]. We assume that each manufacturer produces only one product and maximizes the following objective function

$$Max_{w_i} \pi_i = Ms_i(w_i - c_{mi}) \tag{7.2.8}$$

where M is the potential market size, w is the wholesale price, and c_{mi} is manufacturer i's marginal cost. The first order condition for the manufacturers implies,

$$w - c_m = \left(HQ\right)^{-1}(-s) \qquad (7.2.9)$$

where $H_{ik} = \frac{\partial s_i}{\partial p_k}$ and $Q_{ik} = \frac{\partial p_i}{\partial w_k}$ (i = 1, ... , J, and k = 1, ... , J).

Next, we turn to the retailer's pricing strategy. For the purpose of illustration, we only model a single retailer's pricing behavior even though competition among multiple retailers is possible. One simple rule the retailer can use is to simply charge a fixed markup over wholesale price for each brand, resulting the following specification,

$$p_i = w_i + m_i \qquad (7.2.10)$$

where m stands for the fixed markup. This pricing strategy implies that $\frac{\partial p_i}{\partial w_i} = 1$ and $\frac{\partial p_i}{\partial w_j} = 0$. Substituting those two conditions into equation (7.2.9), we obtain the following pricing equation,

$$p = c_m + m - (H)^{-1}s \qquad (7.2.11)$$

Finally, we can specify the manufacturer and retailer cost c_m as a brand-specific linear function of cost shifters Z, that is

$$c_{mt} = Z_t'\delta_j + \eta_t \qquad (7.2.12)$$

where η_t is the supply side error that we assume a multivariate normal distribution, that is, $\eta_t \sim MVN(0, \Sigma_s)$, or

$$p_t = m + Z_t'\delta_j - H_t^{-1}s_t + \eta_t \qquad (7.2.13)$$

The distribution of observed prices is obtained from the distribution of the supply side error by using change-of-variable calculus. This distribution is of nonstandard form because equation (7.2.13) is implicit in price – that is, price appears on both the left and right side of the equal sign in the terms H and s. The distribution of observed prices is obtained by defining a new variable $r = p + H^{-1}s - m - Z\delta$ that is distributed normal with mean 0 and covariance Σ_s. The likelihood for price is obtained in the standard way as the likelihood for r multiplied by the determinant of the Jacobian ($J = \{\partial r_i/\partial p_j\}$).

7.2.3 Bayesian Estimation

Data augmentation is used to facilitate estimation of the model. We introduce household specific coefficients $\{\theta_i\}$ and supply shock realizations $\{\xi_t\}$ as augmented, latent variables, and use them as conditioning arguments in the model hierarchy. The dependent variables are choice (y_{it}) and prices (p_t), and the model can be written in hierarchical form:

$y_{it} \mid p_t, \theta_i, \xi_t, \epsilon_{it}$ Observed demand

$p_t \mid \{\theta_i\}, \{\xi_t\}, \delta, \eta_t$ Observed prices

$\theta_i \mid \bar{\theta}, \Sigma$ Heterogeneity

$\xi_t \mid \Sigma_d$ Demand shock

$\eta_t \mid \Sigma_s$ Supply shock

ϵ_{it} Extreme value (logit) error

where $\theta_i = (\alpha_i', \beta_i')'$. Observed demand for the ith household is dependent on the household's coefficients (θ_i), the demand shock (ξ_t), the unobserved error (ϵ_{it}), and the explanatory variables, including prices (7.2.3). Observed prices are determined by the set of household coefficients $\{\theta_i\}$, cost shifter coefficients (δ) and the supply shock (η_t), and are set in response to the expected demand across the heterogeneous households (7.2.13). Household coefficients are specified as random-effects, and the demand and supply shocks specified as normally distributed.

Given the household coefficients $\{\theta_i\}$ and demand shocks $\{\xi_t\}$, the joint distribution of demand and prices is obtained by multiplying the conditional (on prices) demand density by the marginal price density. The marginal price density, given $\{\theta_i\}$ and $\{\xi_t\}$, is derived from the supply-side error term, η_t. The conditional demand density, given $\{\theta_i\}$, $\{\xi_t\}$ and prices, is multinomial with logit probabilities (7.2.6). The joint density of all model parameters is then:

$$f(\{\theta_i\}, \{\xi_t\}, \delta, \bar{\theta}, \Sigma_d, \Sigma_s, \Sigma_\theta \mid \{y_{it}\}, \{p_t\})$$

$$\propto \prod_{t=1}^{T} \prod_{i=1}^{I} \Pr(y_{it} \mid \theta_i, p_t, \xi_t) \pi_1(\xi_t \mid \Sigma_d) \pi_2(p_t \mid \{\theta_i\}, \{\xi_t\}, \delta, \Sigma_s) \qquad (7.2.14)$$

$$\times \pi_3(\theta_i \mid \bar{\theta}, \Sigma_\theta) \pi_4(\delta, \bar{\theta}, \Sigma_d, \Sigma_s, \Sigma_\theta)$$

where $\Pr(y_{it} \mid \theta_i, p_t, \xi_t)$ is the logit choice probability for household i at time t, π_1 is the density contribution of the demand error ξ_t, π_2 is the density contribution of the observed prices at time t that depend on consumer preferences and price sensitivities $\{\theta_i\}$, demand errors $\{\xi_t\}$, cost variables and coefficients (Z, δ), and the supply-side error (η_t), π_3 is the distribution of heterogeneity and π_4 is the prior distribution on the hyper-parameters.

Given these augmented variables, estimation proceeds using standard distributions for heterogeneity ($\bar{\theta}$ and Σ) and error covariance matrices (Σ_d and Σ_s). Draws of the augmented variables are obtained from the full conditional distribution for ξ_t and θ_t:

$$[\xi_t \mid *] \propto$$
$$\prod_i \Pr(choice_{it}) \times |J_t| \times (r_t \sim N(0, \Sigma_s)) \qquad (7.2.15)$$
$$\times (\xi_t \sim N(0, \Sigma_d))$$

$$[\theta_i|*] \propto$$
$$\prod_{t=1}^{T} \Pr(choice_{it}) \times \prod_{t=1}^{T} |J_t| \times \prod_{t=1}^{T} \left(r_t \sim N(0, \Sigma_s) \right) \qquad (7.2.16)$$
$$\times \left(\theta_i \sim N(\bar{\theta}, \Sigma_\theta) \right)$$

where

$$\Pr(choice_{it} \neq 0) = \prod_{j=1}^{J} \left(\frac{\exp(V_{ijt})}{1 + \sum_{k=1}^{J} \exp(V_{ikt})} \right)^{I(choice_{ijt}=1)} \qquad \text{(choose one of the brands)}$$

$$\Pr(choice_{it} = 0) = \frac{1}{1 + \exp(V_{it})} \quad \text{(choose outside good)}$$

$$V_{ijt} = \beta_i' x_{jt} + \alpha_i p_{jt} + \xi_{jt}$$

$$|J_t| \text{ is the Jacobian} = \begin{vmatrix} \frac{\partial r_{1t}}{\partial p_{1t}} & \cdots & \frac{\partial r_{1t}}{\partial p_{jt}} \\ \vdots & \ddots & \vdots \\ \frac{\partial r_{jt}}{\partial p_{1t}} & \cdots & \frac{\partial r_{jt}}{\partial p_{jt}} \end{vmatrix}$$

and $r_t = p_t + H_t^{-1} s_t - m - Z_t \delta$

where the jth element of the vector, $H_t^{-1} s_t$, is given by $\dfrac{\sum_{i=1}^{I} s_{ijt}}{\sum_{i=1}^{I} \alpha_i s_{ijt} (1 - s_{ijt})}$

and

$$s_{ijt} = \frac{\exp(V_{ijt})}{1 + \sum_{k=1}^{J} \exp(V_{ikt})}$$

Bayesian analysis does not require the integration of the random-effects (θ_i) and demand shocks (ξ_t) to obtain the marginalized, or unconditional likelihood. Such marginalization is difficult to evaluate because the integral is of high dimension and involves highly nonlinear functions of the model parameters, including the Jacobian needed to obtain the distribution of observed prices. An advantage of using a Bayesian MCMC estimator is that these variables can be used as conditioning arguments for generating other model parameters – resulting in significant simplification in model estimation.

7.3 NON-RANDOM MARKETING MIX VARIABLES

Thus far, we have focused on examples where supply-side behavior has resulted in an "endogeneity" problem in which the marketing mix variables such as price can be functions of unobserved variables. Typically, these ideas are applied to a problem in which the marketing mix variable is being set on a uniform basis for a given market or retailer. However, if firms have access to information regarding the response parameters of a given customer, then the marketing mix variables can be customized to that specific account. This creates a related situation in which the levels of the marketing mix variables are set as a function of the response parameters. Manchanda et al. [2004] explore this problem and offer a general approach as well as a specific application to the problem of allocating sales force in the pharmaceutical industry.

7.3.1 A General Framework

Consider the general setting in which the marketing mix or "x" variables are chosen strategically by managers. The basic contribution is to provide a framework for situations in which the mix variables are chosen with some knowledge of the response parameters of the sales response equation.

Sales response models can be thought of as particular specifications of the conditional distribution of sales (y) given the marketing mix x.

$$y_{it} \,|\, x_{it}, \beta_{it} \qquad\qquad (7.3.1)$$

where i represents the individual customer/account and t represents the time index. For example, a standard model would be to use the log of sales or the logit of market share and specify a linear regression model,

$$\ln\left(y_{it}\right) = x_{it}'\beta_i + \varepsilon_{it} \quad \varepsilon_{it} \sim Normal$$

Here the transform of y is specified as conditionally normal with sales response parameters, β_i. Analysis of (7.3.1) is usually conducted under the assumption that the marginal distribution of x is independent of the conditional distribution in (7.3.1). In this case the marginal distribution of x provides no information regarding β_i and the likelihood factors. If $x_{it}\,|\,\theta$ is the marginal distribution of x, then the likelihood factors as follows

$$\ell\left(\{\beta_i\}, \theta\right) = \prod_{i,t} p\left(y_{it}\,|\,x_{it}, \beta_i\right) p\left(x_{it}\,|\,\theta\right) = \prod_{i,t} p\left(y_{it}\,|\,x_{it}, \beta_i\right) \prod_{i,t} p\left(x_{it}\,|\,\theta\right) \qquad (7.3.2)$$

This likelihood factorization does not occur once the model is changed to build dependence between the marginal distribution of x and the conditional distribution. There are many forms of dependence possible, but in the context of sales response modeling with marketing mix variables a particularly useful form is to make the marginal distribution of x dependent on the response parameters in the conditional model. Thus, our general approach can be summarized as follows:

$$\begin{aligned} y_{it} \,|\, x_{it}, \beta_i \\ x_{it} \,|\, \beta_i, \tau \end{aligned} \qquad\qquad (7.3.3)$$

Equation (7.3.3) is a generalization of the models developed by Chamberlain [1980, 1984] and applied in a marketing context by Bronnenberg and Mahajan [2001]. Chamberlain considers situations in which the x variables are correlated to random intercepts in a variety of standard linear and logit/probit models. Our random effects apply to all of the response model parameters and we can handle non-standard and nonlinear models. However, the basic results of Chamberlain regarding consistency of the conditional modeling approach apply. Unless T grows, any likelihood-based estimator for the *conditional* model will be inconsistent. The severity of this asymptotic bias will depend on model, data, and T. For small T, these biases have been documented to be very large.

The general data-augmentation and Metropolis Hasting MCMC approach is ideally suited to exploit the conditional structure of (7.3.3). That is, we can alternate between draws of $\beta_i \mid \tau$ (here we recognize that the $\{\beta_i.\}$ are independent conditional on τ and $\tau \mid \{\beta_i\}$. With some care in the choice of the proposal density, this MCMC approach can handle a very wide range of specific distributional models for both the conditional and marginal distributions in (7.3.3).

To further specify the model in (7.3.3), it is useful to think about the interpretation of the parameters in the β vector. We might postulate that in the marketing mix application, the important quantities are the level of sales given some "normal" settings of x (e.g., the baseline sales) and the derivative of sales wrt various marketing mix variables. In many situations, decision makers are setting marketing mix variables proportional to the baseline level of sales. More sophisticated decision makers might recognize that the effectiveness of the marketing mix is also important in allocation of marketing resources. This means that the specification of the marginal distribution of x should make the level of x a function of the baseline level of sales and the derivatives of sales with respect to the elements of x.

7.3.2 An Application to Detailing Allocation

Manchanda et al. [2004] consider an application to the problem of allocation of sales force effort ("detailing") across "customers" (physicians). The data is based on sales calls (termed "details") made to physicians for the purpose of inducing them to prescribe a specific drug. In theory, sales managers should allocate detailing efforts across the many thousands of regularly prescribing physicians so as to equalize the marginal impact of a detail across doctors (assuming equal marginal cost which is a reasonable assumption according to industry sources).

The barrier to implementing optimal allocation of detailing effort is the availability of reliable estimates of the marginal impact of a detail. While individual physician level data is available on the writing of prescriptions from syndicated suppliers such as IMS and Scott–Levin, practitioners do not fit individual physician models due to the paucity of detailing data and extremely noisy coefficient estimates obtained from this data. Instead, practitioners pool data across physicians in various groups, usually on the basis of total drug category volume. Detailing targets are announced for each group. Generally speaking, higher volume physicians receive greater detailing attention. Even if detailing had no effect on prescription behavior, volume-based setting of the detailing independent variable would create a spurious detailing effect in pooled data. In addition to general rules which specify that detailing levels are related to volume, it is clear that individual sales force managers adjust the level of detailing given informal sources of knowledge regarding the physician. This has the net effect of making the levels of detailing a function of baseline volume and, possibly, detailing responsiveness. Thus, the independent variable in our analysis has a level that is related to parameters of the sales response function.

Given the need for physician-specific detailing effects, it might seem natural to apply Bayesian hierarchical models to this problem. Bayesian hierarchical models "solve" the problem of unreliable estimates from individual physician models by a form of "shrinkage" or partial pooling in which information is shared across models. A Bayesian hierarchical model can be viewed as a particular implementation of a random coefficients model. If detailing levels are functions of sales response parameters, then standard

Bayesian hierarchical models will be both biased and inefficient. The inefficiency, which can be very substantial, comes from the fact that the *level* of the independent variable has information about the response coefficients. This information is simply not used by the standard approach. We supplement the sales response function by an explicit model for the distribution of detailing which has a mean related to response coefficients. In our application, given that sales (prescriptions) and detailing are count data, we use an Negative Binomial (NBD) regression as the sales response function and a Poisson distribution for detailing. We demonstrate that this joint model provides much more precise estimates of the effects of detailing and improved predictive performance. Rather than imposing optimality conditions on our model, we estimate the detailing policy function used by the sales managers.

7.3.3 Conditional Modeling Approach

A conditional model for the distribution of prescriptions written given detailing and sampling is the starting point for our analysis. Our data are count data with most observations at less than 10 prescriptions in a given month. Manchanda et al. provide some evidence that the distribution of the dependent variable is over-dispersed relative to the Poisson distribution. For this reason, we will adopt the Negative Binomial as the base model for the conditional distribution and couple this model with a model of the distribution of coefficients over physicians. The NBD model is flexible in the sense that it can exhibit a wide range of degrees of overdispersion, allowing the data to resolve this issue. An NBD distribution with mean λ_{it} and overdispersion parameter α is given by

$$\Pr\left(y_{it} = k \mid \lambda_{it}\right) = \frac{\Gamma\left(\alpha + k\right)}{\Gamma\left(\alpha\right)\Gamma\left(k+1\right)}\left(\frac{\alpha}{\alpha + \lambda_{it}}\right)^{\alpha}\left(\frac{\lambda_{it}}{\alpha + \lambda_{it}}\right)^{k} \tag{7.3.4}$$

y_{it} is the number of new prescriptions written by physician i in month t. As α goes to infinity, the NBD distribution approaches the popular Poisson distribution.

We adopt the standard log-link function and specify that the log of the mean of the conditional distribution is linear in the parameters.

$$\lambda_{it} = E\left[y_{it} \mid x_{it}\right] = \exp\left(x'_{it}\beta_i\right) \tag{7.3.5}$$

$$\ln\left(\lambda_{it}\right) = \beta_{0,i} + \beta_{1,i}Det_{it} + \beta_{2,i}\ln\left(y_{it-1} + d\right) \tag{7.3.6}$$

The lagged log-prescriptions term, $\ln\left(y_{it-1} + d\right)$, in equation (7.3.6) allows the effect of detailing to be felt not only in the current period, but in subsequent periods. We add d to the lagged level of prescriptions to remove problems with zeroes in the prescription data. The smaller the number added, the more accurate the Koyck solution is as an approximation. The problem here is that the log of small numbers can be a very large in magnitude that would have the effect of giving the zeroes in the data undue influence on the carry-over coefficients. We choose $d = 1$ as the smallest number which will not create large outliers in the distribution of $\ln\left(y_{it} + d\right)$.

To complete the conditional model, we specify a distribution of coefficients across physicians. This follows a standard hierarchical formulation.

$$\beta_i = \Delta z_i + v_i$$

$$v_i \sim N\left(0, V_\beta\right)$$

The z vector includes information on the nature of the physicians practice and level of sampling (note that we use primary care physicians [PCPs] as the base physician type):

$$z' = (1, SPE, OTH, SAMP)$$

SAMP is the mean (per physician) number of monthly samples divided by 10, SPE is a dummy for a specialist, and OTH is a dummy for other physician types making PCPs base of z and the model in (7.3.4) allows there to be a main effect and an interaction for both physician specialty type and sampling. We might expect that physicians with a specialty directly relevant to the therapeutic class of drug X will have a different level of prescription writing. In addition, detailing may be more or less effective depending on the physician's specialty. Sales calls may include the provision of free drug X samples. The effect of sampling is widely debated in the pharmaceutical industry with some arguing that it enhances sales and others arguing for cannibalization as a major effect. Most believe that sampling is of secondary importance to detailing. Sampling is conditional on detailing in the sense that sampling cannot occur without a detail visit. For this reason, we include the average sampling variable in the mean of the hierarchy that creates an interaction term between detailing and sampling.

7.3.4 Beyond the Conditional Model

The company producing drug X does not set detailing levels randomly. The company contracts with a consultant to help optimize the allocation of their national sales force. The consultant recognizes that detailing targets for the sales force should be set at the physician level and not at some higher level such as the sales territory. According to the consultant, detailing is set primarily on the basis of the physician decile computed by IMS for the quarter prior to the annual planning period. IMS assigns each physician to a decile based on the physician's total prescription writing for all drugs in the therapeutic class. These annual targets are then adjusted quarterly based on previous quarter deciles These quarterly adjustments tend to be minor.

Conditional modeling approaches rely on the assumption that the marginal distribution of the independent variables does not depend on the parameters of the conditional distribution specified in (7.3.6). If total category volume is correlated with the parameters of the conditional response model, then this assumption will be violated. We think it is highly likely that physicians who write a large volume of drug X prescriptions regardless of detailing levels (e.g., have high value of the intercept in equation (7.3.6)) will also have higher than average category volume. This means that marginal distribution of detailing will depend at the minimum on the intercept parameter in equation (7.3.6). This dependence is the origin of the spurious correlation that can occur if higher volume physicians are detailed more.

It is also clear that, although detailing targets are set on an annual basis (and revised quarterly), there is much month to month variation in detailing due to factors outside of the control of the sales force managers. In addition, even though detailing targets are set at a high level in the firm, sales force district or territory managers may change the actual level of detailing on the basis of their own specialized knowledge regarding specific physicians. If sales force managers had full knowledge of the functional form and parameters of the detailing response function, then detailing would be allocated so as to equalize the marginal effects across physicians. Given that the current industry practice is not to compute individual physician estimates, it is unreasonable to assume that firms are using a full information optimal allocation approach.

We adopt a specification of the detailing distribution that allows for some partial knowledge of detailing response parameters. A simple but flexible approach would be to assume that detailing is iid with mean set as a function of the long-run response parameters from equation (7.3.6). Note that the average first order autocorrelation for detailing is less than 0.3. Monthly detailing is a count variable with rarely more than 5 details per month. Detailing is modeled as an iid draw from a Poisson distribution with a mean that is a function of baseline sales and the long-run response to detailing.

The iid model of detailing is as given by the Poisson distribution:

$$\Pr\left(Det_{it} = m \,|\, \eta_i\right) = \frac{\eta_i^m \exp\left(-\eta_i\right)}{m!} \tag{7.3.7}$$

The mean of this Poisson distribution is a function of the (approximate long-run) coefficients as:

$$\ln\left(\eta_i\right) = \gamma_0 + \gamma_1 \left(\beta_{0i} / {}_{(1-\beta_{2i})} \right) + \gamma_2 \left(\beta_{1i} / {}_{(1-\beta_{2i})} \right) \tag{7.3.8}$$

The specification in (7.3.8) allows for a variety of different possibilities. If detailing is set with no knowledge of responsiveness to detailing, then we should expect γ_2 to be zero. On the other hand, if detailing is set with some knowledge of responsiveness to detailing, then we should expect γ_1 and γ_2 to have posteriors massed away from zero. There are a variety of different functional forms for the relationship between the mean level of detailing and the response parameters. We regard our specification as exploratory and as a general linear approximation to some general function of long-run effects.

To summarize, our approach is to enlarge the conditional model by specifying a model for the marginal distribution of detailing. The marginal distribution of detailing depends on conditional response parameters. Using the standard notation for conditional distributions in hierarchical models, the new model can be expressed as follows:

$$y_{it} \,|\, Det_{it}, y_{i,t-1}, \beta_i, \alpha \quad NBD\ Regression \tag{7.3.9}$$

$$Det_{it} \,|\, \beta_i, \gamma \quad Poisson\ Marginal \tag{7.3.10}$$

This dependence of marginal distribution on the response parameters alters the standard conditional inference structure of hierarchical models. In the standard conditional model given only by (7.3.9), inference about the response parameters, β_i, is based on time-series variation in detailing for the same physician and via similarities between physicians as expressed by the random effects or first-stage prior. However, when (7.3.10) is added to the model, inferences about β_i will change as new information is available from the level of detailing. The marginal model in (7.3.10) implies that the level of detailing is informative about responsiveness and this information is incorporated into the final posterior on β_i. For example, suppose $\gamma_2 < 0$ in equation (7.3.8), then detailing is set so that less responsive physicians are detailed at higher levels and this provides an additional source of information which will be reflected in the β_i estimates. Thus, the full model consisting of equations (7.3.9) and (7.3.10) can deliver improved estimates of physician level parameters by exploiting information in the levels of detailing. The model specified is conditional on β_i. We add the standard heterogeneity distribution on β_i.

Another way of appreciating this modeling approach is to observe that likelihood for β_i has two components – the NBD regression and the Poisson marginal model.

$$\ell\left(\{\beta_i\}\right) = \prod_i \prod_t p_{NBD}\left(y_{it}\mid Det_{it}, \beta_i, \alpha\right) p_{Poisson}\left(Det_{it}\mid \beta_i, \gamma\right) \tag{7.3.11}$$

β_i is identified from both the NBD and the Poisson portions of the model. Examination of the form of the likelihood in (7.3.11) and the mean function in (7.3.8) indicate some potential problems for certain data configurations. In the Poisson portion of the model, elements of the γ vector and the collection of β_i values enter multiplicatively. In terms of the Poisson likelihood, $\left[\gamma_1, \{\beta_{1,i}\}\right]$ and $\left[-\gamma_1, -\{\beta_{1,i}\}\right]$, for example, are observationally equivalent. What identifies the signs of these parameters is the NBD regression. In other words, if the signs of the detailing coefficients are flipped, then the NBD regression fits will suffer, lowering the posterior at that mode. This suggests that in datasets where there is only weak evidence for the effects of detailing (or, in general, any independent variable), then there may exist two modes in the posterior of comparable height. Navigating between these modes can be difficult for Metropolis-Hastings algorithms. To gauge the magnitude of this problem, we simulated a number of different data sets with varying degrees of information regarding the effect of the independent variable. We found that for moderate amounts of information, similar to that encountered in our data, the multi-modality problem was not pronounced. However, for situations with little information regarding the $\{\beta_i\}$, there could potentially be two modes.

Manchanda et al. [2004] provide extensive analysis of this model applied to detailing and sales (prescription) data for a specific drug category. They find ample evidence that detailing is set with at least partial knowledge of the sales response parameters. They confirm that detailing is set on the basis of expected volume of prescriptions but also on the responsiveness of sales to detailing activities. Finally, the information in the detailing levels improves the physician level parameter estimates substantially.

8

A Bayesian Perspective on Machine Learning

Abstract

Machine Learning (ML) has emerged in the last few decades as a practical, scalable and impactful discipline that offers practical solutions to estimation problems for large scale data. In this chapter we examine the interplay between ML and the Bayesian paradigm. In particular, we look at the isomorphisms in the two frameworks and discuss how novel ML ideas can be used to enrich the Bayesian toolkit.

8.1 INTRODUCTION

The topic of artificial intelligence (AI) pertains to the idea that machines can emulate, and perhaps at some point exceed, human intelligence. In order for machine to exhibit such intelligence it needs to learn how to accomplish tasks. This aspect of training a machine to learn and perform tasks is broadly defined as machine learning or ML. The ML toolkit is large and often encompasses a number of topics that we would traditionally associate with statistics and optimization. Having said that, there are a number of new constructs that are particular to ML that have had an impact both on academia and on practice. Some of these constructs are broad ideas (e.g., Regularization, Bagging, Boosting) while others are practical toolkits (e.g., Deep Learners, Random Forests). The field of ML is too broad to cover in this chapter so we will focus on a subset of these topics. Our particular goal will be to discuss the interplay between ideas from the Bayesian paradigm and some of these novel ML constructs.

Machine learning and the Bayesian toolkit offer two different philosophies but share a common goal. Both aim to solve problems using data. As such, it might be natural

Bayesian Statistics and Marketing, Second Edition. Peter E. Rossi, Greg M. Allenby, and Sanjog Misra.
© 2024 John Wiley & Sons Ltd. Published 2024 by John Wiley & Sons Ltd.

to think of them as substitutes. This, however, is too simplistic a view of the ML/Bayes relation. There are fundamental differences between the two that are worth articulating and appreciating. First, at its core, ML is primarily about prediction. Essentially, ML seeks to learn patterns in data via algorithms and then generalize that learning to new data. More often than not, ML methods are trained to optimize performance of the model on unseen (validation) data by minimizing some pre-specified loss function. On the other hand, as we have seen in this text, Bayesian methods are focused on inference. Put simply, we want to draw conclusions about the underlying structure, relationships, or causes within a given dataset and articulate the uncertainty with which we hold these conclusions to be true. Even if the ultimate goal is prediction, the path taken by the two frameworks is quite different.

Second, typically Bayes requires the model to be complete (at least the traditional view of Bayes) in that we need to fully specify our assumptions for all components of the model. This can be seen in the requirement that we have a well-defined likelihood without which a number of Bayesian procedures discussed in this text would not work. In contrast, ML models tend to focus more on the data and make the most minimal modeling assumptions. They are more similar to the frequentist paradigm where models can be incomplete as long as the targeted constructs are appropriately calibrated so as to enable prediction. This is not to say that Bayes procedures cannot deal with misspecified or incomplete models, or that ML models cannot be completely specified, indeed they can. In most cases, however, Bayes and ML models differ on this front.

A third, and perhaps more subtle, difference is in the domain the two frameworks connote. Machine learning is an ever changing, imprecisely defined field that can subsume almost any behavior that involves a machine and data. In some sense, all of the Bayesian framework can be seen as a subset of ML. Indeed, when convenient or useful, ML practitioners will co-opt Bayesian methods, ideas, and tools as needed. In contrast, the Bayesian paradigm is a rather strictly defined set of constructs that are organized in a very particular structure.

These differences makes the two frameworks philosophically very distant. Even so, as we will see in what follows, a number of the tools used in the two have similar foundational elements and, in some cases, one can manipulate ML models so they are recast or reinterpreted as Bayesian procedures. There are also cases where the ML tools are truly novel constructs which the Bayes toolkit might benefit from adopting. In either case, engaging in this discussion allows us to appreciate the benefits and shortcomings of each and export the useful ideas across the two.

In this chapter, we will focus on some key ideas in machine learning and show they relate to the Bayesian approach. To be clear, given the breadth of the topic we will not be able to be exhaustive in our coverage. The goal will be to introduce some ideas, outline examples and focus on the manner in which we can use ML and Bayes as complements in our analytic exercises. In the discussion that follows, we will delve into three significant topics in the realm of machine learning and data analysis: Regularization, Ensembles (including Bagging and Boosting), and Flexibility (via Deep Learning). These topics, while distinct are not independent ideas and they often complement each other. Our focus will be on the Bayesian interpretation and usage of these concepts. Throughout our discussion, we will provide examples where appropriate to illustrate these concepts and their applications. We will also explore possible use cases in the field of marketing. For example, regularization can help in building more robust predictive models for customer behavior, ensembles can improve the accuracy of these predictions by combining multiple models, and nonparametric methods can provide the flexibility needed to model complex and nonlinear relationships in marketing data.

8.2 REGULARIZATION

The basic idea behind regularization in machine learning is to impose a set of constraints or restrictions on the parameter space. Typically, these restrictions are chosen to induce some amount of sparsity and shrinkage which then aid the estimation process. A similar role is played by priors in the Bayesian world and, as such, it is not surprising that regularization and the imposition of priors are somewhat isomorphic. To be precise, regularization can be shown to being equivalent to the specification of a form of informative priors on the parameter space. To see the note that we can write the log-posterior for any model as,

$$\ln \pi \left(\theta | D\right) = \ell \left(D | \theta\right) + \ln \pi_0 \left(\theta\right) + C \tag{8.2.1}$$

In the above, ℓ is the log-likelihood and $\ln \pi_0$ is the log-prior. If the (negated) loss function used in the ML routine happens to coincide with ℓ then one can view the log-prior as a regularizer. In what follows, we will examine the celebrated LASSO (Least Absolute Shrinkage and Selection Operator) model and examine this correspondence.

8.2.1 The LASSO and Bayes

Let's start with the linear regression specification,

$$\mathbf{y} = \mathbf{X}\boldsymbol{\beta} + \epsilon \tag{8.2.2}$$

Assume that the set of covariates ($\mathbf{X}$) is reasonably large and we believe that only a subset of them are truly relevant for the regression. The ML approach would then aim to regularize the model by penalizing values of β. The LASSO does exactly this and uses an L1 regularization term to penalize the absolute values of the model parameters. The estimator for the LASSO can be written as:

$$\hat{\beta}_{\text{LASSO}} = \arg \min_{\beta} \left\{ \frac{1}{2} \|\mathbf{y} - \mathbf{X}\boldsymbol{\beta}\|_2^2 + \lambda \|\boldsymbol{\beta}\|_1 \right\} \tag{8.2.3}$$

where λ is the regularization parameter usually described as controlling the degree to which the parameters are shrunk toward zero. If there are K covariates then we note here that the term $\|\boldsymbol{\beta}\|_1 \equiv \sum_k |\beta_k|$. In other words, we are adding a penalty based on the absolute size of the parameters ($\boldsymbol{\beta}$).

In the Bayesian approach, we typically assign a prior distribution to the model parameters $\boldsymbol{\beta}$. In particular, for this case, let's adopt the Laplace distribution as the prior:

$$p(\boldsymbol{\beta} | \gamma) = \prod_{k=1}^{K} \frac{\gamma}{2} e^{-\gamma |\beta_k|} \tag{8.2.4}$$

where, as before, K is the number of model parameters and γ is the scale parameter in the Laplace prior. We will also assume that $\epsilon \sim N\left(0, \sigma^2\right)$ to complete the model. Now, to obtain the maximum a posteriori (MAP) estimate of the model parameters, we need to maximize the conditional posterior distribution of β given the data:

$$\hat{\beta}_{MAP} = \arg \max_{\beta} \left\{ \pi(\beta | \mathbf{y}, \mathbf{X}, \gamma, \sigma^2) \right\} \tag{8.2.5}$$

Using Bayes' theorem, this posterior distribution can be expressed as:

$$p(\beta | \mathbf{y}, \mathbf{X}, \gamma, \sigma^2) \propto p(\mathbf{y} | \mathbf{X}, \beta, \sigma^2) \cdot p(\beta | \gamma) \tag{8.2.6}$$

Taking the negative logarithm of the posterior distribution and dropping the constants, we get:

$$-\log p(\beta | \mathbf{y}, \mathbf{X}, \gamma, \sigma) \propto \frac{1}{2\sigma^2} \|\mathbf{y} - \mathbf{X}\beta\|_2^2 + \gamma \|\beta\|_1 \tag{8.2.7}$$

If we now set $\lambda = \gamma \sigma^2$ then the we can see that maximizing the conditional posterior distribution in the Bayesian approach is equivalent to minimizing the LASSO objective function so that,

$$\hat{\beta}_{MAP} = \hat{\beta}_{LASSO}$$

Both LASSO regression and the Bayesian approach with a Laplace prior lead to similar outcomes, as they encourage sparsity in the estimated model parameters. In other words, some of the parameter estimates are shrunk to zero, resulting in feature selection. Obviously, if we chose to use alternative approaches to fully characterize the joint posterior, say using MCMC, we would get a different set of results and insights. Even so, that fact remains that regularization in ML has very close ties to the idea of imposing informative priors in the Bayes world.

Different forms of regularization can also be shown to have Bayesian prior counterparts. For example, Ridge regression is a linear regression method that uses an $L2$ regularization term to penalize the squared values of the model parameters. The objective function for Ridge regression is:

$$\hat{\beta}_{Ridge} = \arg \min_{\beta} \left\{ \frac{1}{2} \|\mathbf{y} - \mathbf{X}\beta\|_2^2 + \frac{\lambda}{2} \|\beta\|_2^2 \right\} \tag{8.2.8}$$

In this case, it is easy to see that since $\|\beta\|_2^2 = \sum_{k=1}^{K} \beta_k^2$, this corresponds to the a Normal prior distribution on β with mean 0, and variance $\sigma^2 = \frac{1}{\lambda}$. Similarly if we use Gaussian–Laplace mixture prior,

$$p(\beta | \gamma, \sigma, \alpha) = \prod_{k=1}^{K} \left[\alpha \frac{\gamma}{2} e^{-\gamma |\beta_k|} + (1 - \alpha) \frac{1}{\sqrt{2\pi\sigma^2}} e^{-\frac{(\beta_k)^2}{2\sigma^2}} \right] \tag{8.2.9}$$

We would get the equivalent of the Elastic Net (Li and Lin [2010]). For a more detailed discussion of this topic and for other regularization ideas and their link to Bayes, see Polson and Sokolov [2019]. We note here that variable selection has a long tradition in statistics and has seen numerous examples in the marketing domain as well. In that literature, the role of the prior, much like in the LASSO case, is also to induce sparsity.

For example, the Spike and Slab prior proposed by George and McCulloch [1993] can be described as

$$\beta_k \mid \theta, \sigma^2 \sim (1 - \theta)\delta_0 + \theta N\left(0, \sigma^2\right) \tag{8.2.10}$$

A more practical reparameterization is to use $\beta_k = \gamma_k \alpha_k$ where for each parameter index $k \in \{1, \dots, K\}$

$$\gamma_k \mid \theta \sim \text{Bernoulli}(\theta)$$
$$\alpha_k \mid \sigma^2 \sim N\left(0, \sigma^2\right) \tag{8.2.11}$$

Then the complete prior can be written as,

$$p\left(\gamma, \alpha \mid \theta, \sigma^2\right) = \prod_{k=1}^{K}\left[\theta^{\gamma_k}(1 - \theta)^{1 - \gamma_k}\frac{1}{\sqrt{2\pi}\sigma}\exp\left\{-\frac{\alpha_k^2}{2\sigma^2}\right\}\right] \tag{8.2.12}$$

One issue with the Spike and Slab prior is that with large number of parameters the MCMC approach becomes tedious. Alternatively one might consider a MAP estimator and then by using θ and σ as parameters of a penalty function we could treat the optimization as a new ML approach. This is essentially what Rockova and George [2014] implement via an Expectation-Maximization routine that obtain the MAP estimator. In a reversal, the Bayes approach is abandoned for a more practical ML style estimator. The use of the MAP estimator, sampling data to increase speed and efficiency in Rockova and George [2014] are all features borrowed from the ML toolkit.

8.2.2 Discussion: Informative Regularizers

In the above discussion, we have discussed the idea that parameters obtained by minimizing some penalized loss function in ML problems can be reinterpreted as Bayesian MAP estimators. In the Spike and Slab example, we saw that a Bayesian fully specified model could be reduced to designing a new penalty function. One can push this line of thinking to consider the idea that in general, any prior can also be equivalently be thought of as a penalty function or a regularizer when using MAP estimators. Given our discussion above, it stands to reason that this should be the case, and offers some interesting new possibilities. The spike and slab prior is also the inspiration for the setup used by Gilbride et al. [2006]. However, their setup is more involved since they have heterogeneity and a hierarchical setup. That is, they have a spike and slab prior for every unit of analysis. They have to adapt the prior so that the parameters center the Dirac mass at some infinitesimal value rather than zero for practical reasons relating to facilitating their MCMC routine. More generally, one can think hierarchical setups with pre-specified distributions of heterogeneity as priors and consequently as penalties in a regularizer. Since the structure they place on the parameters of interest are now informative we can term them informative regularizers. Implementation of such penalty functions should be rather straightforward. For example, a MAP estimator for the Gilbride et al. [2006] setup could be constructed using ideas in Rockova and George [2014]. More generally, theory based priors such as those presented in Montgomery and Rossi [1999] could also be used as regularizers as could other economically meaning priors that we might wish to place in applied contexts.

8.2.3 Bayesian Inference

Typically, Bayesian procedures need to characterize the entire posterior not just the modes (as in MAP). Given the fact that the LASSO (or for that matter any regularized model) can be seen as being derived from a valid Bayes setup, it should be straightforward to sample from that posterior as well. This would however require us to complete the model and design an appropriate MCMC sampler. The algorithm described below is adapted from Park and Casella [2008] who construct a sampler for the LASSO. As before, the model is a linear regression as described in (8.2.2). The authors reparameterize the problem and express the Laplace prior as a scale mixture of normals using auxiliary variables $\tau_1, \tau_2, \ldots, \tau_p$:

$$p(\beta_k | \lambda, \tau_k) = \frac{\lambda}{2\tau_k} \exp\left(-\lambda \tau_k |\beta_k|\right) \tag{8.2.13}$$

Note that in their setup, each parameter has its own scale parameter τ_k. They then define priors for σ^2 and τ_j. In particular, they use the Inverse-Gamma for σ^2 and an exponential prior for τ_j.

$$p(\sigma^2) \propto (\sigma^2)^{-(\nu_0/2+1)} \exp\left(-\frac{\nu_0 \sigma_0^2}{2\sigma^2}\right) \tag{8.2.14}$$

$$p(\tau_j | \lambda) \propto \exp(-\lambda \tau_j)$$

Under this setup, (Park and Casella [2008]) show that a Gibbs sampler is available that they use to implement their MCMC routine. MCMC samplers for other priors/regularizers are also now available. As we mentioned earlier, the full MCMC route is infeasible for large datasets or applications with large number of model parameters. In those cases, MAP estimators are useful. Later in this chapter, we will explore approximate approaches to characterize the posterior of the parameters by appropriately perturbing the MAP estimators.

8.3 BAGGING

Another powerful and popular tool in the ML toolkit is Bagging or Bootstrap Aggregating. Bagging is an ensemble learning technique that aims to reduce variance and improve the stability of a model by averaging the predictions of multiple base models. The Bagging procedure generates multiple training datasets by sampling with replacement from the original dataset and then trains a base model on each of these new datasets. The final prediction is obtained by averaging the predictions of these base models. Below we focus on the regression case but the ideas can be readily adapted to other contexts.

8.3.1 Bagging for Regression

Consider a nonlinear regression problem where we have a dataset of n observations $\{(\mathbf{x}_i, y_i)\}_{i=1}^{n}$. Let's say we have a model that can be written as,

$$y_i = h(\mathbf{x}_i; \theta) + \epsilon_i \tag{8.3.1}$$

The bagging algorithm for regression can be described as follows:

1. For $m = 1$ to M:

   (a) Generate a bootstrap sample D_m of size n by sampling with replacement from the original dataset.

   (b) Train a base model $h(\mathbf{x}; \hat{\boldsymbol{\theta}}_m)$ on the bootstrap sample D_m.

2. The final (prediction) model is $H(\mathbf{x}) = \frac{1}{M} \sum_{m=1}^{M} h(\mathbf{x}; \hat{\boldsymbol{\theta}}_m)$.

Bagging can be interpreted as a Bayesian procedure by considering the base models as "draws" from a posterior distribution of the model parameters. Above, $\boldsymbol{\theta}$ are the parameters of the base model. Assuming that the ϵ are Gaussian, the likelihood function for the regression problem can be written as

$$p(\mathbf{y}|\mathbf{X}, \boldsymbol{\theta}, \sigma) = \prod_{i=1}^{n} \mathcal{N}(y_i | h(\mathbf{x}_i; \boldsymbol{\theta}), \sigma^2) \tag{8.3.2}$$

where $\mathbf{X}$ is the input matrix, $\mathbf{y}$ is the output vector, $h(\mathbf{x}_i; \boldsymbol{\theta})$ is the prediction of the base model with parameters $\boldsymbol{\theta}$, and σ^2 is the variance of ϵ. For now, let's assume that σ is known or fixed. We will place a prior distribution $p(\boldsymbol{\theta})$ on the model parameters and can denote the posterior distribution of the model parameters, given the data as:

$$p(\boldsymbol{\theta}|\mathbf{X}, \mathbf{y}) \propto p(\mathbf{y}|\mathbf{X}, \boldsymbol{\theta})p(\boldsymbol{\theta}) \tag{8.3.3}$$

In the Bayesian framework, the optimal prediction for a new input $\mathbf{x}^*$ is the posterior predictive distribution:

$$p(y^*|\mathbf{x}^*, \mathbf{X}, \mathbf{y}) = \int p(y^*|\mathbf{x}^*, \boldsymbol{\theta})p(\boldsymbol{\theta}|\mathbf{X}, \mathbf{y})d\boldsymbol{\theta} \tag{8.3.4}$$

If we focused our attention on the mean prediction we would write it as,

$$\hat{y}^* = \int y^* p(y^*|\mathbf{x}^*, \boldsymbol{\theta})p(\boldsymbol{\theta}|\mathbf{X}, \mathbf{y})d\boldsymbol{\theta} \tag{8.3.5}$$

$$= \int h(\mathbf{x}^*; \boldsymbol{\theta})p(\boldsymbol{\theta}|\mathbf{X}, \mathbf{y})d\boldsymbol{\theta} \tag{8.3.6}$$

$$\approx \frac{1}{B} \sum_{b=1}^{B} h(\mathbf{x}^*; \tilde{\boldsymbol{\theta}}_b)$$

Where the second line follows from (8.3.1) and the last line by assuming that we can obtain B (approximate) draws from the posterior $p(\boldsymbol{\theta}|\mathbf{X}, \mathbf{y})$. This description establishes a connection between Bagging and Bayes estimators. The key element left to verify is that by resampling the data and estimating $\hat{\boldsymbol{\theta}}_m$ for each such bootstrap sample we are constructing an approximation to the posterior of $\boldsymbol{\theta}$. In other words, we need to show a mapping from $\hat{\boldsymbol{\theta}}_b$ to $\tilde{\boldsymbol{\theta}}_b$.

To see the connection between Bagging and Bayes we will use the "Weighted Like-lihood Bootstrap" (WLB) introduced by Newton and Raftery [1994]. This approach is similar to the Bayesian Bootstrap proposed by (Rubin [1981]) but uses weights to adjust the likelihood function and combines multiple models to estimate predictive distributions. The basic algorithm is as follows: Assume we have some dataset $\mathbf{D} = \{y_1, y_2, \ldots, y_n\}$ containing n observations.

1. For $b = 1 \ldots B$

   (a) Draw a set of weights $\mathbf{w} = (w_1, w_2, \ldots, w_n)$ from a Dirichlet distribution with uniform hyperparameters $\alpha_1, \alpha_2, \ldots, \alpha_n$, each set to 1.

   (b) Given the weights and a model parameterized by θ, the weighted likelihood function is defined as:

$$L(\theta|\mathbf{w}, \mathbf{D}) = \prod_{i=1}^{n} p(y_i|\theta)^{w_i} \tag{8.3.7}$$

   Here, $p(y_i|\theta)$ is the probability of observing y_i given the model parameter θ.

   (c) For each set of weights, find the value of θ that maximizes the weighted likelihood function. This is denoted as $\hat{\theta}_{\mathbf{w}}$:

$$\hat{\theta}_{\mathbf{w}} = \operatorname*{argmax}_{\theta} L(\theta|\mathbf{w}, \mathbf{D}) \tag{8.3.8}$$

2. Use set $\{\hat{\theta}_{\mathbf{w}(1)}, \hat{\theta}_{\mathbf{w}(2)}, \ldots, \hat{\theta}_{\mathbf{w}(B)}\}$ as approximate B posterior draws.

Newton and Raftery [1994] and other follow-up research show that the collection of maximum weighted likelihood estimators $\{\hat{\theta}_{\mathbf{w}(1)}, \hat{\theta}_{\mathbf{w}(2)}, \ldots, \hat{\theta}_{\mathbf{w}(B)}\}$ can be treated as draws from the approximate (to first-order) posterior distribution of the parameter θ. These draws can be used to calculate various posterior statistics, such as the mean, median, and credible intervals. In particular, in the context of our discussion, the posterior expectation of some function $h(\theta)$ can be approximated as

$$\int h(\theta)\pi(\theta|\mathbf{D})d\theta \approx \frac{1}{B}\sum_{b=1}^{B} h\left(\hat{\theta}_{\mathbf{w}(b)}\right) \tag{8.3.8a}$$

In the case that the function h corresponds to (8.3.5), we see the equivalence between ML and the Bayesian counterparts. In other words, Bagging can be seen as an approximation to the predictive posterior mean. Note that if we simply set the weights all to $1/n$, which corresponds to the nonparametric frequentist Bootstrap, we recover the original Bagging procedure. A more Bayesian flavor of Bagging obtains if we use the weighted likelihood bootstrap.

One should note that the WLB procedure only offers a first-order approximation to the posterior. It sacrifices precision for scalability and tractability and the trade-off is not without consequence. Newton and Raftery [1994] argue for a second stage procedure to resample θ using a sampling-importance-resampling procedure if one needs to refine the procedure to obtain more precise draws form the posterior of interest. This of course reduces the value of the procedure as a fast, scalable and easy to use tool. The main advantage of the weighted likelihood bootstrap is its computational efficiency compared to MCMC methods. While it provides an coarse proxy for the posterior distribution it can, nevertheless, be used for model selection, hypothesis testing, and prediction.

8.3.2 Bagging, Bayesian Model Averaging and Ensembles

Once we view Bagging as a special case of posterior expectations we can also to reinterpret a number of other ML procedures in this light as well. Let's first enlarge the scope of the Bagging procedure to account for model uncertainty. Imagine that we have a set of models such that a particular model $m \in \mathcal{M}$ belongs to that set. Also assume that the model has a set of parameters associated with it θ^m. Then, consider the construct

$$\hat{y}^m(\mathbf{x}^*) = \frac{1}{B}\sum_{b=1}^{B} h(\mathbf{x}^*; \hat{\theta}^m_{\mathbf{w}(b)}) \tag{8.3.9}$$

where the parameters $\hat{\theta}^m_{\mathbf{w}(b)}$ are the MAP estimates under a particular WLB (b) for a randomly drawn model m. As with the WLB if the draws for the model m come from the appropriate posterior model probability $\pi(m|\mathbf{X}, Y)$ then (8.3.9) can be seen as a posterior predictive mean that integrates out the model uncertainty as well. We will refer the reader to Clyde and Lee [2021] for a broader discussion on the topic. Allowing for multi-model approaches within the Bagging paradigm allows us to think of a larger class of ML ensemble models as consistent with Bayesian ideas as well. For example, consider the case of random forests where different sets of covariates ($\mathbf{x}$) are chosen at random along with a Bootstrap sample. and then a decision tree is built for prediction purposes.

1. Define the number of trees in the forest (B).

2. Building Each Decision Tree:

   (a) Bootstrapping: Sample n instances from the training data with replacement, where n is the number of samples in your dataset.

   (b) Creating a Decision Tree: Build a decision tree with a random subset of features (p) using the bootstrap sample.

   (c) Repeat B times.

3. Prediction: Each individual tree gives a numerical prediction $h_b(\mathbf{x}^*)$, and the average of all the trees' predictions is taken as the final prediction of the model.

$$H(x) = \frac{1}{B}\sum_{b=1}^{B} h_b(x^*) \tag{8.3.10}$$

A practical application of this idea is found in Matthew et al. [2015] who use the WLB with trees to construct a Bayesian version of random forests. The authors show how these ideas can be used to scale Bayesian Forests to large datasets in a relatively straightforward manner. There is, obviously, nothing special about trees and the framework can be used to average across any form of weak learners (simple models) to create more powerful predictors. Next we explore an alternative approach to the weak-learner averaging idea called Boosting.

8.4 BOOSTING

Boosting is also an ensemble learning technique that combines multiple base models that are simple (usually referred to as weak learners) to create a more powerful overall model. The main idea behind boosting is to iteratively train a series of simple models, where each new model focuses on correcting the errors made by its predecessor. The final model is formed by taking a weighted sum of the simpler models. Consider a regression problem where we have a dataset of n observations $\{(\mathbf{x}_i, y_i)\}_{i=1}^{n}$ and the model we care about is, as before,

$$y_i = h(\mathbf{x}_i; \gamma) + \epsilon_i \qquad (8.4.1)$$

Let the loss function for the regression be denoted as ℓ then the (gradient-) boosting algorithm can be described as follows:

1. Initialize the model with a constant function, $F_0(\mathbf{x}) = \arg\min_\gamma \sum_{i=1}^{n} \ell(y_i, \gamma)$.
2. For $m = 1$ to M:
   (a) Compute the Residuals: $r_{im} = -\left.\frac{\partial L(y_i, F(\mathbf{x}_i))}{\partial F(\mathbf{x}_i)}\right|_{F(\mathbf{x}_i)=F_{m-1}(\mathbf{x}_i)}$ for $i = 1, \dots, n$.
   (b) Fit a weak learner $h_m(\mathbf{x})$ to the residuals.
   (c) Update the Strong Learner: $F_m(\mathbf{x}) = F_{m-1}(\mathbf{x}) + v h_m(\mathbf{x})$.
3. The final model is $F(\mathbf{x}) = F_M(\mathbf{x})$.

Here, $\ell(y, F(\mathbf{x}))$ denotes the loss function, $F_m(\mathbf{x})$ is the model at step m, $h_m(\mathbf{x})$ is the weak learner at step m, and v is the learning rate. The intuition behind this procedure is as follows. We start with some initial model in step 1. Next, we calculate the residuals and train a new model, not on the original output, but on these residuals. In essence, we're fitting the new model to the negative gradient of the loss function calculated with respect to the prediction of the previous model. This is often called a generalized residual. This process is repeated several times, each time creating a new model that tries to fix the mistakes of the combined preceding models. In the end, to make a prediction, all the models contribute to the output. In the above version, we have used a learning rate to weight/combine the simpler models. Alternative ideas could include other forms of voting, such as a simple majority vote (in classification) or weighted majority vote. In essence, Boosting works by implementing a process that helps to convert a set of simple models into a single strong predictor by focusing on areas where the simple model performs poorly, thus improving the overall prediction accuracy.

8.4.1 Boosting as Bayes

Boosting can also be interpreted as a Bayesian procedure by considering the weak learners as basis functions and the strong learner as a linear combination of these basis functions. The strong learner we described above can, more generally, be described by

$$F(\mathbf{x}) = \sum_{m=1}^{M} \alpha_m h_m(\mathbf{x}) \qquad (8.4.2)$$

where α_m are the weights corresponding to each weak learner $h_m(\mathbf{x})$. As such, it is easy to see how Boosting is also a form of model averaging. Even the Boosting procedure has a very Bayesian flavor. In the initialization phase, we train a weak learner (hypothesis) on the data. This can be seen as the prior belief. We then re-weight or resample the data according to the error rates (or inversely, the performance) of the previous models. This is akin to obtaining new data or evidence. Now, using this "new" data, we train another weak learner. This is similar to updating our beliefs based on the new evidence. The final output of Boosting is a weighted combination of all the weak learners. The weights assigned to each weak learner can be interpreted as their posterior probabilities. The hypothesis with the most weight is considered the most probable.

In the above discussion, we have introduced some ML tools (regularization, bagging, and boosting) that via some re-characterization can be thought of as Bayes procedures. The common theme across these is that they gave up on some notion of precision to achieve some other objective such as scale or complexity. Perhaps the biggest idea that ML has brought to the table is that we can begin to understand complex patterns in a practically viable way. This idea is particularly relevant to a framework that has had a remarkable impact on research as well as society, deep learning. We discuss this next.

8.5 DEEP LEARNING

8.5.1 A Primer on Deep Learning

Deep Learning is a subset of machine learning that employs algorithms to model high-level patterns in data using architectures composed of multiple nonlinear transformations of the input variables. There have been significant developments in the area of deep learning and are numerous varieties of architectures and frameworks that have emerged over the last decade or so. Given space constraints, we cannot go into all of them here. Instead, we will focus our attention on the simplest and most popular deep learning framework, the feedforward neural network.

A feedforward Deep Neural Network (DNN) has an input layer (x), one or more hidden layers (h), and an output layer (y). Each layer is fully connected to the next layer. The network is described as "feedforward" because information travels through the network in one direction: it goes in through the input layer, passes through the hidden layers, and emerges out the output layer. Each layer (h) is comprised of J_l individual neurons or nodes h_{lj}. Each neuron can be defined as (8.5.1)

$$h_{lj} = \phi_l \left(\sum_{i=1}^{J_{l-1}} w_{ij} h_{(l-1)i} + b_{lj} \right) \qquad (8.5.1)$$

In the above,

- h_{lj} is the output of the jth neuron in the lth layer.

- ϕ_l is the activation function used in the lth layer. Common choices for activation functions include the sigmoid, tanh, ReLU[1] (Rectified Linear Unit), and others. The role of this function is to transform the linear input into something possibly nonlinear.

- w_{ij} is the weight associated with the connection from the ith neuron in the $(l-1)$th layer to the jth neuron in the lth layer.

- $h_{(l-1)i}$ is the output of the ith neuron in the $(l-1)$th layer. And the summation $\sum_{i=1}^{J_{l-1}} w_{ij} h_{(l-1)i}$ is the weighted sum of the outputs of the neurons in the $(l-1)$th layer.

- b_{lj} is the bias associated with the jth neuron in the lth layer. The bias term is essentially an intercept, which allows the activation function to be shifted as needed.

- The whole expression (8.5.1) denotes applying the activation function to the weighted sum plus the bias. The operation on the collection of neurons for a layer can be succinctly represented as,

$$h_l = \phi_l \left(w_l h_{l-1} + b_l \right) \tag{8.5.2}$$

Note here that the first layer is simply the input layer so that $h_0 = x$ and the last (output) layer is the prediction of the outcome, that is, $y = h_l$. Putting all the pieces together we can write,

$$y = \sum_{l=1}^{L} \phi_l \left(w_l h_{l-1} + b_l \right) \tag{8.5.3}$$

Collectively, the set of specified elements in the DNN is called the architecture. We can also depict the network (for a given architecture with 3 hidden layers) visually as in Figure 8.1.

In the example, we have three input variables (x), three hidden layers (h), and a scalar outcome (y). Each arrow corresponds to a weight w that links a node in the previous layer to the focal node. Note here that, to avoid visual clutter, we have not depicted the biases (b) in the figure. In what follows, we will use the notation $\theta = \{w, b\}$ to denote all parameters of the DNN and the generative model for y can be written as $y(x, \theta)$. If the outcome is categorical, the last layer is transformed using a Logit kernel. So, for example, if we had a binary outcome, we would use the following as the output layer:

$$y = \left(1 + \exp \left[-\sum_{l=1}^{L} \phi_l \left(w_l h_{l-1} + b_l \right) \right] \right)^{-1} \tag{8.5.4}$$

---

[1] The ReLU activation function is one of the most popular in practice and can be described as the function $\phi(v) = \max\{0, v\}$.

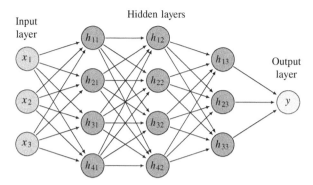

Figure 8.1   Illustration of deep neural network

Given the architecture and a particular set of parameters, the process of evaluating for y using $y(\mathbf{x}_i; \boldsymbol{\theta})$ is called the forward pass. This is essentially what is used for prediction when we have new data, that is, we use $y(\mathbf{x}_{new}; \boldsymbol{\theta})$. The goal of the exercise is then to estimate the parameters $\boldsymbol{\theta}$ by minimizing some pre-specified loss function ℓ. The loss function is dictated by the application at hand and is chosen by the analyst. Typically, in a regression problem the mean squared error is used: $\ell = \sum_i (y_i - y(\mathbf{x}_i; \boldsymbol{\theta}))^2$, while in classification cross-entropy losses are the norm. In the case of the binary outcome, for example, we could use,

$$\ell = \sum_i \left[y_i \ln y(\mathbf{x}_i; \boldsymbol{\theta}) + (1 - y_i) \ln (1 - y(\mathbf{x}_i; \boldsymbol{\theta})) \right] \tag{8.5.5}$$

There are several challenges to estimating the parameters of a DNN. First, the number of parameters are often large and it is therefore common practice to use some form of regularization. In the deep learning literature, this is termed weight decay and is usually a L1 and/or L2 penalty as we saw earlier. The difference is that unlike in LASSO type estimators the penalty parameter is held fixed and not calibrated to the data. A related problem is the high dimensionality of the optimization problem. While the minimization of such an objective function is challenging, the structure of the problem naturally allows for simple gradient based methods to be used. Indeed, the derivative $\frac{\partial \ell}{\partial \theta_k}$ for any parameter θ_k can be obtained by straightforward application of the chain rule. To see this note that,

$$\frac{\partial \ell}{\partial \theta_k} = \frac{\partial \ell}{\partial y} \cdot \frac{\partial y}{\partial \theta_k} \tag{8.5.6}$$

This can be expanded to,

$$\frac{\partial y}{\partial \theta_k} = \sum_{l=1}^{L} \phi_l' (w_l h_{l-1} + b_l) \left[h_{l-1} \frac{\partial w_l}{\partial \theta_k} + w_l \frac{\partial h_{l-1}}{\partial \theta_k} + \frac{\partial b_l}{\partial \theta_k} \right] \tag{8.5.6a}$$

where one can see the recursive structure in these derivatives. This recursion allows us to build out the full gradient map $(\nabla_\theta \ell)$ straightforwardly and use that in our optimization routine as needed. What is even more convenient is that standard deep learning tools such as **Torch** or **Tensorflow** have inbuilt automatic differentiation engines that compute these derivatives exactly, without the need for incremental human effort. The process of constructing the gradient $(\nabla_\theta \ell)$ requires us to go backward through the computational graph and consequently this process is called *backpropagation*.

Another issue in training DNNs is that the datasets are often quite large and evaluating the loss function (and the gradients) repeatedly may consequently be cost prohibitive. To circumvent this issue, it is typical to use *stochastic gradients*. Stochastic gradients are simply the gradient of ℓ evaluated on a subsample S (termed a mini batch). We first define,

$$\tilde{\ell}_S = \sum_{i \in S} (y_i - y(\mathbf{x}_i; \theta))^2 \tag{8.5.7}$$

and then use $\nabla_\theta \tilde{\ell}_S$ as an approximate gradient. As long as this approximate gradient is unbiased (which it is), the optimizer will find a local minima. Optimization methods that use such stochastic gradients are usually termed Stochastic Gradient Descent (SGD). In addition to being practical, there are also other documented advantages of using SGD such as a more through exploration of the loss surface. The SGD procedure is quite simple and can be depicted as

$$\theta^{(t)} = \theta^{(t-1)} - \eta_t \nabla_\theta \tilde{\ell}_{S^{(t)}} \left(\theta^{(t-1)}\right) \tag{8.5.8}$$

Note that at each iteration (or epoch) the mini batch used $\left(S^{(t)}\right)$ is different. To ensure convergence of SGD, the learning rate (η_t) must satisfy two conditions: $\sum_{t=1}^{\infty} \eta_t = \infty$ and $\sum_{t=1}^{\infty} \eta_t^2 < \infty$. The first, ensures that the learning rate doesn't decay too quickly, allowing the algorithm to make progress even with small steps while the second prevents the learning rate from decaying too slowly, ensuring that the steps taken by the algorithm decrease over time. In practice, however, it is common to use a small constant learning rate $\eta_t \equiv \eta$. Repeated application of (8.5.8) will provide estimates of θ upon convergence. There are a number of advances in deep learning theory, architecture design, related optimization methods as well as a variety of nuances, tricks and tools that are relevant for the practical training and implementing of DNNs. These are beyond the scope of this text and we refer the reader to Goodfellow et al. [2016] for an excellent in-depth introduction to these and related topics.

8.5.2 Bayes and Deep Learning

The challenge of implementing a Bayesian DNN arises from the same issues as we have discussed before, namely large parameter sets, complex objective functions and the need to process large amounts of data. Typical Bayesian tools such as MCMC do not scale to the large number of parameters and consequently researchers have taken to constructing various approximate tools. Below we highlight a few connections of stochastic methods used to train DNNs that have a Bayesian flavor.

1. Dropout: Dropout is a regularization technique where random nodes are temporarily "dropped out" during training. The process of "dropping out" nodes refers to a method where during each iteration of the training process, a random selection of

nodes (neurons) in the network are deactivated or "ignored." In other words, they do not contribute to the forward pass nor do they participate in backpropagation for that specific iteration. In addition, during backpropagation, no weight update is calculated for the dropped nodes, so their weights remain as they were for the next iteration. The key idea is that by randomly dropping nodes, the network becomes less reliant on any single node or set of nodes and helps prevent overfitting to the training data. Typically, in dropout, while nodes are ignored during the training process, when making predictions all nodes are included. A variant of dropout, called Monte Carlo (MC) Dropout deviates from this and also applies the random ignoring of nodes during the prediction process. Instead of using a single forward pass, the model is sampled multiple times with dropout enabled, producing different predictions for the same input. By averaging these predictions, the model aims to capture the uncertainty in its predictions.

Dropout, and MC Dropout in particular, can be seen as a practical, approximate implementation of a Bayesian neural network. As we have discussed before, the (posterior) predictive distribution of a model, given some observed data D and a new data point x^*, is given by integrating over the model's parameters, θ:

$$p(y^*|x^*, D) = \int p(y^*|x^*, \theta)p(\theta|D)d\theta$$

Here, $p(y^*|x^*, \theta)$ is the likelihood and $p(\theta|D)$ is the posterior distribution of the parameters. For DNNs, this integral is intractable. Gal and Ghahramani [2016] show that dropout can be interpreted as an approximation to the process of marginalizing over a model's weights, essentially approximating Bayesian inference. We will not go into details here except to point out that the connection between dropout and approximate Bayesian inference is made by recognizing the dropout procedure as a variational approximation. When dropout is applied at prediction time, we generate an ensemble of networks with different "dropout masks," effectively creating a sample of models from the approximate posterior distribution. Predictions are then made by averaging the outputs of these models, as in a standard Monte Carlo estimate:

$$y^* \approx \frac{1}{B}\sum_{b=1}^{B} y_b^*$$

where B is the number of Monte Carlo samples (i.e., dropout masks), and y_b^* is the prediction of the model for the bth dropout mask. This Monte Carlo approximation of the integral gives the predictive distribution, which provides a measure of model uncertainty. Given our discussion earlier about bagging and model averaging the connections to Bayesian inference should be clear.

2. SGD Langevin Processes: An alternative path to Bayesian analysis of DNN's is to exploit the SGD connection to Langevin Dynamics, an idea proposed originally by Welling and Teh [2011]. This approach, called Stochastic Gradient Langevin Dynamics (SGLD), is a variant of SGD where a noise term is added to the update rule, turning the deterministic optimization procedure into a stochastic sampling procedure.

Recall, from (8.5.8), that the original form of the SGD updating rule is,

$$\theta^{(t)} = \theta^{(t-1)} - \eta_t \nabla_\theta \tilde{\ell}_{S^{(t)}} \left(\theta^{(t-1)} \right) \tag{8.5.9}$$

In SGLD, a Gaussian noise term is added to this update rule, which creates a form of diffusion that encourages exploration of the parameter space. The update rule becomes:

$$\theta^{(t)} = \theta^{(t-1)} + \frac{\eta_t}{2} \left(\frac{N}{S^{(t)}} \right) \nabla_\theta \tilde{\ell}_{S^{(t)}} \left(\theta^{(t-1)} \right) + \epsilon_t \tag{8.5.10}$$

Here, $\epsilon_t \sim N(0, \eta_t)$, where η_t retains its interpretation as a learning rate. This noise injection effectively transforms the deterministic optimization into a Markov chain Monte Carlo sampling method. This Markov chain, in the long run, generates samples from the posterior distribution of the model parameters, $p(\theta|D)$. In a Metropolis-Hastings type algorithm, there is an accept–reject step that evaluates proposal candidates before accepting the change in the state. SGLD omits this step and all proposals are simply accepted. Welling and Teh [2011] show that, as $t \to \infty$ and the learning rate $\eta_t \to 0$ at a certain rate, satisfying earlier mentioned conditions ($\sum_{t=1}^{\infty} \eta_t = \infty$, and, $\sum_{t=1}^{\infty} \eta_t^2 < \infty$) the rule in (8.5.10) provides draws from the appropriate posterior even without the accept–reject step. The basic argument underlying SGLD as an approximate posterior sampler is as follows: Since we assume that stochastic gradients are unbiased we can always think of them as the true gradient plus some noise. As $t \to \infty$ and $\eta_t \to 0$ the sampler is more or less centered around the mode of the posterior and the noise term is negligible. As such, the acceptance probability approaches 1 and the need for the accept–reject step is obviated. Consequently, SGLD generates samples from the posterior distribution.

3. Bayes by Backpropagation: Bayes by Backpropagation is a method proposed by Blundell et al. [2015] for performing efficient and scalable Bayesian inference in DNNs. As we discussed earlier, the parameters of a DNN are typically point estimates learned via optimization (like stochastic gradient descent). In contrast, Bayes by Backpropagation represents the parameters by a distribution, typically a Gaussian. Each parameter is associated with a mean and a variance, which are both learned during training.

As before, denote the parameters (weights) of the network as θ, the prior over these parameters as $p(\theta)$, and the observed data as D, we aim to learn the posterior distribution over the parameters given the data, $p(\theta|D)$. Bayes by Backpropagation introduces a variational distribution $q(\theta|\xi)$ that is used to approximate the true posterior. Here, ξ are variational parameters that we optimize to make the variational distribution q close to the true posterior p. This process is known as variational inference (Blei et al. [2017]). The closeness between the variational distribution and the true posterior is measured by the Kullback–Leibler (KL) divergence, which is a measure of the difference between two probability distributions. The goal of the optimization is to minimize this KL divergence:

$$\xi^* = \arg\min_{\xi} KL(q(\theta|\xi)||p(\theta|D)) \tag{8.5.11}$$

However, because the true posterior $p(\theta|D)$ is intractable, we cannot compute this KL divergence directly. Instead, we maximize the Evidence Lower Bound (ELBO), which is equivalent to minimizing the KL divergence:

$$ELBO(\xi) = \mathbb{E}_{q(\theta|\xi)}[\log p(D|\theta)] - KL(q(\theta|\xi)||p(\theta))$$

Here, $\mathbb{E}_{q(\theta|\xi)}[\log p(D|\theta)]$ is the expected log-likelihood of the data under the variational distribution, and $KL(q(\theta|\xi)||p(\theta))$ is the KL divergence between the variational distribution and the prior. The ELBO is optimized using SGD via backpropagation, hence the name "Bayes by Backpropagation." For each minibatch of data, samples of the parameters are drawn from the variational distribution, and these samples are used to compute the gradient of the ELBO with respect to the variational parameters ξ. This gradient is then used to update the variational parameters. Once the variational parameters ϕ are estimated we have a way of sampling from $q(\theta|\xi)$ which provides an approximation to the relevant posterior.

In addition to these, there have been numerous other attempts to design methods that train DNNs that are approximately Bayesian such as variants of the WLB (see e.g., Newton et al. [2021]), Variational approaches and more sophisticated gradient/Hessian-based approximations.

8.6 APPLICATIONS

Our discussion in this chapter has, so far, focused primarily on the methods and tools used in ML and their relation to the Bayes paradigm. The substantive application of Bayes/ML tools in the context of marketing offers exciting opportunities. Elsewhere in this text (Chapter 9) we explore Topic Models using Bayesian techniques. Variable selection approaches are also used in papers such as Gilbride et al. [2006], Joo et al. [2019], and Chandukala et al. [2011] in a variety of application contexts. Generally ML tools can be deployed easily in contexts where the outcome of interest is some function of inputs, in other words when we have prediction related problems where the generative model of interest is $y(x, \theta)$. However, there are a number of other use cases of ML that are relevant to Marketing and to the social sciences more generally. Once such application is the use of ML to accommodate heterogeneity.

Consider the case where the log-likelihood can be written as $\ell(y|t, \theta(x))$. In this setup, the parametric or structural model is encoded in ℓ with parameters $\theta(x)$. y is the outcome variable say purchase or sales, while the t are treatments or, as in our context, marketing mix variables such as price. The x are individual characteristics such as demographics, past behaviors or other individual level descriptors. In particular, the t may be high dimensional and the form of the parameter $\theta(x)$ may or may not be known. This approach is not altogether new and can be thought of as a generalization of the interactions approach to modeling observed heterogeneity. In these scenario, ML tools can be deployed to learn the function θ as well as the elements of x that might be relevant to describing heterogeneity. We will note here that there are no unobserved sources of heterogeneity or that they have been marginalized out. Future research might aim to combine the observed heterogeneity approach with more typical unobserved heterogeneity frameworks discussed in this text. In our discussion below, we will use as a case study a paper by Dubé and Misra [2023], who applied the ideas outlined above in the context of personalized pricing.

8.6.1 Bayes/ML for Flexible Heterogeneity

Dubé and Misra [2023] assume that a prospective, new consumer i with observable features x_i obtains the following incremental utility from purchasing vs not purchasing

$$\Delta U_i = \alpha_i + \beta_i p_i + \varepsilon_i$$

$$= \underbrace{\alpha\left(x_i; \theta_\alpha\right) + \beta\left(x_i; \theta_\beta\right) p_i} + \varepsilon_i \qquad (8.6.1)$$

$$= \tilde{p}_i' \Psi_i \qquad (8.6.2)$$

where $\alpha\left(x_i; \theta_\alpha\right)$ is an intercept and $\beta\left(x_i; \theta_\beta\right)$ is a slope associated with the price, p_i. The notation $\Psi_i = \left(\alpha\left(x_i; \theta_\alpha\right), \beta\left(x_i; \theta_\beta\right)\right)'$ with $\tilde{p}_i = \left(1\ p_i\right)'$ is used for brevity. Note here that the α and β are functions of x and depend on parameters to be estimated. The authors use two different specifications for these functions: For most part, they adopt a linear specification of the functions α and β such that

$$\alpha\left(x_i; \theta_\alpha\right) = x_i' \theta_\alpha$$

$$\beta\left(x_i; \theta_\beta\right) = x_i' \theta_\beta \qquad (8.6.3)$$

We will write $\Psi_i = \Psi\left(x_i; \Theta_{LIN}\right) = \left\{\alpha\left(x_i; \theta_\alpha\right), \beta\left(x_i; \theta_\beta\right)\right\}$ to denote the set of linear functions depicted above. They also present results from a specification where, following ideas in Farrell et al. [2020], they let α and β be outputs of a DNN. In other words, $\alpha\left(x_i\right)$ and $\beta\left(x_i\right)$ are derived as outputs of some DNN which we represent as $\Psi_i = \Psi\left(x_i; \Theta_{DNN}\right)$. Note that, unlike in typical neural networks where the prediction of interest is the outcome y, in this application the DNN outputs the structural parameters of interest $\{\alpha, \beta\}$ as functions of x_i. One can think of this as a nonparametric interactions model as opposed to the linear interactions model we have discussed elsewhere in this text. We will discuss both these specifications below. To complete their model, the authors make the usual assumption that the random utility error ε_i is distributed i.i.d. Logistic and obtain:

$$\mathbb{P}\left(y_i = 1 | p_i; \Psi_i\right) = \frac{\exp\left(\tilde{p}_i' \Psi_i\right)}{1 + \exp\left(\tilde{p}_i' \Psi_i\right)} \qquad (8.6.4)$$

The log-likelihood follows immediately and has the usual form

$$\ell = \sum_i \left[y_i \ln \mathbb{P}\left(y_i = 1 | p_i; \Psi_i\right) + \left(1 - y_i\right) \ln\left(1 - \mathbb{P}\left(y_i = 1 | p_i; \Psi_i\right)\right)\right] \qquad (8.6.5)$$

8.6.2 The Need for ML

Since the number of elements of x could be large, classical approaches to estimation may not be well suited. The authors rely on machine learning methods tailored to the problem so address this issue. In particular, they propose using a LASSO for the linear specification and a DNN to assess robustness to this specification. In both cases, they start by recasting the problem in a Bayesian framework. In particular, using ideas in Bissiri et al. [2016] the authors propose a framework that conducts Bayesian updating using loss functions rather than the typical likelihoods. The goal remains the same – obtaining the posterior distribution of interest. Bissiri et al. [2016] show that for some prior, $h(\Theta)$, data, $\mathbf{D}$, and some loss function $l(\Theta, \mathbf{D})$, the object $f(\Theta|\mathbf{D})$ defined by

$$f(\Theta|\mathbf{D}) \propto \exp\left(-l(\Theta, \mathbf{D})\right) h(\Theta) \tag{8.6.6}$$

is a coherent update of beliefs in the Bayesian sense and consequently represents our posterior beliefs on the parameter vector Θ given the data. For their linear specification, (Dubé and Misra [2023]) use a L_1 penalized (Lasso) negative log-likelihood:

$$l(\Theta, \mathbf{D}) = -\left[\sum_{i=1}^{N} \ell\left(\mathbf{D}_i|\Theta\right) - \lambda \sum_{j=1}^{J} |\Theta_j|\right] \tag{8.6.7}$$

where $\sum_{i=1}^{N} \ell\left(\mathbf{D}_i|\Theta\right)$ is the sample log-likelihood induced by the demand model as described in (8.6.5), and λ is a regularization penalty parameter. They then approximate the posterior $F_\Psi(\Psi|\mathbf{D})$ using a WLB variant of the Bayesian Bootstrap Rubin (e.g., [1981]); Chamberlain and Imbens (e.g., [2003]); Newton and Raftery (e.g., [1994]). In particular, the draw from the posterior distribution of the model parameters using a weighted likelihood bootstrap algorithm (WLB) as outlined in Newton and Raftery [1994]. Since we have discussed the WLB before, we will not repeat details here. Broadly, the procedure draws weights from a standard Dirichlet distribution, assigns them to each observation, and then implements the cross validated LASSO estimator conditioning on these weights. This is repeated B times to obtain an approximate sample from the relevant posterior distribution $F_\Psi(\Psi|\mathbf{D})$ which they denote, for a given Bootstrap sample b as,

$$\hat{\Theta}^b = \arg\max_{\Theta \in \mathbb{R}^J} \left\{\sum_{i=1}^{N} V_i^b \ell\left(D_i|\Theta\right) - N\lambda \sum_{j=1}^{J} |\Theta_j|\right\} \tag{8.6.8}$$

As discussed earlier in this chapter, the draws $\left\{\hat{\Theta}^b\right\}_{b=1}^{B}$ should be interpreted as sample from a first order approximation of the posterior of interest. In the case of using a deep learning setup for $\Psi_i = \Psi\left(x_i; \Theta_{DNN}\right)$, the loss function does not change much. In fact, the same approach outlined above (8.6.8) is used to obtain draws although in this setting the penalty terms are held fixed and no cross-validation is employed (it would be infeasible).

The draws $\hat{\Theta}^b$ obtained from the WLB procedure can be used to ascertain optimal marketing decisions in the usual way. Dubé and Misra [2023] use them to obtain the optimal uniform price as well as optimal personalized prices (conditional on x_i.) This is straightforward since,

$$p_i^* = \arg \max_p (p - c)\, \mathbb{P}\left(y_i = 1 \,|\, p; \Psi_i\right)$$

Other economic quantities are also readily available. For example, the posterior mean surplus across consumers service at some price p^*:

$$\mathbb{E}\left[V(p, x)\,|\,\mathbf{D}, x_i, p = p^*\right] = -\frac{1}{B} \sum_{b=1}^{B} \frac{\log\left(1 + \exp\left(\alpha^b(x_i) - p^* \times \beta^b(x_i)\right)\right)}{\beta^b(x_i)} \qquad (8.6.9)$$

The authors also conduct in-filed validation tests to show that the predicted prices from their procedure deliver profit and sales outcomes in line with predictions.

8.6.3 Discussion

There are a number of points worth noting about the approach in Dubé and Misra [2023]. First, the ML tools adopted were necessary on account of the high-dimensionality of the x vector. Without the regularization, dealing with over a 100 covariates would be tricky to say the least. Further, a full Bayes MCMC procedure such as stochastic variable selection would be expensive. While the regularization deals with variable selection, the DNN approach also allows for the features (x) to be optimally transformed for the problem at hand. Second, the inference procedure in Dubé and Misra [2023] is noteworthy. The proposed algorithm accommodates two different sources of uncertainty. In essence, the approach implements the ideas we discussed in Section 8.3.2 that integrates over the model space (which covariates should be included in the model) hence allowing for model uncertainty. The framework is an model averaging procedure in addition to accommodating the usual parameter uncertainty in a coherent Bayesian fashion. Finally, the approach highlights the value of embedding ML tools in economic models rather than simply using them as predictive devices as expounded in the CS/ML literature. The idea of using flexible ML approaches to incorporate practical nonparametrics is a powerful idea and there could be novel applications of such ideas in marketing that have yet to be explored.

To summarize, this chapter only scratches the surface of the variety of ideas, concepts, and frameworks that has emerged under the ML umbrella. Most of these can be recast or reinterpreted as Bayes procedures and adapted to suit particular inference goals. The more exciting avenues for research, however, lie in the new ideas that have emerged more recently in the ML world that might allow the applied Bayes user to be able to scale Bayes in a way that until now was thought impractical.

9

Bayesian Analysis for Text Data

Abstract

This chapter provides an introduction to Bayesian analysis of text data using the Latent Dirichlet Allocation (LDA) model that models text data in terms of topics and probabilities of words within topics. Textual responses are modeled as a mixture of pure types, or archetypes, that form a convex hull characterizing the distribution of respondent heterogeneity. A topic probability vector characterizes the words provided by each respondent, and can be used to form integrated models of textual response and choice and scaled response. A conjoint dataset is used to illustrate the model. We find that the text data helps clarify the origin of demand.

9.1 INTRODUCTION

Most data available for analysis in marketing is provided as text responses and narratives related to products and their use. These data are useful in understanding precursors to demand that are observed in the marketplace, or in the form of stated preferences arising out of choice experiments. The opinions, beliefs, and perceptions of consumers provide qualitative insight into how consumers find value in the features embodied in a product offering, identify opportunities for product improvement and provide guidance to firms in how best to communicate the benefits of their offerings.

Textual responses comes from many sources including consumer product reviews, transcripts from focus group interviews and open-ended questions in surveys. The advantage of textual responses is that respondents express themselves in a natural manner that is not bound to a set of pre-defined lines of inquiry, which allows for an opportunity to learn what's on people's minds. The challenge of this is in structuring, or supervising the analysis of textual data to be made useful to understanding constructs such as product preferences and demand.

Bayesian Statistics and Marketing, Second Edition. Peter E. Rossi, Greg M. Allenby, and Sanjog Misra.
© 2024 John Wiley & Sons Ltd. Published 2024 by John Wiley & Sons Ltd.

We begin our discussion of with a Latent Dirichlet Allocation (LDA) model for text data (Blei et al. [2003]; Griffiths and Steyvers [2004]), and then examine extensions that avoid its "bag-of-words" assumption. The bag-of-words assumption asserts that the words in a text document appear in an independent and identically distributed (iid) manner. This assumption simplifies the model at the cost of realism in faithfully representing the meaning of a passage. The word "hot" has a different meaning when describing a kettle or a car, and the iid, assumption does not take account of the local phrasing that is present in human speech. We then investigate less restrictive LDA-type models not based on the iid assumption and show how to related a text analysis model to models of other data, such as product reviews (Büschken and Allenby [2016, 2020]).

A feature of the LDA model is that responses from each respondent is modeled as a finite mixture of archetypal responses, and the probabilities associated with this mixture can serve as a mechanism for integrating data from different models. We examine how data arising from a hierarchical model of logit demand can be coupled with models of text to provide a more complete picture of consumer demand. The resulting model is multi-method in that it can produce insights into whom, what and why consumers act as they do.

9.2 CONSUMER DEMAND

We illustrate the LDA model using data on preferences for upgraded seat packages in automobiles and its relationship to driver back pain. Respondents were recruited from a national panel of individuals who reported they experience back pain and have recently purchased or leased an automobile. Respondents were asked to describe any back pain they have and how it was addressed while driving using an open-ended response format. A total of 397 respondents are included in the analysis, and examples of text responses are provided in Figure 9.1. The number of words for each respondent varies from 2 to 47. An average number of words per person is 10.

9.2.1 The Latent Dirichlet Allocation (LDA) Model

The LDA model assumes that words appearing in a document come from a mixture of topics, or archetype responses. A topic may be a set of words expressing an issue like the reliability of a product or the ease of returning a defective product. Each topic comes with a vocabulary that is represented as a probability vector of words appearing in a collection of documents, referred to as a corpus. If a word is strongly associated with a topic, there

---

I make frequent stops, use a lumbar pillow, see my chiro before I leave and when I return.

I get pain in my lower lumbar back when driving for any length of time.

I already have chronic back pain so driving definitely hurts my back. I like to put on the heated seat back to relieve it a bit.

Figure 9.1   Example open-ended responses

Table 9.1   LDA model notation

Variables	Description
N	number of respondents
K	number of topics (archetypes)
T	number of words in each respondent n
S	size of the vocabulary/number of unique words
$w_{n,t}^{LDA}$	respondent n's tth word
$z_{n,t}^{LDA}$	respondent n's topic assignment for tth word
θ_n	K-dim topic probability for respondent n
ϕ_k	S-dim vector representing a unique probability distribution over the S words in topic k

is an increased probability of it appearing, and if a word is rarely used when discussing a topic then the probability is smaller.

Table 9.1 summarizes the notation for the LDA model. The vocabulary for each topic is described by a different probability vector ϕ_k over a fixed set of words. The length of the vector ϕ_k is equal to the number of different words that appear in the corpus (S) and is the same for all topics. The vectors can be combined into a word-topic probability matrix Φ that is of dimension equal to the number of words in the corpus by the number of topics K. Note that document n refers to the text written by respondent n. Therefore, the terms "document n" and "respondent n" are interchangeable.

The goal of text analysis is to characterize each document by its own mixture of topics (θ_n), where each topic is characterized by a discrete probability distribution over words (ϕ_k). That is, the probability that a specific word is present in a text document depends on the latent topic. The tth word appearing in document n, $w_{n,t}$, is thought to be generated by the following process in the LDA model:

- Choose a topic $z_{n,t} \sim$ Multinomial(θ_n),
- Choose a word $w_{n,t} \sim$ from $p(w_{n,t}|z_{n,t}, \Phi)$.

Topics $\{z_{n,t}\}$ and words $\{w_{n,t}\}$ are viewed as discrete random variables in the LDA model, and both are modeled using a multinomial, discrete distribution where the outcomes correspond to the tokenized words as in the example above. We complete the specification of the standard LDA model by assuming a homogeneous Dirichlet prior for θ_n and ϕ_k, which has support on the simplex so the probabilities are constrained to the unit interval and add to one.

$$\theta_n \sim \text{Dirichlet}(\alpha)$$

$$\phi_k \sim \text{Dirichlet}(\gamma)$$

Figure 9.2 displays a plate diagram for the LDA model. The plates indicate replications of documents ($n = 1, \ldots N$), words ($t = 1, \ldots, T_n$), and topics ($k = 1, \ldots, K$). Solid circles represent observed data, shaded circles represent priors and clear circles represent model parameters. Each word $w_{n,t}$ is associated with a latent indicator variable

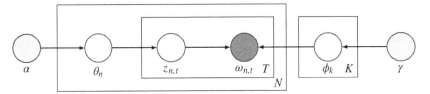

Figure 9.2 Directed acyclic graph for the LDA model

$z_{n,t}$, while the topic probabilities θ_n are assumed constant across words in the document. The LDA model does not impose any structure on the data related to the plates; that is, it assumes that the latent topics $z_{n,t}$ can vary from word to word, sometimes referred to as a "bag-of-words" assumption in the text analysis literature.

Estimation of the LDA model can proceed using a full Gibbs sampler or one that marginalizes, or collapses the draws of the latent indicator variables $z_{n,t}$. We first discuss the full Gibbs sampler, then consider the collapsed Gibbs sampler. The advantage of the full Gibbs sampler is that it simplifies the draws of $z_{n,t}$ but is not so easily generalized. The advantage of the collapsed sampler is that it can be more easily generalized to consider other models with serial dependence.

9.2.2 Full Gibbs Sampler

The full conditional distributions for the model parameters are used to sequentially generate draws as follows:

Step 1. Draw $\{z_{n,t}^{LDA}\}$ for $n = 1, \ldots, N$ and $t = 1, \ldots, T$ given $\{\theta_n\}$, $\{\phi_{s,k}\}$ and $\{w_{n,t}^{LDA}\}$ from a Multinomial distribution:

$$z_{n,t}^{LDA} \sim \text{Multinomial}_K(p_{t,1}, \cdots, p_{t,K}), \quad \text{where} \quad p_{t,k} \propto \theta_{n,k} \phi_{s=w_{n,t}^{LDA},k}$$

Step 2. Draw $\{\theta_n\}$ for $n = 1, \ldots, N$ given $\{z_{n,t}^{LDA}\}$ from a Dirichlet distribution:

$$\theta_n \sim \text{Dirichlet}_K(p_{1,n}, \ldots, p_{K,n}), \quad \text{where} \quad p_{k,n} \propto \sum_{t=1}^{T} I(z_{n,t}^{LDA} = k) + \alpha$$

Step 3. Draw $\{\phi_k\}$ for $k = 1, \ldots, K$ given $\{z_{n,t}^{LDA}\}$ from a Dirichlet distribution:

$$\phi_k \sim \text{Dirichlet}_S(p_{k,1}, \ldots, p_{k,S}), \quad \text{where} \quad p_{k,s} \propto \sum_{n=1}^{N} I(z_{n,t=s}^{LDA} = k) + \gamma.$$

The data $\{w_{n,t}^{LDA}\}$ informs the model parameters in Step 1 of the algorithm, where the multinomial probabilities for the latent indicator $z_{n,t}$ depends on the word–topic probabilities of the observed word.

9.2.3 Processing Text Data for Analysis

The analysis of text data begins with transforming it into a numerical representation. Consider the sentence, "Sometimes I have lower back pain." The text in this sentence can be "tokenized" through a series of commands in the R programming language:

- Read the sentence into R: `mydoc = "Sometimes I have lower back pain."`
- Remove the period: `mydoc=gsub(".","",mydoc,fixed=TRUE)`
- Convert words to lower case: `mydoc=tolower(mydoc)`
- Split the sentence into words: `mydoc = strsplit(mydoc," ")`
- Generate the vocabulary: `w = unlist(mydoc); W=sort(unique(w))`
- Transform words into labels: `W2 = 1:length(W); names(W2) = W`
- Generate word indicators: `for(j in 1:length(mydoc)) {mydoc[[j]] = W2[mydoc[[j]]]}`

These commands result in a numerical representation of the words in the sentence:

- `sometimes i have lower back pain`
- `6 3 2 4 1 5.`

Datasets with multiple respondents can be used to create a set of tokenized documents from a common vocabulary using the above statements. As shown below, additional statements can be added to these commands to strip out white space and remove conjunctions, punctuation and rare words.

9.2.3.1 LDA Analysis   Estimation of the LDA model proceeds as follows. We begin by defining a word-topic count matrix $C_{t,k}^{WT}$ with rows corresponding to words (t) in the vocabulary and columns corresponding to topics (k). Let the (t, k) element indicate the number of times the topic indicator variable $z_{n,t}$ assigns a word t to a topic k across documents n. Similarly, define a topic-document count matrix $C_{k,n}^{TD}$ with rows corresponding to topics (k) and columns corresponding to documents (n), and let the (k, n) element indicate the number of times the topic indicator variable $z_{n,t}$ assigns topic (k) to a document (n) across words (t). The draw of the multinomial topic assignments in Step 1 above proceeds word by word for each respondent by generating a multinomial draw that is proportional to the product of the respondent's topic probabilities θ_n and the word-topic probabilities $\phi_{t,k}$:

- `z[n,t] = which.max(rmultinom(1,1,theta[n] * phi[t,]))`

Draws for Steps 2 and 3 are obtained once the summary count matrices are calculated:

- `C_kn[,n] = table(factor(z[n,1:Data[[n]]$nwords],levels= c(1:K)))}` where `Data[[n]]$nwords` is the number of words provided by respondent n. The term count matrix aggregates the topic assignments across respondents:

- `C_tk = matrix(0,nrow=T,ncol=K)`

```
for (n in 1:N) {
  w = as.vector(Data[[n]])   # Observed words
  zvec = as.vector(z[n, 1:Data[[n]]$nwords])

  # Increment each word in the corpus
    term-count matrix by topic assignment
  for (t in 1:Data[[n]]$nwords) {
    C_tk[w[t], zvec[t]] = C_tk[w[t], zvec[t]] + 1
  }
}
```

Given the summary count matrices, draws of θ_n and ϕ_k are as follows:

- `theta[n] =as.vector(rdirichlet(1, C_kn[,n] + alpha.prior))`

- `phi[,k] = rdirichlet(1, C_tk[,k]+gamma.prior[,k])`

The reason for the simplification is that ϕ_k and θ_n are assumed to be independently distributed across topics (k) and documents (n), and the topic indicators are sufficient statistics for these distributions. Draws of the posterior distribution of model parameters are obtained by the sequential iteration of Steps 1−3.

Table 9.2 displays the 20 highest probability words (ϕ_k) for each of $K = 5$ topics in the car seat dataset. Topic A1 describes physical adjustments made by the driver to relieve their pain while driving, A3 describes a strategy to stop and walk, and topic A5 describes more severe pain associated with a herniated disc. The word list are suggestive of different types of problems faced by drivers and its interpretation is open to much speculation. Part of this is due to the limited number of words in the responses and the bag-of-words assumption of the LDA model. We explore the extension and integration of the LDA model with other data sources below that provide greater context for interpreting the word probabilities ϕ_k.

9.2.4 Collapsed Gibbs Sampler

The collapsed Gibbs sample is an alternative sampling scheme that marginalizes, or integrates out, the topic (θ_n) and word (ϕ_k) probability parameters when generating the draw of the latent indictor $z_{n,t}$. Here we sample each indicator one at a time, conditioning on the remaining indicators. Simplifying notation in the subscripts, we have:

$$\pi\left(z_i|z_{-i}, \alpha, \beta, w\right) = \frac{p\left(z_i, z_{-i}, w|\alpha, \gamma\right)}{p\left(z_{-i}, w|\alpha, \gamma\right)} \propto p\left(z_i, z_{-i}, w|\alpha, \gamma\right) = p\left(z, w|\alpha, \gamma\right)$$

Table 9.2 Highest probability words (ϕ_k) for the LDA model

	LDA Only				
No.	A1	A2	A3	A4	A5
1	adjust	take	stop	use	before
2	seat	break	walk	lumbar	worse
3	posit	before	stretch	support	cause
4	turn	constant	get	pillow	way
5	change	worse	around	cushion	experience
6	keep	mile	bit	low	address
7	can	hurt	hour	behind	disc
8	heat	cause	occasion	short	always
9	driver	move	pull	neck	drive
10	massage	good	long	small	help
11	radiate	way	every	constant	due
12	sometimes	disc	stiff	goe	dont
13	right	ibuprofen	driver	radiate	pain
14	comfort	nothing	really	put	im
15	lean	pill	distance	hip	dull
16	pull	much	minute	ibuprofen	lot
17	area	just	sore	suffer	severe
18	problem	dull	go	upper	turn
19	none	lot	lot	back	period
20	lot	severe	feel	lot	just

where

$$p(z, w | \alpha, \beta) = \iint p(\phi | \gamma) \, p(\theta | \alpha) \, p(z | \theta) \, p(w | \phi_z) \, d\theta \, d\phi$$

$$= \int p(z | \theta) \, p(\theta | \alpha) \, d\theta \int p(w | \phi_z) \, p(\phi | \gamma) \, d\phi$$

Both factors are posteriors of a multinomial likelihood and a Dirichlet prior, and since the Dirichlet distribution is conjugate to the multinomial distribution, each factor is the ratio of beta functions $B(\alpha) = \frac{\prod_k \Gamma(\alpha_k)}{\Gamma(\sum_k \alpha_k)}$.[1] Solving this expression gives:

$$\pi \left(z_{n,t} | z_{-\{n,t\}}, w_{n,t} = s, \alpha, \beta \right) \propto \frac{C_{tk,-nt}^{WT} + \gamma}{\sum_{m'} C_{m't,-nt}^{WT} + W\gamma} \cdot \frac{C_{kn,-nt}^{TD}}{\sum_{t'} C_{t'n,-nt}^{TD} + T\alpha}$$

where $z_{n,t}$ refers to the tth word of document n, and $C_{tk,-tn}^{WT}$ and $C_{kn,-tn}^{TD}$ are the count matrices with the topic assignment for the current word (n, t) excluded. This expression can be used to obtain samples from $z_{n,t}$ conditional on the data (w) and the topic assignments of all other words $z_{-\{n,t\}}$.

---

[1] See for example https://coli-saar.github.io/cl19/materials/darling-lda.pdf

"The hotel was really nice and clean. It was also very quiet. There was a thermostat in each room so you can control the coolness. The bathroom was larger than in most hotels. The breakfast was sausage and scrambled eggs, or waffles you make yourself on a waffle iron. All types of juice, coffee, and cereal available. The breakfast was hot and very good at no extra charge. The only problem was the parking for the car. The parking garage is over a block away. It is $15.00 per day. You don't want to take the car out much because you can't find a place to park in the city, unless it is in a parking garage. The best form of travel is walking, bus, tour bus, or taxi for the traveler. The hotel is near most of the historic things you want to see anyway. I would return to this hotel and would recommend it highly."

Figure 9.3   Example hotel review

9.2.5 The Sentence Constrained LDA Model

One approach to relaxing the iid assumption of the LDA model is to use observed punctuation that is present in the text. Punctuation such as commas and periods are thought to not be informative about the underlying topics used to generate the words, and so they are often pruned from a corpus of words prior to analysis. However, while these articles of speech are prevalent in all topics in a text analysis, they can still provide useful information that signals the change of topics.

Figure 9.3 provides an example. The figure displays an on-line review of a hotel in mid-town Manhattan in New York. The review begins with some general statements about the hotel room, followed by specific comments about the room, breakfast, parking, and connecting transportation before providing a summary evaluation of the hotel by saying, "I would return to this hotel and highly recommend it." Thus, a non-model-based examination of the text suggest that topics are not drawn from the distribution of topics on a word–by–word basis. Instead, topics exhibit a serial dependence that is partially observed in the grammatical structure of sentences.

A sentence-constrained version of the LDA (SC-LDA) model has a generative process described by the following:

- Choose a sentence topic $z_{n,sn} \sim \text{Multinomial}(\theta_n)$,
- Choose a word $w_{n,t} \sim p(w_{n,t}|z_{n,sn}, \Phi)$ for all words t in the sentence sn, $t \in sn$.

These steps appear to be identical to those for the LDA model, except that the latent topic indicator $z_{n,sn}$ is now indexed by the subscript sn for sentence rather than t which denotes the tth word of the document. The words within a sentence are assumed to be generated from a common topic which can change or remain the same from sentence to sentence.

By Bayes theorem, the conditional posterior distribution of the latent indicator variables is given by:

$$\pi\left(z_{n,sn}|z_{-n,sn}, \alpha, \beta, w\right) = \frac{p\left(z_{n,sn}, z_{-n,sn}, w|\alpha, \beta\right)}{p\left(z_{-n,sn}, w|\alpha, \beta\right)} \propto \frac{p(z, w|\alpha, \beta)}{p\left(z_{-n,sn}, w_{-n,sn}|\alpha, \beta\right)}$$

where the expression "$-n, sn$" means "except sentence sn in document n." This expression can be evaluated as the ratio of marginal distributions of words and topics using the expressions (Büschken and Allenby [2016]), appendix A.

The SC-LDA model avoids the bag-of-words assumption and replaces it with a bag-of-sentences assumption where the topic probabilities depend on all the words in a sentence. It introduces contextual effects by conditioning on the sentences provided by the author of the text response. In this sense it is similar to other text model that attempt to introduce contextual effects using machine learning algorithms, such as the word2vec that is available as a R package.[2] The advantage of machine learning algorithms is that they are fast and can process a large volume of text. The disadvantage is that they often not associated with a generative model for the data and do not have a well defined likelihood function. The presence of a generative model becomes important when integrating text with other types of data, as shown below.

9.2.6 Conjunctions and Punctuation

The sentence constrained model is an example of a correlated topic model where words within a sentence are assumed to be generated from the same topic. This assumption may be too restrictive when sentences are long or when there are nuances and complexities to speech that require careful explanation. An example is where a product failure occurs during a specific activity that cases excessive frustration. Modeling textual responses in high involvement settings may require a less rigid autocorrelated structure, possibly related to observed articles of speech such as commas, semi-colons, and prepositions.

An autocorrelated topic model can be developed by allowing topics to carry over from word to word. We can introduce a latent binary indicator variable $\zeta_{n,t}$ equal one to indicate whether the topic assignment to word $w_{n,t}$ is the result of carryover:

$$\zeta_{n,t} \sim Binomial\left(\psi_{n,t}|z_{n,t-1}\right)$$

$$\text{if } \zeta_{n,t} = 1 \quad \text{then} \quad z_{n,t} = z_{n,t-1}$$

$$\text{if } \zeta_{n,t} = 0 \quad \text{then} \quad z_{n,t} \sim Multinomial\left(\theta_n\right)$$

$$\psi_{n,t}|z_{n,t-1} = \frac{\exp\left[\gamma_{0,z_{n,t-1}} + x'_{n,t}\eta\right]}{1 + \exp\left[\gamma_{0,z_{n,t-1}} + x'_{n,t}\eta\right]}$$

where $x_{n,t}$ is a vector of dummy variables that indicate articles of speech such as conjunctions and punctuation, and the baseline carryover probability can be different for each topic. The model specification allows for topic changes at any point in the text, with the probability increasing when observed conjunctions and punctuation are encountered.

The autocorrelated topic model (AT-LDA) avoids the bag-of-words assumption and is found to fit customer review data better for reviews characterized by longer sentences and larger vocabularies (Buschken and Allenby [2020]). The SC-LDA model is shown to fit the data better than the AT-LDA when sentences are shorter and vocabularies are more limited. As reviewers invest greater effort in writing longer reviews, the nuances of speech of found to require more complex models of word generation.

---

[2] https://cran.r-project.org.

9.3 INTEGRATED MODELS

A challenge in the analysis of text data in marketing is the limited number of words that appear in response to open-ended questions and in product reviews. There are typically less than one hundred words unless respondents are specifically asked to provide lengthy answers. Text analysis in marketing therefore greatly benefits from being supervised, or related to some other object of analysis.

Ultimately, the goal of analysis in marketing is to provide insight into the drivers of purchase decisions. The results of the LDA model describe the concerns and interests of consumers, but do not point to specific features of a product they find helpful or useful. In this section, we integrate the LDA model with a conjoint, customer review and fixed-point ratings data to better understand the wants of consumers.

9.3.1 Text and Conjoint Data

Conjoint analysis is a survey-based method used in marketing and economics to estimate demand in situations where products can be represented by a collection of features and characteristics. It is estimated that 20,000 conjoint studies are conducted yearly by firms with the goal of valuing product features and predicting the effects of changes in formulation, price, advertising or method of distribution (Orme [2020]). Conjoint analysis is often the only practical solution to the problem of predicting demand for new products or for new features of products that are not present in the marketplace. A review of the economic foundations of conjoint analysis can be found in Allenby et al. [2019a].

A conjoint survey elicits preferences from respondents by providing them with a set of alternatives and asking which they prefer. Respondents are asked to indicate their most preferred alternative across multiple choice tasks in which the levels of attributes change. The utility provided by the attribute-levels are referred to as "part-worths" in conjoint analysis and can be estimated with a dummy variable regression specification in a logit model of choice. Figure 9.4 shows an example choice task for automobile car seats.

The left side of Figure 9.4 contains a list of product features, or attributes, that change across the choice tasks. A check mark in the choice task indicates the presence of the feature, and a blank indicates that the feature is not offered. The price of each seat grade upgrade package is shown in the bottom row. Explanations of each of the product features are provided in Table 9.3. Since respondents are exposed to multiple choice tasks, the choice data have a panel structure for the estimation of heterogeneous effects.

Notation for the choice model is provided in Table 9.4. A multinomial logit model is used to model the choice data.

$$Pr(y_n = p | \beta_n) = \frac{\exp(x_p' \beta_n)}{\sum_{p=1}^{P} \exp(x_p' \beta_n)}$$

where P denotes the number of choice alternatives and x_p is vector of product attributes for alternative p.

We specify the upper level of the hierarchical Bayes model as a standard model of heterogeneity except that the covariates for the mean are the membership probabilities θ_n:

$$\beta_n = \Gamma \theta_n + \xi_n, \qquad \xi_n \sim N(0, V_\beta)$$

Q1. Which one of the seat packages would you choose?

INTERIOR DETAILS

Seats

Standard features of seats
• 6-way manual front seats
• Cloth seating surfaces

	Package A	Package B	Package C	Standard
6-way power seats		✓	✓	Standard seats are good for me
Upper-back support	✓	✓		
Adjustable lumbar support	✓	✓		
Seat cushion bolsters	✓		✓	
Heated seats			✓	
Seat memory		✓		
Seat massage			✓	
Leather seating surfaces	✓	✓	✓	
Seat package price	$700	$900	$950	$0

Package A   Package B   Package C   Standard Package

Figure 9.4   Seat package choice task

Table 9.3   Glossary of seat package upgrades

Attribute	Description
6-way power seats	Driver and front passenger can electronically adjust the seating position 6 different ways: forward/back, up/down, and angle of recline.
Upper-back support	The upper portion of the seat tilts forward/backward to support the shoulders.
Adjustable lumbar support	The rider can adjust the pressure in the lower back (the lumbar region) for added support.
Seat cushion bolsters	This feature allows the driver to increase or decrease the firmness of the sides of the seat.
Heated seats	This feature provides adjustable temperature heating to the front seat cushions and backs.
Seat memory	This feature offers the option of storing the rider's individual seat settings.
Seat massage	The seat can provide a gentle rolling massage to a lumbar region.
Leather seating surfaces	Leather instead of a cloth covering is used for the vehicle's seats.
Seat package price	The cost of the seat packages is determined by the combination of attributes.

Table 9.4   Choice model notation

Variables	Description
H	Number of choice tasks for each individual n
P	Number of alternatives in each choice task
M	Number of attribute levels in each choice task
y_n	H-dim vector of choices for respondent n
β_n	M-dim vector of part-worths for respondent n
Γ	$M \times K$ matrix of regression coefficients for the mean of the random effects distribution of heterogeneity
V_β	$M \times M$ covariance matrix of random effects distribution of heterogeneity

where the kth column of Γ is the average part-worth estimate of the kth archetype. A standard conditional prior on the random-effects parameters is assumed:

$$V_\beta \sim IW(v, V)$$

$$\Gamma | V_\beta \sim N\left(\overline{\Gamma}, V_\beta \otimes A^{-1}\right)$$

We note that since the membership vector θ_n is constrained to sum to one, the interpretation of the Γ coefficients is nonstandard. This constraint does not invalidate the use of regression analysis in the model, only in the interpretation of the parameters. In a standard regression model without constraints on the covariates, regression coefficients can be interpreted as the expected change in the dependent variable for a change in an independent variable, holding fixed the other variables in the model. In our case, the other variables cannot be held fixed because the elements of θ_n sum to one and an increase in one element must be accompanied with a decrease in one or more of the others. The correct interpretation of the coefficients in Γ is in terms of the membership profiles formed as a linear combination of columns given θ_n. We can observe the difference is the expected value of β_n for $\theta'_n = (1,0, \dots ,0)$, for example, vs any other vector on the simplex and interpret the change using the predicted $\Gamma'\theta_n$ coefficients.

Figure 9.5 displays the DAG for the integrated choice model. From the DAG, we can see that the proposed model is integrated through the membership vector θ_n and is therefore informed by two data sources: text w_n^{LDA} and the choice data y_n. The joint posterior distribution of the LDA+Choice model is:

$$p(\Omega | Data) \propto \left[\prod_{n=1}^{N} \left(\prod_{t=1}^{T} p\left(w_{n,t}^{LDA} | z_{n,t}^{LDA}, \phi_{1:K}\right) p\left(z_{n,t}^{LDA} | \theta_n\right) \right) \right.$$

$$\times p\left(y_n | X_n, \beta_n\right) p\left(\beta_n | \theta_n, \Gamma, V_\beta\right) p\left(\theta_n | \alpha\right) \Bigg]$$

$$\times \left[\prod_{k=1}^{K} p\left(\phi_k | \gamma\right) p\left(\lambda_k | \tau\right) p\left(\Gamma | V_\beta, \overline{\Gamma}, A\right) p\left(V_\beta | v, V\right) \right]$$

where $\Omega = \left\{ z_{n,t}^{LDA}, \{\theta_n\}_{n=1}^{N}, \{\beta_n\}_{n=1}^{N}, \Gamma, V_\beta, \{\phi_k\}_{k=1}^{K} \right\}$.

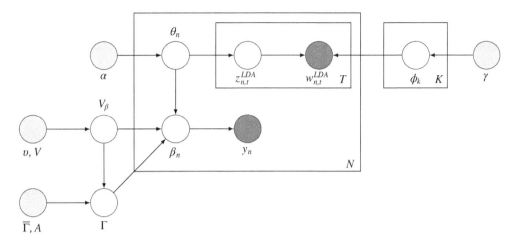

Figure 9.5 Directed acyclic graph for the LDA+choice model

Estimation of the integrated model requires modifications to the draws of θ_n and β_n:

Step 1. Draw $\{\theta_n\}$ for $n = 1, \dots, N$ given $\{z_{n,t}^{LDA}\}$, $\{z_{n,j}^{GoM}\}$ and $\{\beta_n\}$ via the following Metropolis-Hastings (MH) algorithm proposed by Chib and Greenberg [1995c].

(a) Generate a candidate $\theta_n^{new} \sim \text{Dirichlet}_K(p_{1,n}, \dots, p_{K,n})$, where $p_{k,n} \propto \sum_{t=1}^{T}$ $\text{I}(z_{n,t}^{LDA} = k) + \alpha$ as in the LDA model.

(b) Accept/reject θ_n^{new} based on the Metropolis ratio

$$\alpha = \frac{f(\beta_n|\theta_n^{new}, \Gamma, V_\beta)}{f(\beta_n|\theta_n^{old}, \Gamma, V_\beta)},$$

where $f(\cdot)$ is the density of the multivariate normal distribution.

This is necessary since β_n makes a likelihood contribution to the draw of $\{\theta_n\}$. Since θ_n^{new} is generated from the posterior of the LDA model, all elements in the Metropolis acceptance ratio α cancel out, except for the likelihood component of the regression model.

Step 2. Generate β_n^{new} for $n = 1, \dots, N$ given $\{\theta_n\}$, Γ, V_β via the random-walk Metropolis-Hastings (RWMH) algorithm:

(a) Draw candidate $\beta_n^{new} \sim N(\beta_n^{old}, s_\beta \cdot V_\beta)$. In RWMH algorithm, β_n^{old} is the previous value of β_n and s is the step size.

(b) Accept β_n^{new} with following probability:

$$\text{Pr(accept)} = \min\left[1, \frac{\ell_n(\beta_b^{new}|\{y_n\}, X_n) \cdot f(\beta_b^{new}|\theta_n^{old}, \Gamma, V_\beta)}{\ell_n(\beta_b^{old}|\{y_n\}, X_n) \cdot f(\beta_b^{old}|\theta_n^{old}, \Gamma, V_\beta)}\right],$$

where $f(\cdot|\Gamma, V_\beta)$ is the density of the multivariate normal distribution with mean $\Gamma'\theta_n^{old}$ and variance V_β.

Step 3. Generate Γ and V_β using $B = \Gamma'G + \Xi$ where G is a matrix with each θ_n^{old} as a row vector, B is a matrix with each β_n as a row vector, and $\Xi \sim N(0, V_\beta)$.

Figure 9.6 displays the distribution of membership probabilities $\{\theta_n\}$. Respondents are assigned to one of archetypes (i.e., segment) corresponding to their highest membership probability. The plot reveals how respondents live on a simplex defined by the five archetypes as well as how likely it is that each individual can be characterized by one of the archetypes. Some respondents are positioned near the vertices of the plot, indicating that they are well described by those archetypes, whereas others are positioned near the center and are better characterized as a mixture of archetypes. The classification of respondents into segments is useful when considering respondent-level actions that might be taken (e.g., direct marketing), or when determining the size of segments based on economic criteria in addition to a purely statistical measure. For example, cost considerations in responding to each archetype would change the highest-probability assignment rule to one based on expected profitability.

Parameter estimates for the LDA+Choice model are reported in Tables 9.5 and 9.6. Table 9.5 displays the 20 highest probability words for each archetype ϕ_k. Archetype A1 reports using medication to alleviate their back pain. Archetype A2 has problems with the hip and lumbar, and Archetype A3 describes chronic pain and driving long distances. Archetype A4 relies on the use of heat to relieve back pain, and Archetype A5 prefers to deal with pain by stopping their car and stretching.

Table 9.6 reports the regression coefficients Γ. The integration of choice data into the model provides insight into what respondents with different ailments want to purchase.

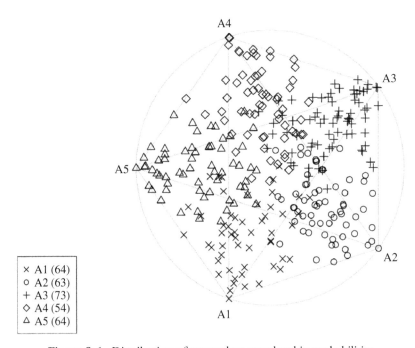

| × A1 (64) |
| o A2 (63) |
| + A3 (73) |
| ◇ A4 (54) |
| △ A5 (64) |

Figure 9.6   Distribution of respondent membership probabilities

Table 9.5 Highest probability words (ϕ_k) for the LDA+choice model

	LDA+Choice Model				
No.	A1	A2	A3	A4	A5
1	take	constant	experi	heat	stretch
2	like	support	suffer	none	hour
3	alev	behind	dont	adjust	stop
4	hurt	pillow	short	posit	minut
5	break	put	low	chang	everi
6	becaus	use	period	seat	feel
7	medic	small	address	turn	walk
8	start	lumbar	caus	shoulder	ach
9	just	back.	way	lean	go
10	sharp	neck	distanc	lot	stiff
11	problem	side	cushion	keep	stretch.
12	noth	right	massag	possibl	get
13	ibuprofen	surgeri	long	can	around
14	mile	due	im	shift	sore
15	rest	hip	chronic	tri	trip
16	befor	spine	car	longer	occasion
17	move	upper	time	littl	need
18	pill	alway	sit	help	dull
19	sometim	one	drive	comfort	driver
20	littl	also	good	often	pull

Table 9.6 Upper-level regression estimates (Γ) for the LDA+choice model

Attributes	A1	A2	A3	A4	A5
6-way power seats	0.97 (0.50)	−0.12 (0.44)	**1.04** (0.38)	**1.13** (0.43)	**2.07** (0.57)
Upper-back support	**1.25** (0.59)	0.49 (0.43)	**1.34** (0.36)	−0.02 (0.53)	0.67 (0.47)
Adjustable lumbar support	**1.69** (0.55)	**3.08** (0.57)	**2.19** (0.52)	0.62 (0.58)	**2.43** (0.64)
Seat cushion bolsters	**1.07** (0.43)	−0.19 (0.41)	**0.89** (0.34)	−0.02 (0.50)	0.85 (0.49)
Heated seats	**1.11** (0.49)	−0.10 (0.51)	**1.40** (0.40)	**2.19** (0.49)	**1.49** (0.50)
Seat memory	0.75 (0.45)	−0.72 (0.40)	**1.61** (0.29)	0.47 (0.45)	**1.07** (0.53)
Seat massage	**1.59** (0.50)	−0.21 (0.44)	**1.30** (0.35)	0.40 (0.57)	0.90 (0.55)
Leather seating surfaces	0.49 (0.56)	**−1.49** (0.45)	0.52 (0.43)	**1.18** (0.53)	**1.30** (0.52)
Seat package price	**−0.43** (0.21)	0.00 (0.18)	**−0.60** (0.16)	−0.19 (0.21)	**−0.63** (0.23)

- * The marginal posteriors with means in bold have 95% credible intervals that do not contain zero.
- * Standard deviations are in parentheses, ().

For example, respondents who stop and stretch (Archetype 5) during their drives find value in having six-way power seats and an adjustable lumbar support. Individuals represented by A2 do not have strong positive preferences for any of the product features except lumbar support, and actually would prefer to not to purchase a car seat with a leather seating surface. Respondents represented by A1 deal with their pain by taking medication and prefer most of the car seat features in the study. The Archetypes differ in what they want and their sensitivity to price, with the text data providing insight into the reasons for respondent preferences.

Table 9.7 provides an integrated summary of the text and choice data. The advantage of the integrated model structure is that these data elements are organized around archetypal characterizations that provide greater insight into the nature of demand than standard models of heterogeneity that describe the mean of random-effect distributions. We find that respondents characterized as having severe back pain (A2) and who use heat as a pain remedy (A4) generally place low value on car seat features. Similarly, those with chronic pain place high value on nearly all of the features. The individual-level membership probabilities $\{\theta_n\}$ allows for flexibility in characterizing the distribution of heterogeneity. We find that the text and scaled response data helps clarify the origin of demand in our analysis.

9.3.2 R Code for Text and Conjoint Data

Processing data for text and choice analysis requires three major steps. The raw data first needs to be read into the R session and the choice data needs to be organized in a list structure that includes respondent choices (y) and explanatory variables (X). In our analysis, we employ a data-based screen to identify respondent data with low in-sample fit. We engage in respondent screening if we find that their choice responses are less than that would be expected by chance. The second step of analysis involves converting text data to a numeric representation and integrated into the list structure. The third step is to estimate the joint text and choice model.

9.3.2.1 Reading in the Data: Import.R

The raw data for our study is contained in the Excel spreadsheet "N415.csv" and "Design Matrix.csv." The first file contains the survey responses described in Kim and Allenby [2022]. There were two versions of the conjoint exercise and the corresponding design matrices are contained in the second file. We begin by reading in the data:

```
dat = as.matrix(read.csv("N415.csv"))
```

and the design matrices:

- ```
  DesignX = as.matrix(read.csv("Design Matrix.csv",
  header = F))
  ```
  ```

Table 9.7 Archetype profiles for the LDA+choice model

	Taking medicine (A1)	Severe back pain (A2)	chronic (A3)	Heat and adjust (A4)	Stop and stretch (A5)
Text	Medic, sharp, ibuprofen, pill	Constant, pillow, lumbar, surgeri	Address, cushion, massag, chronic	Heat, adjust, posit, chang	Stretch, stop, walk, occasion
Choice	• upper-back support • adjustable • seat cushion bolsters • heated seats • seat massage	• adjustable • leather seating surfaces (negative)	• 6-way power seats • upper-back support • adjustable • seat cushion bolsters • heated seats • seat memory • seat massage • price sensitive	• 6-way power seats • heated seats • leather seating	• 6-way power seats • adjustable • heated seats • seat memory • leather seating • price sensitive
Market size	0.20	0.20	0.23	0.17	0.20

The list structure for the data is then created following normal convention used in the *bayesm* package:

```
lgtdata = NULL
  for(hh in 1:nhh){
     nver = as.numeric(dat[hh,"choice"])    # two versions of the
                                                design matrix
     y = Y[hh,]
     XX = X[,,nver]
     lgtdata[[hh]] = list(X = XX, y = y, nver = nver)
}
```

Next, we estimate a hierarchical Bayes logit model using the data:

```
Mcmc1 = list(keep=10, R=20000)
Data1 = list(lgtdata=lgtdata, p=alt)
Prior1 = list(ncomp=1)
out = rhierMnlRwMixture(Data=Data1, Prior=Prior1, Mcmc=Mcmc1)
```

We employ an in-sample screen to remove respondents who do not appear to be paying attention to the survey by first estimating a threshold value for respondent log likelihood:

```
for(resp in 1:nresp){
  bstari = out$betadraw[resp,,]
  for(rep in (burnin+1):ncol(bstari)){
    lli[rep-burnin] = llmnl(bstari[,rep], y = lgtdata[[resp]]$y,
                            X = lgtdata[[resp]]$X)
  }
  loglike[resp] = mean(lli)
}
lowlike = loglike < task*log(.25)   # Resp with low log-likes
```

where task is the number of choice tasks, each with four choice options. The likelihood per choice task must be above 0.25 for the respondent to not appear to be randomly guessing. We remove respondents with log likelihood values below the limit:

```
lgtdata_cen = NULL
i = 1
for(resp in 1:nresp){
  if(!lowlike[resp]) {
    lgtdata_cen[[i]] = lgtdata[[resp]]
    i=i+1}
}
```

The choice data is now organized as a list for analysis with 403 potentially available respondents.

9.3.2.2 Pre-processing the Text Data: Text.R   The text data come from two open-ended questions in the survey prompting respondents to write about their back pain and how they deal with it while driving. The text data are contained in the R file "dat" and combined for analysis:

```
mydoc = NULL
for (n in 1:nobs.all){
  mydoc = append(mydoc, paste(as.character(dat[,"N2"][n]),
                            as.character(dat[,"N5"][n])))
}
```

We next define sets of words that are removed from analysis because they are not specific to a topic:

```
Conjunctions = c("for","and","nor","but","or","yet","so" …
Punctuation = c(",",".",";",":", …
```

and remove numbers, whitespace and stem words to avoid trivial differences such a plural versions of a word:

```
for(i in 1:length(mydoc)){
   mydoc[[i]] = removeWords(mydoc[[i]],Conjunctions)
   mydoc[[i]] = removeWords(mydoc[[i]],Punctuation)
   mydoc[[i]] = gsub("-","",mydoc[[i]],fixed=TRUE)
   mydoc[[i]] = mydoc[[i]][which(mydoc[[i]]!="")]
   mydoc[[i]] = stemDocument(mydoc[[i]], language = "en")
     .
     .
     .
}
```

Additional error checks are present in the R script that ensures there is actual content beyond a single word in the text responses. Once we have an acceptable set of words, we develop a vocabulary for analysis:

```
w = unlist(mydoc)        # Updated vocabulary (W)
W = sort(unique(w))      # 125 words
W2 = 1:length(W)         # Transforms words into labels for a
                           vector 1:length(W)
names(W2) = W            # W2 is the nominal words
                           vector with labels
```

Next we generate a nominal set of word indicators for each document:

```
mydoc_nom = mydoc
for(i in 1:nobs.all){
   for(j in 1:length(mydoc[[i]])){mydoc_nom[[i]][[j]] =
      W2[mydoc[[i]][[j]]]
   }
}
```

9.3.2.3 Estimating the Joint Model: Estimate.R
The first step in the analysis of the joint text and choice data is to integrate the two datasets into the list format. The variable "new_insample" is defined in the R script as the size of the calibration dataset comprising 80% of the original data:

```
Data = NULL
for(h in 1:length(new_insample)) {
   hh = new_insample[h]
   X.h = regdata[[hh]]$X
   y.h = regdata[[hh]]$y

   Word = as.numeric(mydoc_nom[[hh]])
   N = nw[[hh]] # number of words in the text written by the
                  individual h
   Data[[h]] = list(X = X.h, y = y.h, Word = Word, N = N)
}
```

Storage arrays and initial values are then assigned to all of the model parameters. The MCMC chain for estimating the joint model is similar to the model for hierarchical logit model except for generating the draw for θ. Following Step 1 from the integrated model algorithm above, we generate θ_n using the MH algorithm three proposed by Chib and Greenberg [1995c] where a candidate draw is generated from a portion of the conditional posterior distribution and accepted with probability determined by the remainder of the posterior. The candidate draw is from a Dirichlet distribution and the acceptance probability is determined by the multivariate normal density of heterogeneity:

```
cand.theta = as.vector(rdirichlet(1, C_dk[hh,] + alpha.prior))
# theta

# Likelihood of theta WRT beta
cand.post = dmvn(oldbetas[hh,],t(cand.theta)%*%oldGamma,oldVbeta,
            log=TRUE)
old.post = dmvn(oldbetas[hh,],oldG[hh,]%*%oldGamma,oldVbeta,
            log=TRUE)

ldiff = cand.post - old.post
aalpha = min(1, exp(ldiff))
if (aalpha < 1){unif=runif(1)} else{ unif=0}
if (unif <= aalpha){oldG[hh,] = cand.theta} else {rej_g = rej_g + 1}
```

The other parameters of the joint model are drawn from conditional distributions that are the same as in the standard models. Given θ_n, the matrix of coefficients Γ and covariance matrix V_β are from a standard multivariate regression model and the draw of β_n can be obtained from a random-walk MH algorithm. The draw of ϕ_k is similarly unaffected.

Estimates of the high probability words that characterize the archetype responses are obtained using the commands:

```
Phi.summary = round(apply(Phidraw[,,burnin:end],1:2,mean), 3)

prob_list <- list()
topN = 20
topN_list = matrix(NA, topN, K)
## Loop through the columns of Phi.summary and calculate the order
for (i in 1:K) {
  prob_list[[i]] <- order(Phi.summary[, i], decreasing = TRUE)
  topN_list[,i] <- W[head(prob_list[[i]], topN)]
}
```

9.3.3 Text and Product Ratings

The topic probabilities for a document θ_n provide a low-dimensional representation of the words in a document that can be associated with a product's rating using a censored regression model:

$$r_n = f \quad if \quad c_{f-1} \le \tau_n \le c_f$$

and

$$\tau_n \sim N\left(\theta'_n \delta, \sigma^2\right)$$

where the cut-points $\{c_f\}$ are used to censor the latent continuous variable τ_n to produce a discrete, fixed point rating r_n. We assume that rating is measured with f fixed-point categories (e.g., $f = 5$ indicates a fixed point rating scale with five categories). We also assume that document topic probability vector θ_n serves as covariates in the latent regression, and so θ_n is also informed by both the ratings r_n and the words w in the review.

The estimation of the regression model requires three additional steps beyond the simple Gibbs sampler discussed above, and a modification to the draw of θ_n. The draw of θ_n changes because it is also appears in the expression for the mean of the latent variable τ_n. Again, we use the Metropolis-Hastings algorithm for the new draw of θ_n by modifying the original draw of theta:

1. Generate a candidate θ_n^{cand} from the Dirichlet distribution as described above, for example, from the full Gibbs sampler.

2. Accept/reject θ_n^{cand} based on the Metropolis ratio:

$$\alpha = \frac{p(\tau_n | \delta, \theta_n^{cand}, \sigma_\varepsilon^2, c)}{p(\tau_n | \delta, \theta_n, \sigma_\varepsilon^2, c)},$$

which is the ratio of truncated univariate normal distributions. Since we generate the candidate θ_n^{cand} from the posterior of the LDA model, all elements in the Metropolis acceptance ratio α cancel, expect for the likelihood component of the regression model:

```
cand.theta = as.vector(rdirichlet(1, C_dk[hh,] + alpha.prior))
# theta

# Likelihood of theta WRT tau_n
tau_cand = dot(cand.theta, delta)
tau_old = dot(oldG[hh,],delta)

y1 = c_(f-1) # lower cutpoint
y2 = c_f     # upper cutpoint

cand.post = pnorm(y2-tau_cand,sigma,log.p=TRUE)
                 - pnorm(y1-tau_cand,sigma,log.p=TRUE)
old.post = pnorm(y2-tau_old,sigma,log.p=TRUE)
                 - pnorm(y1-tau_old,sigma,log.p=TRUE)

ldiff = cand.post - old.post
aalpha = min(1, exp(ldiff))
if (aalpha < 1){unif=runif(1)} else{ unif=0}
if (unif <= aalpha){oldG[hh,] = cand.theta} else {rej_g = rej_g + 1}
```

The remaining parameters of the regression model are estimated as an ordinal probit regression model with cut-points c_0 fixed to $-\infty$ and $c_f = +\infty$. The ordinal probit model is discussed in Section 4.1 and implemented in the `rordprobitGibbs` routine in `bayesm`. Cut-points are estimated using a Metropolis-Hastings algorithm for candidate cut-points, and given the cut-points the remaining parameters are estimated as a Bayesian regression model.

The vector of coefficients δ in the latent regression model provide a sense of the importance and valence of the relationship between topics and ratings. A regression coefficient of zero indicates that rating do not vary with the prevalence of the topic in a review or document, and large regression coefficients indicate that small chances in a topic probability are associated with relatively large changes in ratings. The estimates of the cut-points $\{c_f\}$ provide additional information in terms of the change in the latent rating τ_n needed for a change in the observed discrete rating r_n. If the estimates of $\{c_f\}$ are close to one another, then relatively small changes in the latent rating are needed to expect changes in the observed discrete rating, while estimate of the cut points that are further apart require larger changes in the latent rating to expect to see changes in r_n.

An advantage of the LDA-ratings model is that it simultaneously reduces the dimensionality of the words in a corpus of reviews and provides an indication of the valence of the topic probabilities to the ratings. A negative regression coefficient implies a negative topic in that the more prevalent it is in a product rating, the less likely the product will receive a favorable review. Similarly, a positive regression coefficient implies a higher rating for higher prevalence of topics associated with positive regression coefficients. Without the availability of the topic probabilities, analysts would need to relate the words in the reviews to the rating through either a dummy variable regression model, or a regression model with covariates that summarized certain aspects of the review such as its valence or sentiment. The advantage of using an LDA model for the words is that it provides access to the original reviews through the word-topic probability estimates ϕ_t. Further details of these models can be found in Büschken and Allenby [2016].

9.3.4 Text and Scaled Response Data

People differ in why they want products, and there are many reasons for the diversity of preferences and sensitivities that exist in a market. Some reasons are best understood with open-ended responses where respondents write about their experience with the product and their lives. Other reasons are better understood using fixed-point rating scales where the researcher queries respondents for specific information. We now consider an integrated model of text and scaled responses that combines data from both data formats.

The Grade of Membership (GoM) model is a mixed membership model useful for the analysis of fixed-point ratings scale data. It differs from LDA model where each response, or word, is an iid draw from a distribution of words in a vocabulary for each topic. Instead, it assumes there are J questions measured on a discrete scale of possibly different dimensions. The simplest GoM model is for pick-all-that-apply data in which a vector of binary responses to J questions are provided by respondents. The J questions might be related to past behaviors (e.g., indicating which from a set of brands respondents have purchased in the past year) or possibly respondent opinions measured in terms of an agree/disagree 2–point rating scale. We will use the following notation for the GoM model (Table 9.8).

The GoM model differs from the LDA model where each word is thought to arise from a multinomial draw from a latent topic defined by a set of probabilities for each word. In the GoM model, the response to each of the J scaled questions are assumed to be generated from a latent response profile. For a two-point rating scale, this means that

Table 9.8 GoM model notation

Variables	Description
J	Number of categorical questions
L_j	Number of categorical responses for question j
$w_{n,j}^{GoM}$	Respondent n's categorical response for jth statement
$z_{n,j}^{GoM}$	Respondent n's archetype assignment for jth statement
$\{\lambda_{j,k}\}_{j=1}^{J}$	Array of probability distributions $\lambda_{j,k}$ over the L_j response options for each question j and archetype k

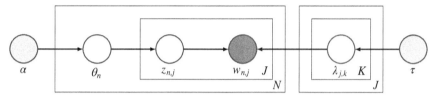

Figure 9.7 Directed acyclic graph for the GoM model

the multinomial draws in the LDA model for each word are replaced by J draws from a binomial distribution.

Figure 9.7 displays a plate diagram for the GoM model. The elements of θ_n are the probabilities that respondent n is represented by, or belongs to, each of K types. The types are called topics in an LDA model, and each document is modeled as a mixture of topics. The types are called exemplars, or archetypes, in a GoM model and each respondent's answers to the scaled response questions are modeled as a mixture of archetypes. Thus, the membership probability vector θ_n represents the location of the respondent (or document) within the convex haul of the types $\Lambda = \{\lambda_k\}$ or $\Phi = \{\phi_k\}$.

The word-topic matrix of word probabilities Φ is replaced in the GoM model by an array of response probabilities Λ with J rows, one for each scaled response question. For binary response data, responses are characterized by one of the two response probabilities since $\Pr(\text{Yes}) = 1 - \Pr(\text{No})$, and the array of response probabilities takes the form of a two-dimensional matrix with J rows and K columns, with each entry indicating the probability of a "Yes" response to question j and archetype k. The GoM model is similar to the LDA model in that the vector of membership (topic) probabilities θ_n characterize responses for the nth individual (review), but different in the model used to generate responses. Data generation for the GoM model takes the form:

For each of J questions:

- Choose an archetype $z_{n,j} \sim \text{Multinomial}(\theta_n)$,
- Choose a scaled response $w_{n,j} \sim p(w_{n,j} | z_{n,j}, \Lambda)$

where the scaled responses have a discrete distribution with probabilities $\lambda_{k=z_{n,j}}$.

Estimation of the GoM model is similar to the LDA model. The full conditional distributions for the model parameters are used to sequentially generate draws as follows:

Step 1. Draw $\{z_{n,t}^{GoM}\}$ for $n = 1, \ldots, N$ and $j = 1, \ldots, J$ given $\{\theta_n\}$, $\{\lambda_{j,k}\}$ and $\{w_{n,j}^{GoM}\}$ from a Multinomial distribution:

$$z_{n,j}^{GoM} \sim \text{Multinomial}_K(p_{j,1}, \cdots, p_{j,K}), \quad \text{where} \quad p_{j,k} \propto \theta_{n,k}\lambda_{j,k}(w_{n,j}^{GoM})$$

Step 2. Draw $\{\theta_n\}$ for $n = 1, \ldots, N$ given $\{z_{n,j}^{GoM}\}$ from a Dirichlet distribution:

$$\theta_n \sim \text{Dirichlet}_K(p_{1,n}, \ldots, p_{K,n}), \quad \text{where} \quad p_{k,n} \propto \sum_{j=1}^{J} \text{I}(z_{n,j}^{GoM} = k) + \alpha$$

Step 3. Draw $\{\lambda_{j,k}\}_{j=1}^{J}$ for $j = 1, \ldots, J$ and $k = 1, \ldots, K$ given $\{z_{n,t}^{GoM}\}$ from a Dirichlet distribution:

$$\lambda_{j,k} \sim \text{Dirichlet}_{L_j}(p_{k,1}, \ldots, p_{k,L_j}), \quad \text{where} \quad p_{k,l} \propto \sum_{n=1}^{N} \text{I}(z_{n,j}^{GoM} = k \text{ and } w_{n,j}^{GoM} = l) + \tau$$

The scaled response data $\{w_{n,j}^{GoM}\}$ inform the model parameters in Step 1 and Step 3 of the algorithm.

R code for these draws are as follows:

```
# Create GoM storage arrays:
#    nhh = number of respondents;
#    J = number of fixed point rating questions;
#    Lj number of response categories per question;
#    K archetypes
oldZ.gom = matrix(NA,nhh,J)         # Latent archetype
oldtheta = matrix(NA,nhh,K)         # Membership probability vector
oldLambda = array(NA,c(J,Lj,K))     # Response profile
Niz = matrix(NA,nhh,K)              # Counts of archetype assignments
```

Step 1. Draw $\{z_{n,t}^{GoM}\}$:

```
for (hh in 1:nhh) {
  for (j in 1:J) {
    pr = oldLambda[j,respdata[hh,j],] * oldtheta[hh,]
    oldZ.gom[hh,j] = which.max(rmultinom(1,1,pr))
  }
}
```

Step 2. Draw $\{\theta_n\}$:

```
for (hh in 1:nhh) {
  Niz[hh,] = table(data.table(factor(oldZ.gom[hh,],
                 levels=c(1:K))))
  thetanew = as.vector(rdirichlet(Niz[hh,] + alpha.prior))
}
```

Step 3. Draw $\{\lambda_{j,k}\}_{j=1}^{J}$:

```
# Obtain counts across respondents
Njk = array(NA,c(J,Lj,K))
for(k in 1:K){
  for(j in 1:J){
    Njk[j,,k] =
      table(data.table(factor(respdata[oldZ.gom[,j]==k,j],
      levels=c(1:Lj))))
  }
}
# Generate draw of lambda
for(k in 1:K){
  for(j in 1:J){
    oldLambda[j,,k] = rdirichlet(Njk[j,,k] + tau.prior)
  }
}
```

In both the LDA and GoM models, the model parameter θ_n is a discrete probability distribution that characterizes the nth respondent. In the LDA model it represents vector of probabilities that summarizes the topics in the text document. In the GoM model it is a vector of probabilities indicating the respondents "grade" or likelihood of membership in each of the K archetypes. Parameter estimates of (ϕ_k) and (λ_k) from these models describe archetypes that characterize the responses of individuals, with each individual characterized as a mixture of archetypes.

An integrated mixed membership model views the text (i.e., open ended responses) and scaled preference responses of an individual as represented by the same probability vector θ_n. Figure 9.8 displays the DAG for the integrated model. The parameters z in the LDA and GoM models are augmented variables that simplify the estimation of model parameters. For the LDA model, $z_{n,t}^{LDA}$ is drawn for each of the T words in the text, and in the GoM model $z_{n,j}^{GoM}$ is drawn for each of the J scaled items in the survey.

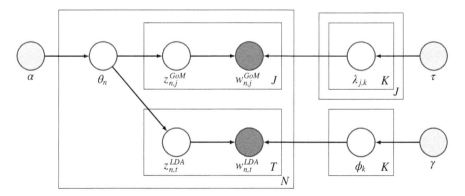

Figure 9.8 Directed acyclic graph for the LDA+GoM model

Estimation of the integrated LDA+GoM model involves minor changes to the previous estimation algorithms:

Step 1. Draw $\{z_{n,t}^{LDA}\}$ for $n = 1, \ldots, N$ and $t = 1, \ldots, T$ given $\{\theta_n\}$, $\{\phi_{s,k}\}$ and $\{w_{n,t}^{LDA}\}$ from a Multinomial distribution:

$$z_{n,t}^{LDA} \sim \text{Multinomial}_K(p_{t,1}, \cdots, p_{t,K}), \quad \text{where} \quad p_{t,k} \propto \theta_{n,k}\phi_{s=w_{n,t}^{LDA},k}$$

Step 2. Draw $\{z_{n,t}^{GoM}\}$ for $n = 1, \ldots, N$ and $j = 1, \ldots, J$ given $\{\theta_n\}$, $\{\lambda_{j,k}\}$ and $\{w_{n,j}^{GoM}\}$ from a Multinomial distribution:

$$z_{n,j}^{GoM} \sim \text{Multinomial}_K(p_{j,1}, \cdots, p_{j,K}), \quad \text{where} \quad p_{j,k} \propto \theta_{n,k}\lambda_{j,k}(w_{n,j}^{GoM})$$

Step 3. Draw $\{\theta_n\}$ for $n = 1, \ldots, N$ given $\{z_{n,t}^{LDA}\}$, $\{z_{n,j}^{GoM}\}$ from a Dirichlet distribution:

$$\theta_n \sim \text{Dirichlet}_K(p_{1,n}, \ldots, p_{K,n}), \quad \text{where} \quad p_{k,n} \propto \sum_{t=1}^{T} I(z_{n,t}^{LDA} = k) + \sum_{j=1}^{J} I(z_{n,j}^{GoM} = k) + \alpha$$

Step 4. Draw $\{\phi_k\}$ for $k = 1, \ldots, K$ given $\{z_{n,t}^{LDA}\}$ from a Dirichlet distribution:

$$\phi_k \sim \text{Dirichlet}_S(p_{k,1}, \ldots, p_{k,S}), \quad \text{where} \quad p_{k,s} \propto \sum_{n=1}^{N} I(z_{n,t=s}^{LDA} = k) + \gamma.$$

Step 5. Draw $\{\lambda_{j,k}\}_{j=1}^{J}$ for $j = 1, \ldots, J$ and $k = 1, \ldots, K$ given $\{z_{n,t}^{GoM}\}$ and $\{w_{n,j}^{GoM}\}$ from a Dirichlet distribution:

$$\lambda_{j,k} \sim \text{Dirichlet}_{L_j}(p_{k,1}, \ldots, p_{k,L_j}), \quad \text{where} \quad p_{k,l} \propto \sum_{n=1}^{N} I(z_{n,j}^{GoM} = k \text{ and } w_{n,j}^{GoM} = l) + \tau$$

The algorithm for the integrated model is the same as the component models except for the draw of the membership probabilities θ_n where the latent topic assignment variable $z_{n,t}^{LDA}$ and the latent membership indicator archetype assignment variable $z_{n,j}^{GoM}$ combine to inform the conditional posterior distribution:

```
thetanew = as.vector(rdirichlet(Niz[hh,] + C_dk[hh,] + alpha.prior))
```

9.4 DISCUSSION

The advantage of integrating text and choice data is being able to learn about the context of consumption and the motivations for consumer preferences for product features.

Text responses offer respondents the ability to identify reasons for product use in their own language. Consumer choices are ultimately of interest to researchers in marketing and economics, and the ability to relate these data to each other provides a more complete picture of consumer demand than that available with models that are not integrated. Moreover, the membership probabilities θ_n provide a means of relating multiple sets of data from text, scaled response and choice data.

The ability to estimate an integrated model is dependent on specifying a joint likelihood for the data. The advantage of Bayesian methods is that they are likelihood based and embrace the likelihood principal, where all information from the data about the model parameters is expressed in the likelihood function. Modern machine learning methods are often not associated with a well-defined likelihood function, making it difficult to generate joint models of behavior where information from text data can be combined with other forms of data.

10

Case Study 1: Analysis of Choice-Based Conjoint Data Using A Hierarchical Logit Model

Abstract

In this case study, we consider a workhouse model in marketing and economics, the random coefficient logit model. One popular application of this model is for the analysis of conjoint survey data.[1] Conjoint survey data has a panel format in which respondents choose among a small number of alternative products specified by specific product features (attributes) and specific values of these features (levels). Each respondent in the conjoint survey makes a relatively small number of choices (usually between 8 and 16) and these responses are analyzed using a random coefficient logit model. In `bayesm`, we implement Bayesian inference for a random coefficient logit model with both a mixture of normal components as well as the ability to impose sign constraints on the random coefficients in the routine, `rhierMnlMixture`.

10.1 CHOICE-BASED CONJOINT

The most popular form of conjoint survey analysis is called choiced-based conjoint (CBC). At least 10,000 CBC surveys are conducted per year in a wide variety of commercial and academic contexts. A frequent commercial application is to forecast demand (either in sales or market share) for new products or new product features. Here the hypothetical nature of the survey context is useful to make forecasts of the demand for products, features, or configurations which are not currently in the

---

[1] See also Sections 5.4 and 5.5.

Bayesian Statistics and Marketing, Second Edition. Peter E. Rossi, Greg M. Allenby, and Sanjog Misra.
© 2024 John Wiley & Sons Ltd. Published 2024 by John Wiley & Sons Ltd.

marketplace. Conjoint surveys can provide one input into firm decisions regarding which new products are likely to be successful prior to introduction in the marketplace.

The best way to view a CBC conjoint study is as a simulation of purchase decision involving choice among competing products. For example, a consumer is in the market for a class of cars, performs some sort of research and makes an informed choice among a somewhat limited consideration set of cars. In conjoint analysis, we assume that products can be meaningfully described by a finite and relatively small number of key features or attributes. Figure 10.1 shows a typical conjoint choice set or "choice task" involving choice between four hypothetical digital cameras, described by six features. The "none of these" choice option is sometimes called by economists the "outside" option in the sense that the consumer is comparing the utility afforded by each of the four cameras and contrasting that with the utility that can be obtained "outside" the choice set by purchasing some other, perhaps, completely unrelated product.

Conjoint analysis assumes a specific consideration set and assumes that, given a consideration set, all information on the relevant product features is readily available to potential consumers. In the example of choice of a car, it is assumed that not only has the consumer narrowed their consideration set to a relatively small set of alternative cars but that the set of product features relevant to that consumers decision is relatively small. This is not to say that cars and many other products don't have a very large set of potential attributes but that the conjoint study can meaningfully focus on a relatively small set of features (typically no more than 10 features) without fully specifying the set of all important product features. It is standard to instruct respondents that all other features (not displayed in the choice tasks) are to be held "constant" or assumed to be the same across all choice alternatives.[2] Given the instructions that all other attributes are to be held constant,

Scenario 1 of 16	Camera 1	Camera 2	Camera 3	Camera 4	
Brand	Canon Powershot	Panasonic Lumix	Sony Cyber-shot	Nikon COOLPIX	
Megapixels	16	16	16	16	
Optical Zoom	10x	4x	10x	4x	
Video	Full HD Video (1080p) with Stereo Microphone	HD Video (720p)	HD Video (720p)	Full HD Video (1080p) with Stereo Microphone	None of these
Swivel Screen	No	Yes	No	Yes	
Wifi	Yes	Yes	No	Yes	
Price	$79	$279	$379	$179	
Which of these digitial cameras do you prefer?	◉	◉	◉	◉	◉

As a reminder, you can hover over the features to view the definitions.

Figure 10.1 Camera survey choice task

---

[2] Whether or not this is a reasonable assumption depends critically on the quality of the conjoint design. In particular, if there is a wide variation in prices and the survey omits many important features, then respondents tend to ignore the instruction to "hold everything constant") and assume that higher priced alternatives have some hidden positive feature. This compromises the external validity of the conjoint survey and causes the researcher to conclude that price sensitivity is very low, accentuating the so-called "hypothetical" bias in a survey.

it is difficult to interpret the random error term in the RUM (Random Utility Model) formulation of the logit model (10.1.1) as the utility afforded by unobservable attributes.

The response variable in a conjoint survey is a multinomial or categorical response with the product features as inputs. It should be noted (see Chapter 13) that the CBC framework can be extended to continuous demand (nonunit demand). A natural statistical model would be a multinomial logit model in which each of the vector of multinomial choice probabilities is linked to the regression function which is simply a linear combination of product attributes. The coefficients in a standard MNL are sometimes called "partworths" in that they can be interpreted as utility weights corresponding to product features. This approach can be rationalized as a random utility model in which the utility of each choice can be expressed as a linear combination of product attributes (the X or independent variables) with a random error term that reflects choice error.

$$U_j = a'_j\beta + \varepsilon_j, \quad j = 1, \dots J \tag{10.1.1}$$

Here there are J choices and the vector, a_j, is the levels of each of the attributes included in the CBC study with β providing the weights (sometimes called "part-worths") that accord utility to each attribute-level combination. It is important to note that, in a utility-based interpretation of the MNL, this equation provides the utility of a specific respondent. Utility is a scale that is individual-specific and is not strictly comparable across respondents as it is well-known that any monotone function of a utility leaves demand (in this case choice) constant. Each element of the β vector provides the contribution to utility afforded by the specific attribute-level combination in the a_j vector. For example, if the attribute is price, then the coefficient on price can be interpreted as the negative of the marginal utility of income. The assumption of linearity in the utility function rules out any interactions between attributes. This can be relaxed by adding interaction terms or other functions of the attributes, but this is seldom done.

If some attributes (such as price) are continuous, it is typical to design the conjoint survey with at least three different "levels" of the attribute and create dummy variables for each level, thereby conferring a "non-parametric" flavor to the analysis. If we assumed that the random utility error term has as specific distribution then the standard MNL model can be derived:

$$Pr(y = j) = \frac{\exp\left(a'_j\beta\right)}{\sum_j \exp\left(a'_j\beta\right)}$$

If there is a "none of the above" or "outside" alternative in the analysis, then it is standard to normalize with respect to the outside alternative and write:

$$Pr(y = j) = \frac{\exp\left(a'_j\beta\right)}{1 + \sum_j \exp\left(a'_j\beta\right)}$$

Here the values of the a_j vector (including price) are assumed to be zero so that the a_j reflect the difference between the attribute value for each of the "inside" goods and the outside good.

10.2 A RANDOM COEFFICIENT LOGIT

Clearly, the MNL model cannot be fruitfully applied to conjoint survey data without some further assumptions regarding the variation of the β coefficients across respondents. A long history in marketing and economics suggests all consumers are different in the relative utility weight they assign to various product features. One approach would be to treat each respondent in the survey as different and attempt to fit MNL models to each respondent. Clearly, this is folly as there are typically only a handful of observations for each respondent (8–16). A maximum of the MNL likelihood may not even exist. The Bayes estimator will be saved by an informative prior. One such prior can be constructed by the assumption of a random coefficient model. That is, we assume there is some commonality among the respondents in the sense that each of their β_i vectors (i is the index of the respondent) is an iid draw from a common distribution:

$$y_i | A_i, \beta_i$$
$$\beta_i \sim N\left(\mu_\beta, V_\beta\right)$$
(10.2.1)

Here A_i is the matrix of the attributes for each of the choice alternatives in each of the choice tasks, stacked up. If there are four choice alternatives (excluding the outside option) and 16 choice tasks, then A_i is a 4×16 or p (number of choice alterantives) by *nvar* (the number of attribute-level combinations used in the survey) matrix. The second equation above is the random coefficient model, which (in this case) is a normal model. Clearly, this is not a useful model unless some way of assigning or estimating the hyperparameters, μ_β, V_β, is provided.

As we saw in the chapter on hierarchical models, the Bayesian approach is to put a prior on these normal distribution parameters and conduct Bayesian inference on all parameters using an MCMC approach derived from a hybrid Gibbs sampler. The basic model above can be extended to the case of a mixture of normals model for the first stage of the prior. This allows for important flexibility in that a mixture of normals can easily approximate any continuous distribution including ones with multi-modality, skewness, and bent or "banana-shaped" contours. For ease of exposition, we will consider just a simple one-component mixture of normals and we will also not allow for the possibility that there are observable characteristics of respondents that might drive the mean of the random coefficient distribution. All of these enhanced capabilities are available

R in *bayesm*.

10.3 SIGN CONSTRAINTS AND PRIORS

The full Bayesian approach adds priors on the random coefficient distribution:

$$\mu_\beta | V_\beta \sim N\left(\overline{\mu}, V_\beta \otimes a_\mu^{-1}\right)$$
$$V_\beta \sim IW\left(v, V\right)$$
(10.3.1)

These are standard conjugate priors and can easily be assessed if it is desired to be relatively uninformative with respect to the parameters of the normal distribution of het- R erogeneity. The default settings in `bayesm` are to designed to spread out the prior on the covariance matrix by using a barely (by 3 df) prior IW prior and a very diffuse normal prior. Specifically, by setting

$$\bar{\mu} = 0, a_{\mu} = .01, v = nvar + 3, V = vI \tag{10.3.2}$$

This works well if the partworths are on roughly the same scale (be careful of continuous variables as attributes – for example, if price is quoted in \$ or in 1,000\$ this will change the partworth dramatically). In general, it is a wise practice to scale the attribute variables so that they have roughly unit std deviation (note: most variables in conjoint are dummy variables and do not require scaling).

Not only does the normal distribution assumption have limitations in terms of flexi-bility (note that the classical random coefficient literature not only almost always uses a normal distribution but often limits the distribution to subsets of the β vector such as only the price coefficient), there is a real problem when we have a priori sign information. For example, in order to be considered a valid demand system, the price coefficient must be negative for all respondents. With a normal random coefficient distribution, there will always be a right tail that puts probability mass on positive price coefficients. This prob-lem is exacerbated by consumer heterogeneity. Most conjoint survey respondents display considerable price sensitivity while others may display little sensitivity to price. Given the rather small amount of sample information per respondent, a model with a normal distri-bution of heterogeneity will be inclined to center the distribution over a negative value but also show a large variance in order to accommodate those respondents who appear to be price insensitive. Any calculation which is based on assuming that the conjoint data can be treated as valid demand data, that is, more like revealed preference data than stated preference data, will be highly sensitive to the presence of posterior mass on economically irrational values of price sensitivity. For this reason, most conjoint practitioners imposes some sort of sign restriction on the price partworth. If prices are entered as a series of dummy variables, then a price "monotonicity" constraint is desired – the partworths in price must be decreasing in the price level they represent.

One simple way to impose a positive or negative sign constraint is to reparametrize the model in terms of a "deeper" set of part worths which are unconstrained. For the example of a price coefficient, we have

$$\beta_p = -e^{\beta_p^*} \tag{10.3.3}$$

Here β_p^* is unconstrained, $\beta_p^* \in (-\infty, \infty)$, but β_p is constrained to be negative. Several aspects of this reparameterization should be noted. First, there is now an inverse rela-tionship between the "deep" parameter, β_p^*, and price sensitivity. Secondly, the scale of the parameters is now different – β_p^* is on the log-scale, while β_p is not. If we assume that β_p^* is normal, then β_p will follow the negative of a log-normal distribution. What is dis-tinctive about the log-normal distribution is that the density declines to zero as ordinate gets closer to zero. In other words, a log-normal distribution imposes the prior view that price coefficients are not only nonpositive but that there is an interval near zero, which is near zero probability. This contrasts with a truncated normal distribution that does not decline to zero as the ordinate gets closer to zero. On these grounds, one could argue that the log-normal distribution is more appropriate than truncated normal distribution for partworths that are strictly less or greater than zero.

However, the price coefficient is not the only partworth that might be considered for sign constraints. Many attributes in conjoint studies are positive in the sense that we expect that as the attribute increases, utility increases as well. Thus, there is an argument for imposing sign constraints on these partworths as well. The problem here is that if you use the exponential re-parameterization in (10.3.3) except without the negative sign, then you are imposing the constraint that the partworth for this attribute is strictly positive. This rules out the possibility that you can determine from your data that respondents don't really assign any appreciable utility to this attribute. For this reason, we do not recommend imposing sign constraints without first checking the unconstrained model fit. If there is a considerable mass on implausible values of partworth coefficients, it is possible that this may result from respondents who are not sufficiently diligent completing the survey or because of implausible designs that call into question the fundamental assumption/ instruction that all other aspects of the hypothetical products are the "same" across choice alternatives.

The most popular, by far, software used for conjoint survey design and analysis is Sawtooth Software. In the CBC implementation in Sawtooth, there are options to impose sign constraints. However, these options are implemented with an incoherent and non-Bayesian "tying" procedure that violates Bayes theorem and produces results which are unusable for an serious application in which a truly sign constrained model is desired.

From a technical point of view, a sign constraint parameterization is trivial, just rewrite the likelihood function and use the same prior on the "deep" parameters instead of the actual partworths. When draws are complete, simply transform back to the original parameterizations.[3] However, the implied prior on the sign-constrained part-worths should be examined to see if it reflects the investigators prior information. In particular, standard diffuse priors essentially imply that you think very large positive and negative values of the "deep" parameter are possible, inducing a very informative prior on the transformed, original partworths.

To understand this problem, let's consider the consequences of using a standard prior on the partworth vector. A standard "diffuse" prior (see 10.3.2) on the partworth coefficients is designed to center the normal distribution of the partworths on zero and center the var–cov matrix of the partworth vector on the identity matrix. More importantly, these settings impose barely proper and extraordinarily diffuse priors. To illustrate this, Figure 10.2 displays draws from the deep parameter, β_p^*, as well as the induce prior (induced via reparameterization) from (10.3.3). Here we consider a priors where the

---

[3] Given the use of a custom-tuned Metropolis RW chain, the Jacobian of the transformation from the deep parameters to the original part worths must be computed and used.

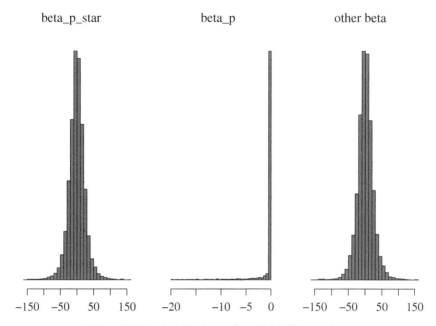

Figure 10.2 Implications of standard diffuse priors

partworth vector is of length 10 and we impose a sign restriction only on the "price coefficient." The leftmost graph shows the prior distribution of β_p^*, which is centered at zero and very diffuse. The rightmost figure shows the prior distribution of any other non-sign constrained element of the partworth vector. In the middle, we see the implied or induced prior distribution on the price coefficient after transformation to impose negativity. Clearly there is a huge mass near zero which comes from the 50% of the prior draws on β_p^*, which are over large negative values. This is a highly informative prior which would shrink respondent price partworths toward zero.

Clearly we do not want to impose such strong prior information on our analysis. In order to avoid this problem, we implement a new default prior in *bayesm*, which avoids this problem by lowering the diffuseness of the prior and moving the prior on the β_p^* parameter to be centered over a small positive value.

$$\bar{\mu} = \bar{\mu}^*, a_\mu = 0.1, v = nvar + 15, V = vDiag(d). \tag{10.3.4}$$

Here $\bar{\mu}_i^* = 0$ if unconstrained, and $\bar{\mu}_i^* = 2$ if constrained. $d_i = 4$ if unconstrained and 0.1 if constrained. Figure 10.3 shows the same prior distributions for each of constrained and unconstrained parameters. The modified diffuse prior retains a very noninformative view on the unconstrained parameters (see rightmost histogram) while imparting a much more reasonable but still diffuse prior on the price partworth. Obviously, whether or not this prior is appropriate depends to some real extent on the scaling of the price variable and some experimentation may be necessary.

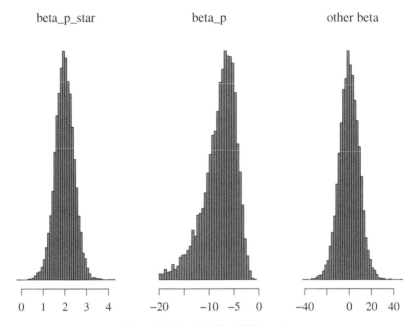

Figure 10.3   Modified diffuse prior

10.4 THE CAMERA DATA

To illustrate the fitting of a hierarchical logit model to conjoint data, we consider the
R digital camera dataset. This dataset is included in *bayesm*, accessible with the R command
`data(camera)`. We designed a conjoint survey to estimate the demand for features in the
point and shoot submarket. We considered the following seven features with associated
levels:

1. Brand: Canon, Sony, Nikon, Panasonic

2. Pixels: 10, 16 mega-pixels

3. Zoom: 4x, 10x optical

4. Video: HD (720p), Full HD (1080p) and mike

5. Swivel Screen: No, Yes

6. WiFi: No, Yes

7. Price: $79–279

We focus on evaluating the economic value of the swivel screen feature, which is illustrated in Figure 10.4. The conjoint design was a standard fractional factorial design in
which each respondent viewed 16 choice sets, each of which featured four hypothetical
products. A dual response mode was used to incorporate the outside option. Respondents were first asked which of the four profiles presented in each choice task was most
preferred. Then the respondent was asked if they would buy the preferred profile at the

LCD screen tilts and can be swung away from the camera body.

270°

175°

Figure 10.4   Swivel screen attribute

stated price. If no, then this response is recorded as the "outside option" or "none of the above." Respondents were screened to only those who owned a point and shoot digital camera and who considered themselves to be a major contributor to the decision to purchase this camera.

The survey was fielded to the Sampling Surveys International internet panel in August 2013. We received 501 completed questionnaires.[4] We recorded time to complete the conjoint portion of the survey. The median time to complete is 220 seconds or about 14 seconds per conjoint task. The 25th percentile is 151 seconds and the 75th percentile is 333 seconds. To check sensitivity to time spent on the survey, we conducted analyses deleting the bottom quartile of the respondents and found little change. It is a common and well-accepted practice to remove respondents who "straight-line" or always select the same option (such as the left most choice). The idea is that these "straight-liners" are not putting sufficient effort into the choice task. Of our 501 complete questionnaires, only two respondents displayed straight-line behavior and were eliminated. We also eliminated six respondents who always selected the same brand and two respondents who always selected the high price brand. Our reasoning is that these respondents did not appear to be taking the trade-offs conjoint exercise seriously. We also eliminated 23 respondents who always selected the outside option as their partworths are not identified without prior information. This leaves was 468 respondents out of an original size of 501. As is customary, we eliminated who appeared to be guessing. If someone is simply randomly picking one of the 5 choice options (including the outside outcome, then their choice probability would be 1/5 and corresponding log-likelihood would be $16 * log(1/5) = -26$. Respondents whose log likelihood was close to this figure were eliminated,[5] leaving a total of 332 respondents. It should be noted that in these days of large internet panels of unknown quality, investigators should be even more concerned than ever before regarding the quality of their sample respondents. Many respondent simply rush through surveys of all types, leaving data of questionable quality. Conjont surveys are particularly demanding in that they involve the repetitive task of examining very similar (in format) conjoint choice tasks screens and a larger number of product attributes. There are two ways to deal with this problem: 1. to screen out respondents based on a combination

---

[4] This study was part of a wave of four other very similar conjoint studies on digital cameras each with the same screening criteria. For all studies in the wave, 16,185 invitations were sent to panelists, 6,384 responded. Of those who responded to the invitation, 2,818 passed screening and of those passing screening 2,503 completed the questionnaire. Thus, the overall completion rate is 89% which is good by survey standards.

[5] A respondent who guess have a hit rate of 20% (1/5). We required respondents to have a hit rate of at least 40%, which means a threshold log-likelihood value of $16*log(0.40) = -14.66$.

of straight-lining, speeding, or guessing and 2. to institute a nontrivial reliability test in the screener portion of the survey. In recent work, we have found asking respondents to input the exact model of their electronic device and checking this against an internal table of model numbers (note a device such as the iPhone has a multiple digit "model number" which is not the same as simply iPhone 14 or some well-known version number). We have found that 20–30% of respondents are screened out when using such a reliability test.

The digital camera survey was designed to evaluate the importance and value from adding a tilt-screen feature to the camera as shown in Figure 10.4. There are many ways to consider the value of a feature but certainly the fundamental exercise involves the comparison of cameras that are otherwise identical but for the swivel screen feature. Given that partworths are not comparable across respondents unless they are normalized by a price or other variable that has a ratio scale, simply describing partworths provides almost no information regarding value. In the companion case study, we will consider using the conjoint survey as the basis for a valid demand system and directly answering such important questions as what are consumers willing to pay for a swivel feature and what can firms charge for the addition of this feature, two completely different questions.

First, let's take a look at the camera dataset and familiarize ourselves with the format R that *bayesm* want to see all panel data.

10.4.1 *Panel Data in* bayesm

Panel data is widely used in all of the social sciences. Panel data can be obtained by simply observing a "panel" or set of cross-sectional units (households, consumers, markets, key accounts, ...) over time. In the case of a conjoint survey, the "panel" is a subset of respondents who meet the appropriate screening criteria and are observed to make choices in a small number of choice tasks. Unless there is a severe problem with noncompletion,

```
> data("camera")
> str(camera,list.len=2)
List of 332
 $ :List of 2
  ..$ y: int [1:16] 1 2 2 4 2 2 1 1 1 2 ...
  ..$ X: num [1:80, 1:10] 0 1 0 0 0 0 1 0 0 0 ...
  .. ..- attr(*, "dimnames")=List of 2
  .. .. ..$ : chr [1:80] "1" "2" "3" "4" ...
  .. .. ..$ : chr [1:10] "canon" "sony" "nikon" "panasonic" ...
 $ :List of 2
  ..$ y: int [1:16] 3 5 5 3 1 2 5 4 5 1 ...
  ..$ X: num [1:80, 1:10] 1 0 0 0 0 0 0 0 0 0 ...
  .. ..- attr(*, "dimnames")=List of 2
  .. .. ..$ : chr [1:80] "81" "82" "83" "84" ...
  .. .. ..$ : chr [1:10] "canon" "sony" "nikon" "panasonic" ...
 [list output truncated]
```

Figure 10.5 Contents of camera dataset

conjoint survey panel data is a "balanced" panel in the sense that there an equal number of choices/observations made for each respondent. Other panel datasets often have a different number of observations for each panel member as some members leave and others join the panel. In addition, the order of observations is important in many panels as the observations for any panel unit are observed over time. In a conjoint survey, the time ordering of responses is typically ignored.

The list object in R is an ideal container for panel data. Each of the panelists' data is stored as an element of the panel's list. Each panelists' data is, in turn, also stored as a list. Thus, the complete panel dataset is stored as a list of lists. In the case of the camera dataset, we have a list structure of 332 elements. Each element contains the choices (in the multinomial vector y) and a matrix which stores the attributes for each choice alternative is each of the choice tasks, denoted X. X is a matrix with nvar columns (each column corresponds to an attribute-level combination) and 4×16 rows. Note that in most software packages, panel data is stored in large arrays with a respondent identifier. This sort of data storage is prone to errors in data manipulation and requires the number of X variables to be the same for each respondent.

Let's take a look at what's in the camera data:

The str function displays information about any R object. Camera is a list with 332 elements, one per respondent. Each respondent's data is also a list of two elements, y and X. y records the multinomial choices (1–4 for hypothetical products and 5 for "none of the above"). The X is a matrix that provides information on the attributes, which define each choice alternative. Let's take a look at part of the X data for the second respondent:

```
> X_2 = camera[[2]]$X[1:5,]
> rownames(X_2)=c(rep("",5))
> X_2
  canon sony nikon panasonic pixels zoom video swivel wifi price
    1    0    0       0        1     1     1      0     1   0.79
    0    0    0       1        1     0     0      1     1   1.79
    0    1    0       0        1     1     0      0     0   2.29
    0    0    1       0        1     0     1      1     1   1.29
    0    0    0       0        0     0     0      0     0   0.00
```

Figure 10.6   Top of X matrix for second respondent

We use the indexing operator for lists, [[]], to access the second respondent. It is important to realize that camera[[2]] is a list of two elements – the data on choices and attributes for the second respondent. We access the X matrix and choose the first five rows which describes the first choice task for this respondent. There are 10 attributes that include brand, pixel (high or low), zoom (4x/10x), video capability (720/1040), swivel screen (yes/no), and price in $100.00. This is the result of a conjoint design exercise in which combinations of attributes are typically sampled at random from the set of all possible combinations and then trimmed for balance by attribute level.

With the data already in the "list of lists" format and with the correct names (note: the choices and the attribute matrix must be labeled "y" and "X" or the *bayesm* routine will not work, we can procedure to "run the model."

R

10.5 RUNNING THE MODEL

To perform Bayesian inference for the camera dataset, we will have to select a model that includes the likelihood and prior. Clearly, this data is well-suited for a hierarchical Multinomial Logit model. To complete the specification, we must choose the form of the first stage prior (sometimes known as the random coefficient distribution), for this illustration we will use a one component normal mixture, or simply a normal model. We note that this is a complete distribution of heterogeneity that allows all partworths to vary across respondents.

R   In working the *bayesm*, we must also choose three sets of quantities:

1. **Prior**: This includes the form of the prior, any desired sign constraints, and prior hyper-parameters

2. **Data**: what object is the data stored in

3. **MCMC** Parameters: How many draws should be drawn and with what frequency should these draws be saved.

We will use a simple normal model, default choices for prior hyperparameters and we will impose a sign constraint that the price partworth must be negative. It is important to understand that Hybrid Gibbs samplers can have relatively high autocorrelation and, therefore, we should use a large number of draws. Fortunately, even laptop computers are able to make $100,000$ s of draws even in complicated models well with in the "coffee break" period. We will set $R = 50,000$ and keep every 10th draw just reduce memory usage (though even this is hardly a constraint any more). This will take about 2 minutes on garden variety equipment.

```
> ncomp = 1 # normal distribution of "deep" coefficients
> nvar = 10
> Prior1 = list(ncomp=1,SignRes=c(rep(0,9),-1)) # constrain price coef to be < 0
> R=50000
> Mcmc1= list (R=R,keep=10)
> Data1 = list(p=5,lgtdata=camera)
> out=rhierMnlRwMixture(Data=Data1,Prior=Prior1,Mcmc=Mcmc1)
Z not specified
Table of Y values pooled over all units
ypooled
   1    2    3    4    5
1100  936 1035  898 1343


Starting MCMC Inference for Hierarchical Logit:
   Normal Mixture with 1 components for first stage prior
   5  alternatives;  10  variables in X
   for  332  cross-sectional units
```

Figure 10.7   Running rhierMnlRwMixture

```
> str(out,list.len=4)
List of 4
 $ betadraw: 'bayesm.hcoef' num [1:332, 1:10, 1:5000] 2.7769 -4.1669 4.1603 1.745
 ...
 $ nmix    :List of 3
 ..$ probdraw: num [1:5000, 1] 1 1 1 1 1 1 1 1 1 1 ...
 ..$ zdraw   : NULL
 ..$ compdraw:List of 5000
 .. ..$ :List of 1
 .. .. ..$ :List of 2
 .. .. .. ..$ mu   : num [1:10] 0.5376 0.319 0.479 -0.0293 1.0199 ...
 .. .. .. ..$ rooti: num [1:10, 1:10] 0.377 0 0 0 0 ...
 .. ..$ :List of 1
 .. .. ..$ :List of 2
 .. .. .. ..$ mu   : num [1:10] 0.763 0.599 0.482 0.294 0.86 ...
 .. .. .. ..$ rooti: num [1:10, 1:10] 0.352 0 0 0 0 ...
 .. ..$ :List of 1
 .. .. ..$ :List of 2
 .. .. .. ..$ mu   : num [1:10] 0.72 0.376 0.52 0.147 0.998 ...
 .. .. .. ..$ rooti: num [1:10, 1:10] 0.332 0 0 0 0 ...
 .. ..$ :List of 1
 .. .. ..$ :List of 2
 .. .. .. ..$ mu   : num [1:10] 0.9 0.661 0.622 0.306 0.925 ...
 .. .. .. ..$ rooti: num [1:10, 1:10] 0.324 0 0 0 0 ...
 .. .. [list output truncated]
 ..- attr(*, "class")= chr "bayesm.nmix"
 $ loglike : num [1:5000, 1] -2669 -2907 -3153 -3268 -3380 ...
 $ SignRes : num [1:10, 1] 0 0 0 0 0 0 0 0 0 -1
```

Figure 10.8 "out" object

The first few lines above sets the number of normal components to 1 (ncomp), sets the number of X variables or attribute variables to 10 and provides the list argument for prior settings. The key part is the argument, SignRes, which tells the routine which partworths to constrain to be positive (1) or negative (−1) or to leave unconstrained. Next we tell *bayesm* where to find the data and how many choice alternatives there are (5). We capture all output from the MCMC run in the object "*out*." The *bayesm* routine starts by shows us how many choices for each alternative was made over all respondents. In a well-designed conjoint, we should see approximately the same number of choices for each choices 1–4. Choice 5 is the outside option. If we have designed very unattractive products, we could see the outside option (5) as very large and we would have to redesign the survey. *Bayesm* then echoes the model and prior setting used as well as the MCMC parameters and provides (assuming nprint not set to 0) a running tab on the iterations. At the end, the object out is created.

Let's take a brief look at what is in "*out*."

Of course, "*out*" is another of these lists. The first list element is a three dimension array, "betadraw," these are the draws for each respondent. The first dimension indexes the respondent, the second dimension the attribute, and the third the draws. Note that there are only 5000 draws stored for each respondent in an effort to reduce memory

```
> BurnIn=1000 # discard first 10,000 draws
> names=c("Canon", "Sony", "Nikon", "Panasonic", "pixels",
+          "zoom", "video", "swivel","wifi", "price")
> plot(out$betadraw,names,burnin=BurnIn)
```

Figure 10.9   Plotting respondent draws

consumption. Thus, `betadraw[5,2,1]` is the first draw of the second partworth for the fifth respondent. If you want all the draws for a specific respondent, simply write "`betadraw[5,,]`" this will be a 10×5000 array for that respondent. These betadraws can be used to make inferences regarding the posterior distribution of any particular respondents partworths.

The next element of out is called "`nmix`" and this is a list itself that provide draws of the common normal mixture parameters. Since we have only a single component mixture of normals this structure is perhaps more cumbersome that needed. "`nmix$probdraw`" are draws of the mixing probabilities for each of the ncomp mixture components. Since we only have a single component this matrix is just a long vector of ones. The parameters of each of the ncomp normal mixtures are stored in `compdraw`, again as a list of lists. In this case, there is only one component so `compdraw[[1]]` is a list of two elements – the mean and Cholesky root at the first MCMC iteration. `compdraw[[1]][[1]]$mu` is the first draw of μ and `compdraw[[1]][[1]]$rooti` is the first draw of the inverse of the Cholesky root of $V_\beta = R^t R$. We can use the draws in compdraw to make inferences about the distribution of the partworths across respondents.

R   Many users of `bayesm` are somewhat frustrated that the routines simply end with the creation of a draw object and the user must decide how to use this simulation output to make the appropriate inferences and predictions.

10.6 DESCRIBING THE DRAWS OF RESPONDENT PARTWORTHS

R There are plot methods associated with `bayesm` that are very useful to plot and examine the distribution of coefficients across different members of the panel. Let's try these out and learn a bit about how the partworths vary across respondents.

Here we set a "burn-in" of 10,000 draws (remember we are storing only every 10th draw). The "plot" generic function understands that we are desiring a plot of draws **R** in a hierarchical setting. The plot method written for `bayesm` randomly selects 30 or so respondents and plots the posterior distribution of each partworth on a separate page. At the end, the plot method will plot a histogram of the posterior means for each partworth and respondent. Let's take a look at the partworths on the swivel screen first.

We can see that respondents have different views regarding the swivel screen. Some appear to have a higher utility than others. Of course, it is not strictly kosher to compare raw partworths across respondents but there is some hope that the shrinkage inherent in the hierarchical model may force more comparability. In the next chapter, we will use only metrics that avoid utility comparisons that are not strictly comparable across consumers. The boxplots in Figure 10.10 show the spread of the posterior is fairly

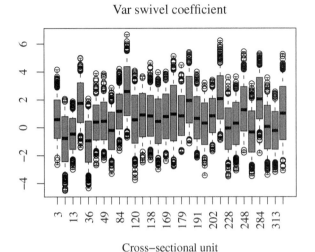

Figure 10.10   Swivel screen partworths for various respondents

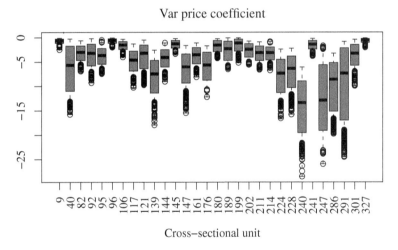

Figure 10.11   Price partworths for various respondents

comparable across respondents, which is a direct reflection of the balanced conjoint orthogonal- design.

There are even larger differences among respondents in their price partworths as shown in Figure 10.11. Some respondents are fairly insensitive to price while the majority are highly sensitive. Figure 10.12 provides the histogram of the posterior means (Bayes estimates) of selected partworths, which confirms this finding.

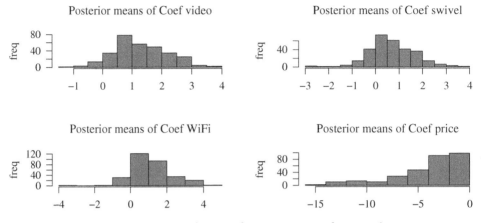

Figure 10.12 Distribution of posterior mean of partworths

10.7 PREDICTIVE POSTERIORS

10.7.1 Respondent-Level Partworth Inferences

Posterior inferences about specific respondent's partworths are useful. As we have seen, it is a simple task to compute the posterior means of the partworth vector and report aspects of the distribution of these Bayes estimators across respondents. Typically, we will see a table of average partworth estimates as well as some measure of the degree of heterogeneity or spread in the distribution of partworth estimates such as the standard deviation or quantities across respondents. As we have explained, these distributions are of questionable value due to the uniqueness of the utility scale for each respondent. That is, β_i is a vector of nvar elements which are expressed with a utility scale unique to the ith respondent. For this point of view, it is meaningless to average or summarize in anyway the raw partworths across respondents. This means, among other things, that it is conceptually incorrect to talk about the "average" or "representative" respondent by using any summary statistic of a single partworth alone. For example, if you compute the median of the posterior mean estimates of partworths across respondents, this is not, as might seem without further reflection, the "median" or representative respondent.

However, if you divide each of the attribute partworths by the price coefficient then you will have converted respondent-specific utility into the money metric as the price coefficient should be interpreted as the marginal utility of money or income. Many, if not the majority of conjoint practitioners, will call this ratio the willingness to pay (WTP) for the attribute for that particular respondent. As we will see in Chapter 11, this is not true. For this reason, we call this method a pseudo WTP computation or, pWTP. For example, the pWTP of the ith respondent for the swivel-screen (SS) attribute is given by:

$$pWTP_{SS,i} = \frac{\beta_{SS,i}}{-\beta_{p,i}} \tag{10.7.1}$$

Obviously, we can use the draws from the posterior for the ith respondent to build up the posterior distribution of pWTP for any partworth and any respondent. The problem with this point of view is that it does ignore an important aspect of the model. If we truly believe that the "deep" partworths are distributed as a mixture of normals (in this case simply normal), then we are not using that part of the model to its fullest advantage.

10.7.2 Posterior Predictive Distributions

If the model specifies that $\beta \sim N(\mu_\beta, V)$, then we could compute the expected pWTP (or any other function of β) as a summary statistic.

$$E\left[pWTP_{SS}|\mu_\beta, V_\beta\right] = \int \frac{\beta_{SS}}{-\beta_p}\phi\left(\beta|\mu_\beta, V_\beta\right) d\beta \qquad (10.7.2)$$

Here ϕ is the multivariate normal density function. Of course, this is a function of the normal distribution parameters which we do not know. As Bayesians we have been taught that, instead of plugging in some "consistent" estimators of these parameters, we should integrate over the appropriate posterior distribution. This results in the predictive posterior distribution of whatever quantity we desire. That is, for each posterior draw of (μ_β, V_β), we calculate the integral in 10.7.2, this gives us one draw from the posterior predictive distribution of $E\left[pWTP_{ss}\right]$. Of course, we can simply use simulation to compute the integral. More formally, we consider a function of the β vector, $f(\beta)$, and we are simulating from the posterior predictive distribution of this function. For example, the probability that this function lies in a set A is given by:

$$Pr\left(f(\beta) \in A\right) = \int_A f\left(\beta|data\right) d\beta$$

$$= \int_A \int f(\beta)\phi\left(\beta|\mu_\beta, V_\beta\right) p\left(\mu_\beta, V_\beta|data\right) d\mu_\beta dV_\beta d\beta \qquad (10.7.3)$$

This double integral essentially means a simulation within a loop. As we loop over all posterior draws of the normal distribution parameters, we compute the expected pWTP for those normal parameters by making a large number of draws from the appropriate normal distribution. We then proceed to the next posterior draw of the normal distribution parameters. In the case of the camera data, we have (after burn-in and "thinning") 4000 draws from the posterior distribution, $\left(\mu_{beta}, V_\beta|Data\right)$, for each of these draws we can make a large number of draws from the associated normal distribution and compute the simulation-based approximation to (10.7.2). The result of this procedure will be a vector of 4000 draws from the posterior predictive distribution of $E\left[pWTP_{SS}\right]$.

Let's look at the code to do this with the "out" object we created. This code is shown in Figure 10.13.

We start by creating arrays to store various computed quantities including each draw from the posterior predictive distribution of pWTP. The loop is over each valid draw from the posterior of the model parameters (after the BurnIn period). Recall that draws from the mixture of normal distributions of the β parameter are stored is a list of lists called **R** "nmix." We use a *bayesm* utility routine to draw 4000 draws from each of the posterior draws of the normal distribution parameters. Recall that the model assumes the "deep" partworth parameters are distributed as a mixture of normals. For this reason, we must transform the "deep" price parameter back to the utility scale. For each simulation

```
> NIter=length(out$nmix$compdraw)
> pWTP=matrix(double((NIter-BurnIn)*nvar),ncol=nvar)
> betastar=matrix(double((NIter-BurnIn)*nvar),ncol=nvar)
> beta=matrix(double((NIter-BurnIn)*nvar),ncol=nvar)
> pWTP=double(NIter-BurnIn)
>
> for(iter in (BurnIn+1):NIter){
+    betastar=rmixture(n = 4000, pvec = out$nmix$probdraw[iter],
+                      comps = out$nmix$compdraw[[iter]])$x
+    beta= cbind(betastar[,1:9],-exp(betastar[,10])) # transform price coef
+    pWTP[iter-BurnIn]= mean(100*beta[,8]/-beta[,10])
+ }
> quantile(pWTP,probs=c(.025,.05,.25,.50,.75,.95,.975))
    2.5%       5%      25%      50%      75%      95%    97.5%
28.38797 31.04509 38.89811 44.94529 51.51996 61.57908 65.17830
> mean(pWTP)
[1] 45.42567
```

Figure 10.13 Computing the posterior predictive distribution of pWTP

inside the loop, we compute the average pWTP for the swivel screen feature (note prices are in $100 s). We store each of these means into the pWTP vector which holds 4000 draws.[6] Now that we have 4000 draws from the posterior predictive, we can summarize these draws in whatever way we choose. The posterior predictive distribution of pWTP is centered around $45 with a rather wide range. This shows that a sample size of only 300–400 is apt not to be adequate for precise estimation of quantities such as pWTP as can be seem from the quantiles presented in Figure 10.14. A 95% posterior interval

pWTP($)

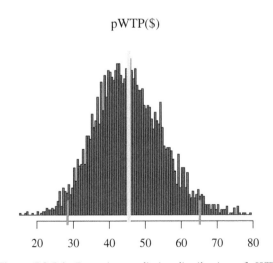

Figure 10.14 Posterior predictive distribution of pWTP

---

[6] The number of draws from the normal inside the loop is fairly arbitrary. Since these are iid draws, we may not need as many as posterior MCMC draws, but in an abundance of caution we choose 4000 draws.

would extend from \$28.39 to \$65.18. The Figure shows the histogram of the draws from the posterior predictive distribution of $E[pWTP]$ along with the posterior mean (vertical light grey line) and the 95% posterior interval shown by the grey lines.

We should strongly emphasize that it is inappropriate to interpret these pWTP results as providing a valid measure of actual consumer WTP. Even worse, these computations have little or no bearing on what a firm in a competitive market can actually charge for a digital camera with the swivel screen feature. Case Study 2 in Chapter 11 will address both concerns by providing a valid measure of WTP and indicating how a supply side analysis can be constructed to allow for the computation of equilibrium prices.

10.8 COMPARISON OF STAN AND SAWTOOTH SOFTWARE TO BAYESM ROUTINES

There are numerous implementations of MCMC approaches to models used in marketing and econometrics. `bayesm` is unusually broad in coverage and has been re-written in C++ for improvements in speed of execution. However, several other implementations are worthy of notice.

10.8.1 Comparison to STAN

For general-purpose of Bayesian modeling, the STAN project has achieved a great deal of success and has strong advocates. There many advantages of STAN but a few deserve comment:

1. STAN includes a modeling language that allows the researcher to specify models in terms of graphs or hierarchies of distributions. Most common univariate and multivariate distributions are implemented and, of course, the user can add their own distributions if desired.

2. STAN can, in principle, handle any combination of likelihood and prior distributions that is available within the STAN environment. Priors receive very little emphasis in most applied work. But in STAN, the analyst is largely freed from constraints on selection of priors. Priors can be chosen for their ability to represent prior information rather than simply for convenience. There is, however, a disturbing tendency of some in the STAN community to specify models without priors or with improper priors. This, in our view, abandons much of the value of the Bayesian approach to inference and opens the possibility of undefined posterior distributions, particularly in the context of hierarchical applications. However, the developers of STAN cannot be responsible for misuse of their software.

3. A combination of the "No-U-Turn" sample and some clever re-parameterizations offer the promise of very low autocorrelation in the MCMC sampler used by STAN.

There are also some disadvantages to STAN.

1. STAN cannot handle parameter spaces that destroy differentiability of the log-posterior in the parameters.

2. STAN imposes a substantial computational penalty in that the gradient of the log-posterior must be evaluated at every MCMC iterations (even though this gradient is evaluated with analytical results generated by automatic differentiation software).

R   In this section, I will compare STAN to the relevant *bayesm* routine. When comparing different MCMC methods, the relevant metric is effective sample size per minute of computation. That is what is the equivalent iid sample size that is computed per minute of clock time? For example, even though *bayesm* is blazing fast compared to many other implementations, it may well be that the draws from the *bayesm* implementation are much more autocorrelated than other algorithms, effectively erasing the speed differential. This is apt to be particularly true with respect to comparisons with STAN whose MCMC method plus reparameterizations in the hierarchical setting may run slower (fewer draws per minute of computing time) but may more than make up by a much lower autocorrelation and larger effective sample size.

I will use a rather generic and low-powered CPU, a machine using the Intel i7-9750H CPU with 2.6 GHz clock speed and running Windows 10. All comparisons are between *bayesm* version 3.1-4 and rstan version 2.21.2. Barring parallelization (which, in

R principle, can be applied to both *bayesm* and STAN), there is no reason to believe than the relative performance of *bayesm* and STAN-based models will change materially in future versions.

R   I will use the camera dataset to illustrate the relative performance of the *bayesm* routine, rhierMnlRwMixture, and STAN. Identical priors and identical values of prior hyperparameters were used. The STAN implementation is based on an example in the STAN user guides and uses the STAN functions, inv_wishart, multi_normal, and categorical_logit. Comparison of the simulated univariate distributions of individual level and common parameters showed close correspondence between STAN and *bayesm* (via QQplots). A 10% burn-in period was used with 10,000 draws (note: any computing time does NOT include the compilation step that STAN requires to compile the automatically generated code to evaluate the gradient vector of the log-posterior).

R   While *bayesm* produces 9874 draws per minute of computing, STAN only produces 17.96, making STAN impractical for even this small data set unless the investigator is willing to wait more than an hour for the results. On the other hand, the effective sample size for STAN is 796.44 per 1000 draws while the *bayesm* produces a much more highly autocorrelated, with only a 28.81 effective sample size per 1000 draws. However, the dramatic difference speed still means that *bayesm* produces 19.89 times the number of effective draws per minute of computing time.

It also appears that much of the reduction in autocorrelation achieved by the STAN sample is due to a reparameterization of the hierarchical model and not the exact MCMC algorithm.

With current computing equipment, the STAN implementation is so slow that analysis of even moderate size data sets with hierarchical models is not practical unless the analyst is willing to wait several hours for the results.

I also checked other common models including a nonhierarchical logit model, binary

R probit, and ordinal probit models. The ratio of *bayesm* to STAN effective sample size per minute is 27.9 for the multinomial logit, 2.07 for the binary probit, and 3.16 for the ordinal probit. I should emphasize that all of these results are averages over 100

runs for the same dataset. The multinomial logit was run on the same camera data with no hierarchy, that is, a pooled MNL model with about 3000 observations. The binary probit and ordinal probit were examined with simulated data and moderate sample sizes of 10,000 or 3000, respectively.

10.8.2 Comparison with Sawtooth Software

Sawtooth Software Inc (SSI) offers a hierarchical Bayes implementation to analyze MNL data collected from conjoint surveys. Sawtooth specializes in this area and is the leading vendor for software to analyze conjoint data. Sawtooth does not pretend to offer a comprehensive suite of software, instead SSI specializes in the conjoint area. In addition, SSI offers a whole set of tools that a survey researcher can use to design and field a conjoint survey as well as to undertake demand and market forecasts which are often incorrectly called "market-simulations."

R   Using the same camera dataset that is in `bayesm`, we compared the `bayesm rhierMnl-RwMixture` implementation to the Lighthouse Studio available from SSI, version 9.8.1. We used the academic teaching license that limits conjoint datasets to a maximum of 250 respondents (not observations). We attempted to set identical prior specifications and used a run length of 30,000 draws with a burn in of 10,000. `bayesm` dominates SSI software producing both more draws per minute of computation and much less autocorrelated draws. This combines to afford `bayesm` a large advantage over SSI with 488.86 effective sample size per minute of computation vs 161.43/minute for SSI. On the basis of effective sample size per minute of computing, `bayesm` has a 3:1 advantage over SSI. Given that `bayesm` is free and offers a wealth of other extensions such as a mixture of normals for the distribution of partworths in the population, there seems little to recommend SSI as simply method for analysis of CBC conjoint data. However, the convenience of other features such as the design and fielding of conjoint surveys may compensate for these short-comings.

It should be emphasized that, given the very high autocorrelation of SSI draws, a careful investigator should use more than 100,000 draws if simulation error is to be held to an acceptable minimum.

11

Case Study 2: WTP and Equilibrium Analysis with Conjoint Demand

Abstract

In this chapter, we will consider how conjoint data can be used as a valid measure of demand and how to compute a valid Willingness to pay (WTP) measure based on a conjoint study which is properly designed to measure demand. As is well known, WTP is not a market price and we consider what additional data and assumptions are required to use a conjoint survey as part of a market price calculation. This will allow us to provide a true measure of what a firm can change for a product feature and what sort of profits can be generated by the addition of this feature–something that has eluded the conjoint literature until very recently.

In this case study, we will extend our analysis of conjoint data by regarding the conjoint study as a method of demand measurement, in particular, a method for properly assessing the demand for products defined by a set of features. We will begin by providing some of the economic foundations for interpreting conjoint data as valid demand data. We will also discuss some of the advantages and disadvantages of conjoint data relative to observation demand data.

We will define a proper measure of WTP for products and product features and discuss the relationship between this measure and other ad hoc approaches that are widely used by conjoint practitioners.

Clearly, demand is only one half of the story. If a properly specified set of competitors, costs and equilibrium definition are used, then a valid supply analysis can be added to the demand system and equilibrium profit, prices and quantities can be computed. Ultimately, this will address the more fundamental question in marketing – what is the market value of a product or a product feature? Remarkably, the answer to this question has eluded virtually the entire academic and commercial world of conjoint analysis for more than 30 years. We will illustrate this approach to fully specifying and calculation market equilibrium with the digital camera data discussed in case study 1 in Chapter 10. Much of the material in this chapter is derived from Allenby et al. [2014a, 2014b].

11.1 THE DEMAND FOR PRODUCT FEATURES

The demand schedule facing any firm's product can be constructed from the distribution of preferences across the population of consumers. We take a characteristics view of products; that is, products are simply bundles of features or characteristics. Firms pick a particular bundle of features which can be thought of as a point in the space of possible product configurations. The competitive structure of the industry is defined not only by the number of competing products but also by the positions of those products in the characteristics or product feature space. Given a set of products with associated demand, a competitive equilibrium can be computed, enabling computation of the incremental profits flowing to the firm offering a product with focal features.

We use a standard model for the demand for differentiated products. Consumers choose only one product at any one purchase occasion and may elect to not purchase any product. A firm introduces a product into the marketplace by announcing a particular configuration of product features/characteristics and a price. The demand schedule facing this firm is the expected sales of the product given the price, configuration and the distribution of preferences across consumers. For example, a digital camera product might consist of a brand name (e.g., Sony), particular resolution (measured in megapixels), a particular level of zoom performance, a style of display (swivel screen or not), etc. The firm announces a price for this product. Consumers evaluate this product and compare it to other products in the marketplace. Each consumer makes a choice and these choices add up to total demand for the product. Each consumer is represented by a specific utility function which is specified by a set of utility weights on the product characteristics. A choice model can predict the probability that a consumer will pick a specific product from a set of products based on the characteristics of each product and the weights which represent that consumer's preferences for features.

11.1.1 The Standard Choice Model for Differentiated Product Demand

The consumer faces a marketplace that consists of J products each specified by a characteristics vector, x_j, and a price, p_j. If there are k characteristics, then the characteristics vector specifies the specific values of each of the k characteristics. In our example, $x_j = $ (Sony, 16 mp, 10× zoom, without swivel screen). Many characteristics have only a discrete set of possibly unordered values. For these characteristics (like the presence or absence of a swivel screen display), we would introduce a set of binary indicator or dummy variables into the characteristics vector. For example, if there are four major brands in the market place, then four indicator (0/1) variables[1] would be included in the characteristics vector.

---

[1] Unlike standard regression models, a characteristics model does not have an intercept.

The utility derived from the purchase and use of any product is modeled via what has become a standard random utility model (McFadden [1981]) for choice applications. The observed characteristics and price enter the utility for the jth brand in a linear fashion. An error term is introduced to account for unobserved factors which influence choice. The error term is often interpreted as representing unmeasured characteristics which influence choice-specific utility.

$$u_j = \beta' x_j - \beta_p p_j + \varepsilon_j \qquad (11.1.1)$$

x_j is a $k \times 1$ vector of attributes of the product, including the feature that requires valuation. For mathematical convenience, it is assumed that the error terms have an extreme value type I or Gumbel distribution. Consumers are assumed to know the realization of the error term and simply choose the alternative with maximum total utility.

To derive the standard choice model, we must calculate the probability that the jth alternative has the maximum utility, employing the assumption that the error terms have a Gumbel distribution with a scale parameter of 1. That is, since we do not observe the error terms but only the deterministic portion of utility, there is an entire region of possible errors that are consistent with the choice of a specific alternative. To derive the choice probability, we have to integrate over this region of possible values of the errors using the joint distribution of the errors. The fact that the error terms have positive support (can take on, in theory, any value in the interval $(-\infty, \infty)$) means that any choice alternative is possible even if the deterministic portion of utility for that choice alternative is small relative to all other alternatives. The Gumbel distribution produces a standard logit model for the choice of products.

$$\Pr(j) = \frac{\exp\left(\beta' x_j - \beta_p p_j\right)}{\sum_{j=1}^{J} \exp\left(\beta' x_j - \beta_p p_j\right)} \qquad (11.1.2)$$

It should be noted that utility is measured only on an interval scale with an arbitrary origin. That is, the same number can be added to the utility of all J alternatives without altering which alternative has maximum utility or the choice. This property is nothing more than the statement that only relative utility matters in choice models. We can arbitrary assign one of the product as the "base" alternative and express utility relative to this base alternative. In most situations, it is reasonable to assign the "outside" option as the base alternative. That is, one of the possibilities is that consumer decides not to purchase one of the J products and spends his or her money on other types of goods. For example, not everyone has a point and shoot digital camera. This is because the base utility for nonpurchase is higher for some than the utility they would obtain net of price from purchasing in the product category. It is common, therefore, to assign a utility of 0 to the "outside" option and, therefore, the utility for each of the J products is expressed relative to the outside option of nonpurchase. Thus, a model with the outside option or possibility of nonpurchase of any of the J products can be written

$$\Pr(j) = \frac{\exp\left(\beta' x_j - \beta_p p_j\right)}{1 + \sum_{j=1}^{J} \exp\left(\beta' x_j - \beta_p p_j\right)}$$

$$\Pr(no\ purchase) = \frac{1}{1 + \sum_{j=1}^{J} \exp\left(\beta' x_j - \beta_p p_j\right)} \qquad (11.1.3)$$

Of course, most firms do not observe the choice decisions of individual consumers. Instead, firms face the aggregate demand for their products which is the sum of the quantities demanded of each possible consumer. In most product markets, there are a very large number of potential customers. It is well documented that consumers differ dramatically (Allenby and Rossi [1999]) in their preferences. In the context of our choice model, this means that consumers have very different utility weights, β. For example, some consumers are very price sensitive and regard all digital cameras as roughly equivalent in terms of performance/features. For these consumers, choice is a matter of locating the lowest priced alternative. Others may be less price sensitive and more loyal to specific brands. In terms of product features, some consumers assess little or no value to a given feature, while others may accord it substantial value. The best way to think about this is that there is a continuum of preferences for any one characteristic rather than only a small number of "segments" or types of consumers. In this situation, the consumer population is best thought in terms of the distribution of preferences. The fact that there are a very large number of different types of consumers (each with different preferences) means that we can think of there being a continuous distribution of consumer preferences. This continuous distribution is represented by a density, $p\left(\beta, \beta_p\right)$. To "sum" over all possible consumers to create aggregate demand is the same as averaging the choice probabilities with respect to the density of consumer preferences.

$$Sales_j = M \times \int \Pr\left(j|\beta, \beta_p\right) p\left(\beta, \beta_p\right) d\beta d\beta_p \tag{11.1.4}$$

M is the total size of the market (the number of total potential customers). The choice probability averages over the distribution of preferences is the market share that the firm can expect to attain given the configuration of the product characteristics and price. In this model where consumers only buy at most one product, market share is equivalent to unit sales.

Product features have value to the extent to which consumers obtain utility from the feature. For example, if the first product characteristic is a focal product feature, then demand for that feature depends on the distribution of the first element of the β vector over the population of potential customers. If most consumers place positive and "large" weight on this feature, then the firm can increase sales by adding the feature while keeping price constant or maintain sales with a higher price. Economic valuation of any product feature will require estimation of the demand system given by (11.1.3) and (11.1.4).

11.1.2 Estimating Demand

We have seen that valuation of a product feature must be driven on the demand side by the utility weight accorded that feature and the distribution of these utility weights across individuals. Thus, a critical, but not the only, component of feature valuation is to estimate the demand system. In particular, we must estimate the distribution of utility weights for the product feature which requires observing demand for products with and without this particular feature. We should start by delineating the set of features that determine demand for this type of products or product category. This list might be very extensive. For example, new versions of a smartphone may incorporate changes to over

1000 different features, only some of which we interested in valuing. If we could specify a list of the product features or characteristics, we would then, in an ideal world, conduct controlled experiments in which product features and prices are varied in according to a randomized orthogonal design. We would also want to observe individual demand or choice decisions. With this panel data structure (multiple observations for the same consumer), we could estimate the utility weights for a sample of consumers. This would provide an estimate of the distribution of preference parameters and allow us to construct an estimate of aggregate demand. From this aggregate demand estimate, we could then undertake some type of equilibrium analysis to compare firm profits with and without the focal feature.

From a practical point of view, the estimation of demand for features may require data not available to us. Typically, we have access only to aggregate sales or market share data for a small number of time periods. With a great deal of aggregate share data, it is possible, in principle, to estimate the distribution of utility weights but, with only small number of observations, these estimates are apt to be unreliable. More importantly, we typically do not have access to data in which the same product (in terms of brand and other relevant characteristics) is sold both with and without a given feature. Without variation in the inclusion of the focal feature, no amount of data, at either the aggregate or consumer level, will allow us to estimate the utility weights accorded the feature.

In some cases, we can observe aggregate sales of a product in a time series in which the feature is added at some point during the time series. Again, we might consider using this time series variation as a way of identifying the value of the feature. However, such analysis must also control for other changes in the market such as the feature composition of competing brands. Moreover, for many consumer products, there is very little time series variation in prices. Without variation in prices, we will not be able to determine how consumers might trade off the utility they obtain for the focal feature against an increase in price. For example, consider the addition of the 5G feature to smartphones. There is a times series of aggregate sales for smartphones both before and after the introduction of the 5G feature. However, the prices of smartphones do not vary much over time, making it difficult to determine what sort of price premium the 5G feature could command.

In observational data, we also face the problem of omitted product characteristics and price endogeneity (see, for example, Berry et al. [1995]). No matter what set of characteristics we are able to measure and add to our demand model, one can always make the argument that there are omitted or unmeasured product characteristics which drive demand. If this is the case, then we might expect that firms set price, in part, with reference to these omitted characteristics. This means that prices are not exogenous and typically the utility weight on price, β_p, would be biased downward. In general, there may be unobserved drivers of demand that are correlated with price and thus we cannot use all of the variation in price (however small in the first place) to estimate the price coefficient.

In summary, standard observational data is inadequate for valuation of features for four reasons: 1. we often do not observe the same product with and without the feature, 2. we only have a short time series of aggregate data, 3. there is often little or no price variation, and 4. what price variation we observe may be confounded with unobserved or unmeasured product characteristic, resulting in endogeneity biases.

To overcome the shortcomings of observational data, the only real option would be to collect experimental demand data at the consumer level. In many contexts, such market experiments might be either infeasible or too costly. Instead, we can employ a survey-based experimental method called conjoint analysis to conduct what amounts to controlled, but simulated, experiments on the consumer level. With modern Bayesian statistical methods, this data can be used to obtain estimates of the distribution of consumer preferences and aggregate demand that can then be used to compute a profit-based measure of product value.

11.2 CONJOINT SURVEYS AND DEMAND ESTIMATION

While it may be impractical to conduct field experiments in which features are added and removed from products to assess demand, it is possible to devise a survey-based simulation that achieves the same end. In a choice-based conjoint survey, products are described as bundles of characteristics. A survey respondent is recruited and confronted with a sequence of choice tasks. In each task, the respondent is asked to choose from a menu of hypothetical products described by their characteristics. Figure 10.1 shows a typical choice task from a survey designed to measure demand for digital cameras. The respondent is asked to choose between four hypothetical cameras. Each hypothetical camera is described by seven characteristics. Each characteristic is described by a number of different values or levels. For example, price has four possibilities that are spread over the range from $79 to $379, while there are only two possibilities for the swivel screen feature (present or not). In addition, the respondent was given the option of electing not to purchase any of the products offered in each choice task screen.

The respondent is faced with a small number of choice tasks (in this survey there were 16 choice tasks or "screens"). The conjoint choice task is designed to simulate a marketplace in which products are described by their features. In making choices, respondents reveal their preferences for each of the product characteristics (including the focal feature) in the same way as they would in the marketplace. Thus, conjoint data is designed to measure preferences in the spirit of revealed preference theory. That is, we are not asking each respondent directly – what is the maximum you would be willing to pay for this camera or how important is feature A relative to feature B? Instead, we are deducing what these preferences are via simulated choice behavior. The assumption is that the preferences respondents reveal in the conjoint exercise are the same as the preferences which dictate behavior in the marketplace. There is a long history of successful application of conjoint methodology and surveys in the forecasting of demand for new products and in forecasting customer response to price changes (see Orme [2009] for many examples).

11.2.1 Conjoint Design

To undertake a rigorous conjoint survey for the purpose of estimating a valid demand schedule, the following criteria must be met:

- A representative sample must be used.
- Major product features must be included in the design.
- Product features/characteristics must be described in terms meaningful to the respondents.

- A proper randomized experimental design must be used to select the combinations of characteristics and levels used in the conjoint survey.

- The "outside" or "no choice" option must be included.

- The major set of competitive brands must be included.

11.2.1.1 Sampling Procedure In any survey, we are extrapolating or projecting from the sample to the relevant population. In the application of conjoint surveys to product valuation, the relevant population is the population of potential customers for the product category. The best way to insure representativeness for this population is to start with a representative sample of all consumers in the US and then screen this sample to those who are in the market for the products in consideration. The only sampling procedures that can guarantee representativeness are random sampling procedures. For example, if we have a list of all US residents, we would select respondents at random (each respondent is equally likely) for the screening portion of the survey.

Unfortunately, much survey work today is done with internet panels that are not random samples. An internet panel is a group of potential respondents who have indicated a willingness to take surveys using web-based methods. Invitation emails are sent to a random sample of the internet panel with a link to the survey. The problem with most internet panel providers is that the internet panelists are "harvested" via display ads and email lists. This is what statisticians call a convenience sample, in that the sample is collected by any means necessary and is not designed to be representative of any population. The fact that only a random sample of the entire internet panel is invited does not mean that the final sample of screened respondents is a random sample of the population of potential customers. Yeager et al. [2011] document that nonrandom internet panels can yield biased estimates of population quantities. There are internet providers who attempt a close approximation to random samples. If at all possible, higher quality providers should be used.

If the survey sample is not obtained via high quality, screened random samples, then it is still possible to use the sample but only if checks for representativeness are made. The typical check for representativeness is to compare the demographic characteristics of the pre-screened sample (this requires asking demographic questions prior to screening) with census-based demographics. This assures representativeness only if the product preferences of consumers are highly correlated with demographics. For example, we could find that our sample collected by nonscientific methods is representative on the basis of the gender demographic variable. That is, the sample proportion of women is the same as the population proportion. This doesn't, however, guarantee that the views of our sample with respect to a specific product feature are representative of the relevant population. This would only be true if preferences for the product feature are closely related to the gender of the consumer. Clearly, for most products, demographics cannot explain very much of the variation in brand and feature preferences. Our recommendation is that variables more closely related to the product category be used. For example, if we were doing a survey of smartphones, we might insert questions on ownership of smartphones by make or model and compare the market shares of our survey with those known in the US market.

Considerations of sample representativeness are critical to the reliability and generalizability of any survey, conjoint or otherwise. No survey evidence should be considered relevant unless evidence of representativeness is provided.

Finally, since the objective of the conjoint survey is an estimation of a valid demand system, the target population for the sampling procedure must consist of all potential customers for the products in the survey. Basically, the survey researcher must screen for consumers in the market for the set of products that is the subject of the survey. It is a mistake to focus on only those consumers who have purchased specific products or brands in the past. These considerations are much the same as the consideration which inform the relevant market from an anti-trust point of view.

11.2.1.2 Inclusion of Product Features   The heart of the demand for product features is a specification of the relevant product characteristics. For many products, the number of features is so large it would seem impractical to ever attempt to partial out the portion of utility or demand which can be ascribed to any one feature. One reaction is that this makes any characteristics approach to demand impractical. However, the logit choice model of demand (11.1.3) has an important property called the Irrelevance of Irrelevant Alternatives (IIA) (see, for example, Train [2003]). This property implies that any set of characteristics which are constant across choice alternatives drop out of the choice probabilities. For example, partition the characteristics vector into two parts, (x_0, x_1). x_0 is varies across choice alternatives while x_1 does not.

$$\Pr(j) = \frac{\exp\left(\beta_0' x_{0,j} + \beta_1' x_1\right)}{\sum_j \exp\left(\beta_0' x_{0,j} + \beta_1' x_1\right)} = \frac{\exp\left(\beta_0' x_{0,j}\right)}{\sum_j \exp\left(\beta_0' x_{0,j}\right)} \tag{11.2.1}$$

The IIA property of logit[2] means that if consumers assume that only a subset of the characteristics are varied while all other characteristics are constant across choice alternatives, then we can construct valid demand estimates by testing only a subset of relevant product characteristics. This greatly enhances the power of a choice-based conjoint survey. In theory, as long as we assume the logit model of demand is correct, we only have to examine a subset of product features. We will have to instruct respondents to assume, in making their choices, that all other features are constant. This is common in conjoint surveys.

In contrast, aggregate demand modeling with observational data does not hold the unobserved characteristics constant across products. If we do not observe important product features, then we might find that prices are determined, in part, by the unobserved features, creating a so-called endogeneity bias problem. We would have to find variables, called instruments, that move prices but are uncorrelated with the unobserved characteristics. It can be difficult to identify such variables. In theory, this is not a problem in a conjoint survey because the respondents are specifically instructed to assume that all features/characteristics other than those varied in the survey are constant across alternatives.

---

[2] It should be noted that although we are assuming that IIA holds at the individual consumer level, this does not mean that IIA holds as a property of the aggregate demand system.

Taken literally, the IIA property means we could conduct a conjoint survey with only two characteristics – the focal feature and price. In this extreme case, we are constructing a very unrealistic simulation of the marketplace.[3] For example, if one of the hypothetical products in the choice task has a very high price, it is difficult to expect that the respondents will "hold constant" other features. Their natural inclination is to assume that, perhaps, the high price indicates that this hypothetical product has very important features missing from other alternatives. This violation of the "hold constant" instruction is much less likely if other important features are included in the conjoint survey design. This means that, even though we may only wish to test a small number of features, we must include many of the other important features in the product. This, of course, does not mean you have to include all features of the product, a "generic" criticism of conjoint methods. All scientific models are abstractions which attempt to capture the important aspects of the problem. For example, when *Consumer Reports* provides comparisons of cars, smartphones, televisions, or other consumer products, they do not list all features, but, instead, concentrate on the important features. It is important to undertake research prior to the conjoint survey design to determine what are the major and most important features of the product.

The strong consensus is that a valid conjoint design must include important product features in addition to the focal features (the features which are being valued). Failure to include at least some of the most important features will typically bias the results toward a higher value of product features and a diminished price sensitivity which will ultimately lead to unreasonable equilibrium prices. It should be recognized that there is fundamental tension in the design of conjoint choice task exercises. The tension is between the desire to inject variation in both price and product features to learn more from a given size sample and the extent to which the survey offers realistic choices to respondents. If one were to design a conjoint survey with few or none of the important price features and a great deal of price variation, it is likely that respondents will violate the "spirit" of the conjoint experiment which is to "hold all other features not specified constant" or to assume that there are no unobservable features. For example, if a conjoint survey to study the demand for mid-sized sedans does not include important reliability, performance, and style features and has large price variation, then respondents will be confronted with two (or more) hypothetical sedans with vastly different prices but no feature differences specified of any important feature. Respondents will be tempted to ignore the ceteris paribus instruction and assume that the higher-priced product as some desirable unobservable characteristic. This will serve to reduce the measured sensitivity to price in much the same way as endogeneity afflicts demand estimation with nonexperimental data.

11.2.1.3 Description of Features
A general principle of conjoint survey design is that the product characteristics must be defined in terms that are understandable and meaningful to the respondents. For example, smartphone battery capacity should be specified in terms of battery life in use rather than in units of capacity such as milliamperes.

Features and functionality must be described to the respondent in simple and meaningful terms. This may involve the use of graphics and video descriptions. In order to construct a meaningful survey to value these features, descriptions of the features must be

---

[3] As Orme [2009] puts it, "realism begets better data."

understandable by the survey respondents. This certainly means that careful pre-testing will be required (for a discussion on surveys and pre-testing see Sheatsley [1983]).

11.2.1.4 Experimental Design   Conjoint surveys are properly viewed as experiments in which the products considered in the simulated choices are designed for maximum discrimination between features. Once a set of characteristics/features (called attributes in the conjoint literature) are selected, then the levels of these attributes must be selected. For example, if we include megapixels as a characteristic or attribute of digital cameras then we need to select the specific values (e.g., 10 or 16 mp). Given the set of attributes and the possible levels for each attribute, the conjoint design task then consists of "creating" hypothetical products as bundles of these attributes and specific levels. While this is a highly technical subject (see, for example, Box and Draper [1987], chapters 4 and 5), the central intuition is that we must vary each attribute independently of the other attributes in order to learn about the utility weights for each attribute. For example, we can't always have those products with the focal feature be priced more than those hypothetical products without the feature.

There are well-known and reliable ways of automatically generating choice tasks that will yield maximally informative and balanced choice task designs. It is this experimental design that makes conjoint surveys especially valuable as these designs create hypothetical products with configurations that are designed for the purpose of revealing survey respondents preferences. In the marketplace, we tend to see less independent variation of product features and price.

There is some controversy in the conjoint literature as to whether or not fully orthogonal designs should be used. For example, if we were to design a survey on the demand for sports cars and an orthogonal design was used, then price might be uncorrelated with key attributes such as acceleration or handling. This is a disconnect to the real world which may cause some respondents to be confused. Again, careful and rigorous pretesting is the way to see if an orthogonal design approach is too extreme and requires modification to build-in some of the correlations seen in the real world.

11.2.1.5 The Outside Option   In the real marketplace, not all potential customers purchase one of the available products. For example, although the penetration of digital cameras is high, not everyone has one. This means that, in order to be realistic, the conjoint study must include the "outside option" or "none-of-the-above" to allow respondents to opt-out of purchase. This is especially important in the evaluation of new product features. Firms invest in the development of new patented features in order to compete for customers not only with existing products but also with the hope of attracting new customers into the market. For example, the design features incorporated into the first iPad greatly expanded the market for tablet computers. If the outside option is not included in the conjoint study, then demand can only come at the expense of competing products with no growth in the overall market.

Practitioners of conjoint have found that it makes a difference how the "outside" option is included in the conjoint study. Just adding a column for "none of these" has been found to cause respondents to be overstate their purchase intentions (Brazell et al. [2006]). One possible explanation for this "overoptimistic" behavior is that respondents don't pay sufficient attention to the price attribute. Another is that respondents sometimes feel awkward rejecting products they believe the conjoint survey designer has a personal stake in. It is common to use what is called a "dual response" mode of incorporating the outside option. In the choice task, respondents are forced to choose one option which is their preferred option. Then the respondents are asked explicitly if they would purchase their preferred product at the stated price.

11.2.1.6 The Set of Competing Brands   Our method of valuing a product feature is to compute the incremental profits/price that the firm would earn from including this feature in their product. This depends on the structure of competition in the market. Competition is defined both by the number of competitors but also by the position of their products in the marketplace. Unless competing brands are included in the conjoint analysis, it will be impossible to make realistic estimates of the profits that can be realized from addition of the focal feature.

In summary, conjoint surveys when properly designed, pre-tested and applied to representative samples can be used to estimate industry demand for product features. However, a proper valuation of product features does not end with the production of conjoint data. This data must be analyzed to produce reliable estimates of aggregate demand for the relevant firms and we must undertake equilibrium calculations. Case Study 1 demonstrated how a hierarchical MNL can be fit to conjoint data using the **R** appropriate `bayesm` routine and provide posterior inferences regarding individual respondent preference parameters as well as the distribution of preference over the population which is the key constructing market demand.

11.3 WTP PROPERLY DEFINED

11.3.1 Pseudo-WTP

In Case Study 1, we defined pseudo-WTP simply as a way of normalizing partworths to convert utility into a \$ metric which can be meaningly compared across respondents and across product attributes. Another interpretation is that pseudo-WTP is the amount by which price can be increased while leaving the utility of products with and without a feature the same. What is called "WTP" in the conjoint literature is one attempt to convert the partworth of the focal feature, f, to the dollar scale. Using a standard dummy variable coding, we can view the partworth of the feature as representing the increase in *deterministic* utility that occurs when the feature is turned on.[4] If the feature part worth is divided by the price coefficient,[5] then we have converted to the dollar scale, which is a ratio scale. Some call this a WTP for the product feature.

$$WTP \equiv \frac{\beta_f}{\beta_p} \tag{11.3.1}$$

This WTP measure is often justified by appeal to the simple argument that this is the amount by which price could be raised and still leave the "utility" for choice alternative

---

[4] For feature enhancement, a dummy coding approach would require that we use the difference in partworths associated with the enhancement in the "WTP" calculation.

[5] We have defined the price coefficient such that this is always positive. See (11.1.3).

J the same when the product feature is turned on. Others define this as a "willingness to accept" by giving the completely symmetric definition as the amount by which price would have to be lowered to yield the same utility in a product with the feature turned off as with a product with the feature turned on. Given the assumption of a linear utility model and a linear price term, both definitions are identical.[6] In the conjoint literature (Orme [2001]), WTP is sometimes defined as the amount by which the price of the feature-enhanced product can be increased and still leave its market share unchanged. In a homogeneous logit model, this is identical to (11.3.1).

The WTP measure is properly viewed simply as a scaling device. That is, WTP is measured in dollars and is on a ratio scale so that valid inter and cross respondent comparisons can be made. As such, WTP should properly be interpreted as an estimate of the change in WTP from the addition of the feature.

$$\Delta WTP = WTP_{f*} - WTP_f \qquad (11.3.2)$$

Here WTP_{f*} is the WTP for the product with the feature and WTP_f is the WTP for the product without the feature. The measure of WTP described here is what is commonly used by conjoint practitioners.

11.3.2 Pseudo WTP for Heterogenous Consumers

Even in the case of homogeneous customers, we have seen that pWTP should not be regarded as a proper measure of economic value. In the case of heterogeneous consumers, additional problems are associated with the pWTP concept. In almost all choice-based conjoint settings, Hierarchical Bayes methods are used to estimate the choice model parameters. In the Hierarchical Bayes approach, each respondent may have different logit parameters, β and β_p, and the complete posterior distribution is computed for all model parameters, including individual respondent level parameters. The problem, then, becomes how to summarize the distribution of pWTP which is revealed via the HB analysis. The concept of pWTP provides no guidance as to how this distribution should be summarized. One natural summary would be the expectation of pWTP where the expectation is taken over the distribution of model parameters.

$$\mathbb{E}\left[pWTP\right] = \int \frac{\beta_f}{\beta_p} p\left(\beta_f, \beta_p | Data\right) d\beta_f d\beta_p$$

However, there is no compelling reason to prefer the mean over any other scalar summary of the distribution of pWTP. Some propose using a median value of pWTP instead. Again, there are no economic arguments as to why the mean or median or any other summary should be preferred. The statistical properties of various summaries (e.g., mean vs median) are irrelevant as we are not considering the sampling performance of

---

[6] In practice, reference price effects often make pWTA differ from pWTP, see Viscusi and Huber [2012] but, in the standard economic model, these are equivalent.

an estimator but rather what is the appropriate summary of a population distribution. A proper economic valuation will consider the entire demand curve as well as competitive and cost considerations. Equilibrium quantities will involve the entire distribution via the first order conditions for firm profit-maximization. These quantities cannot be expressed as a function of the mean, median or any other simple set of scalar summaries of the distribution of pWTP.

However, it is possible to provide a rough intuition as to why the mean of pWTP may be a particularly poor summary of the distribution for equilibrium computations. It is the marginal rather than the average consumer that drives the determination of equilibrium prices. Exactly where, in the distribution of pWTP, will the marginal customer be is determined by nature of the distribution as well as where supply factors that "slice" into the distribution of pWTP. It is possible to construct cases where the average pWTP vastly overstates the pWTP of the marginal customer. If the bulk of the market has a low value of pWTP and there is a small portion of the market with extremely high pWTP, then a profit maximizing firm may set price much lower than average pWTP so as to sell to the majority of potential customers who have relatively low pWTP. There are situations where the greater volume from low pWTP consumers outweighs the high margins that might be earned from the high pWTP segment. In these cases, mean pWTP will vastly overstate the price premium a firm will charge over cost for a product. It is more difficult, but possible, to construct similar scenarios for median pWTP.

One of the major problems with using any measure of the central tendency of the distribution of pWTP (either mean, median, or mode) is that this includes consumers whose pWTP is insufficient to be in the market. That is, our surveys should qualify respondents to be in the "market" for the products (with a screening question such as "do you plan to buy a digital camera in the next six months?"). However, simply averaging pWTP over all survey respondents averages in those whose pWTP for the product as a whole is below the market price of any products and, therefore, would not purchase in the product category. This is a downward bias for pWTP computations.

11.3.2.1 Pseudo-WTP and "Market Simulators"

In the case of homogeneous consumers, it is easy to see that the pseudo-WTP can be interpreted as the amount by which you can reduce the price of the "disadvantaged" product (the product without the feature) and maintain the same utility level. What this means is that if you define the "market" as a set of M homogeneous consumers, then pseudo-WTP is the amount that you can reduce the price of the product without the feature and have consumers indifferent between the two products – one with the feature and the other, an otherwise identical product, without the feature. This is closer to the proper economic analysis of pWTP as defined by the amount that you need to compensate consumers to accept an inferior product.

In the case of heterogenous consumers, it is clear that $E\left[pWTP\right]$ is not the right amount to define this indifference point. Orme [2001, 2021] outline the standard approach in the case of heterogeneous consumers. Define "market" demand as the sum of demand (sum choice probabilities across all consumers). Take two products each identical in all other features and turn the focal feature on for one and level it off for the other. Find the price reduction required to make the market demand equal for both products. Let a be the vector of the attributes of the product with the focal feature not present (i.e., one of the elements of a is set to zero) and let a^* be a vector that is identical in all respects except that the focal feature is now set to 1. Let p^* be the price of the product with the feature. $pWTP$ is the solution to the following:

$$\int \frac{\exp\left(\beta' a - \beta_p p^*\right)}{1 + \exp\left(\beta' a - \beta_p p^*\right)} p\left(\beta, \beta_p | data\right) d\beta d\beta_p$$

$$- \int \frac{\exp\left(\beta' a^* - \beta_p p^*\right)}{1 + \exp\left(\beta' a^* - \beta_p(p^* - pWTP)\right)} p\left(\beta, \beta_p | data\right) d\beta d\beta_p = 0 \qquad (11.3.3)$$

In most implementations, the integral is approximated by replacing the integration with respect to the posterior predictive with a simple sum over each of the respondents with the β vectors set to the posterior mean for that respondent. This the exact procedure described in Orme.

The unfortunate term "Market Simulator" has been coined for this process. It is clear that there is no market of any kind here (there are only two hypothetical products, typically of the same brand and there is no supply side) and this is simply an alternative to averaging the ratio of partworths that has no economic justification. Orme [2001] has pointed out a number of problems with this approach to computing pWTP and has warned conjoint practitioners not to misinterpret these quantities as what a firm can charge for a feature enhancement. Orme often uses his "Gilligan's Island" example to illustrate why pWTP is not a market price. There can be a consumer (Mr. Howell in the example) who has an extremely high pWTP to leave the island. Mr. Howell does not have to pay this high price if there is regular ferry service off the island. This statement has been misinterpreted by Orme to be an argument that pWTP depends on competition. pWTP does not depend on competition. pWTP is an inherently individual characteristic that does not change as a function of actions of competition. It is true that Mr. Howell will be a lot happier if there is a cheap ferry service (he makes a surplus of the difference between his pWTP and the ferry price) but that does not mean that competition changes utility. This fundamental confusion about the difference between utility and competition persists in Orme's more recent writings (Orme [2021]). Here he advocates using a richer choice set with multiple competitive alternatives in the "market simulator" and observes that adding competitive alternatives tends to reduce the pWTP estimates. This is true but not because the enhanced pWTP computation moves closer to the market price but because Orme is groping toward better definition of pWTP which depends on the choice set. This has been present in the economics literature for decades.

11.3.3 True WTP

To develop a measure of WTP that is based in economic analysis, we need to return to the idea that WTP is the outcome of an indifference experiment in which the utility of an enhanced choice-set is compared to that of a reduced choice set, expressed in $ terms.

WTP is a measure of social welfare derived from the principle of compensating variation. That is, WTP for a product is the amount of income that will compensate for the loss of utility obtained from the product; in other words, a consumer should be indifferent between having the product or not having the product with an additional income equal to the WTP. Indifference means the same level of utility. For choice sets, we must consider the amount of income (called the compensating variation) that must be paid to a consumer faced with a diminished choice set (either an alternative is missing or diminished by omission of a feature) so that consumer attains the same level of utility as a consumer facing a better choice set (with the alternative restored or with the feature added). Consumers evaluate choices *a priori* or before choices are made. Features are valuable to the extent to which they enhance the attainable utility of choice. Consumers do not know the realization of the random utility errors *until* confronted with the specific choice tasks and the description of the choice alternatives. Addition of the feature shifts the deterministic portion of utility or the mean of the random utility. Variation around the mean due to the random utility errors is equally important as a source of value.

The random utility model was designed for application to revealed preference or actual choice in the marketplace. The random errors are thought to represent information unobservable to the researcher. This unobservable information could be omitted characteristics that make particular alternatives more attractive than others. In a time series context, the omitted variables could be inventory which affects the marginal utility of consumption. In a conjoint survey exercise, respondents are explicitly asked to make choices solely on the basis of attributes and levels presented and to assume that all other omitted characteristics are to be assumed to be the same. It might be argued, then that there role of random utility errors is different in the conjoint context. Random utility errors might be more the result of measurement error rather than omitted variables that influence the marginal utility of each alternative.

However, even in conjoint setting, we believe it is still possible to interpret the random utility errors as representing a source of unobservable utility. For example, conjoint studies often include brand names as attributes. In these situations, respondents may infer that other characteristics correlated with the brand name are present even though the survey instructions tell them not to make these attributions. One can also interpret the random utility errors as arising from functional form mis-specification. That is, we know that the assumption of a linear utility model (no curvature and no interactions between attributes) is a simplification at best. We can also take the point of view that a consumer is evaluating a choice set prior to the realization of the random utility errors which occurs during the purchase period. For example, consider the value of choice in the smartphone category at some point prior to a purchase decision. At the point, the consumers knows the distribution of random utility errors which will depend on features not yet discovered or from demand for features which is not yet realized (i.e., the benefit from a better browser is not known with certainty prior to choice). When the consumer actually purchases a smartphone, he or she will know the realization of these random utility errors.

To evaluate the utility afforded by a choice set, we must consider the distribution of the maximum utility obtained across all choice alternatives. This maximum has a distribution because of the random utility errors. For example, suppose we add the feature to a product configuration that is far from utility maximizing. It may still be that, even with the feature, the maximum deterministic utility is provided by a choice alternative without the feature. This does not mean that feature has no value simply because the product it is being added to is dominated by other alternatives in terms of deterministic utility. The alternative with the featured added can be chosen after realization of the random utility errors if the realization of the random utility error is very high for the alternative that is enhanced by addition of the feature.

More formally, we can define WTP from a feature enhancement by using the indirect utility function associated with the choice problem. Let A be a matrix that defines the set of products in a choice set. A is a $J \times K$ matrix, where J is the number of choice alternatives and K is the number of attributes which define each choice alternative (other than price). The rows of the choice set matrix, a_j, show the configuration of attributes for the jth product in the choice set. That is, the jth row of A defines a particular product – a combination of attribute levels for each of K attributes. If the kth attribute is the feature in question, then $a_{j,k} = 1$ implies that the feature has been added to the jth product. Let A denote a set of products that represent the marketplace without the new feature and A^* denotes the same set of products but where one of the products has been enhanced by adding the feature. We define the indirect utility function for a given choice set as

$$V\left(p, y | A\right) = \max_{x} \ U\left(x | A\right) \quad \text{subject to } p'x \leq y \tag{11.3.4}$$

WTP is defined as the compensating variation required to make the utility derived from the "feature-poor" choice set, A, equal to the utility obtained from the feature-rich choice set, A^*.

$$V\left(p, y + WTP | A\right) = V\left(p, y | A^*\right) \tag{11.3.5}$$

As such, WTP is a measure of the social welfare conferred by the feature enhancement expressed in dollar terms. The choice set may include not only products defined by the K product attributes but also an outside option which is coded as row of zeroes in the feature matrix and a price of 1.0. Thus, a consumer receives utility from three sources: 1. observed characteristics of the set of products in the market, 2. expenditure on a possible outside alternative, and 3. the random utility error.

For the logit demand system, the indirect utility function is obtained by finding the expectation of the maximum utility (see, for example, McFadden [1981]).

$$V\left(p, y | A\right) = E\left[\max_{j} U_j | A\right]$$

$$= \beta_p y + \ln \sum_{j=1}^{J} \exp\left(a'_j \beta - \beta_p p_j\right) \tag{11.3.6}$$

To translate this utility value into monetary terms, we divide by the marginal utility of income. In these models, the price coefficient is viewed as the marginal utility of income. We can then transform the utility value in (11.3.6) into monetary terms, resulting in the "social surplus" (Trajtenberg [1989]).

$$W\left(A | p, \beta, \beta_p\right) = y + \ln\left[\sum_{j=1}^{J} \exp\left(\beta' a_j - \beta_p p_j\right)\right] \Big/ \beta_p \tag{11.3.7}$$

We can then solve for WTP using the (11.3.5).

$$WTP = \ln\left[\sum_{j=1}^{J} \exp\left(\beta' a_j^* - \beta_p p_j\right)\right] \bigg/ \beta_p - \ln\left[\sum_{j=1}^{J} \exp\left(\beta' a_j - \beta_p p_j\right)\right] \bigg/ \beta_p \quad (11.3.8)$$

It is straightforward to see that the true WTP measure developed above will often be less than pseudo-WTP. The pseudo-WTP is based on the marginal utility of adding the feature and is calculated from the deterministic part of utility (the part of utility that depends on observable product features/attributes). In an additive utility model, this incremental utility is the same no matter what the values of the other features (the other elements of the a_j vector) are. The "market simulation" approach (11.3.3) is very similar but it should be noted that the nonlinearity of the logit probability locus used to form aggregate demand means that although the utility function is additive in features there is some difference as to which alternative the focal feature enhancement is added to. However, if two choice alternatives have similar market shares in the "base" configurate without the feature turned on (a_j), then the pWTP calculated using the "market" simulator approach will be very similar.

The true WTP measure does not suffer from this undesirable property in that the value of a feature enhancement depends critically on the total set of products. Consider a situation in which there are only two products but one of the products has a much higher utility and market share in the base configuration without the feature enhancement. If you add the feature enhancement to the product with lower market share, it will not have the same effect as adding the enhancement to the product with high base-line market share as measured by WTP. Adding a small feature enhancement will only result in a higher achievable maximum utility for those draws of the random utility errors that actually change the base-line utility of the inferior product. On the other hand, if the feature enhancement is added to the high baseline utility product, this will tend to increase the maximum achievable utility by more as there is a much larger set of utility errors that are consistent with the popular product having the highest utility. Similarly, the addition of alternatives to the choice set will tend to reduce WTP as there is a higher probability that random utility errors will result in other alternatives than the alternative with the feature enhancement achieving the maximum utility.

It should also be noted that the WTP measure developed here can be applied not just to the enhancement of the choice set by altering the product features in existing products but also by adding an entirely new product or choice alternative. Of the original applications of choice-based modeling was to value in terms of increased attainable utility the addition of the BART rapid transit system as an alternative to driving or taking a bus in the Bay area. Here the BART system is a new alternative in with different values of cost, speed and ease-of-use as compared to existing systems. In principle, WTP calculations can be used to assess the social value of the BART system as offering an enhanced commuting choice set and this value could be compared to the capital costs of the system to undertake a social return calculation. The problem with this sort of use of the WTP measure is that each choice alternative provides a source of value because of the assumption of an iid error term with unbounded support. As you add even redundant alternatives the random utility error terms add a source of value that some believe is highly unrealistic. In the application here we are considering the enhancement or reconfiguration of existing products so that we avoid this troublesome drawback to the standard logit model (sometimes this problem is called the red bus–blue bus problem).

11.3.4 Problems with All WTP Measures

The basic problem with all WTP measures is that these measures depend only on consumer utility and, as such, are demand side only. The supply side must be considered. Adding the supply side means that marginal costs of providing each alternative must be considered (including, perhaps, the increase in marginal cost that may derive from the addition of the feature). A proper supply analysis does not end with a measure of marginal costs but also includes the specification of a set of relevant competing products and allowing for the possibility of price adjustment. If competitors are allowed to adjust their price after the focal feature is added, we also need to make assumptions about how industry equilibrium prices are set. One can imagine that if a valuable new feature is added to one product in a competitive marketplace, then we can expect that the firm that owns the focal product will want to raise price and, feeling a stiffer competition from the focal product, other firms will want to lower their prices. A new equilibrium is achieved as a point of stability in which there are no profit incentives for any firm to alter deviate from the equilibrium price vector. In the next section, we will develop such an approach to this problem. In a nutshell, any WTP computation, no matter how well done, cannot answer the question of what a firm can charge for feature enhancement to their products or more fundamentally, what increment in profits (if any) can be expected from the feature enhancement. It seems that, ultimately, the entire conjoint literature should be focused on profits that can be obtained from product configurations and should recognize that, with conjoint survey data alone, it is impossible to make profit calculations or compute market prices.

11.4 NASH EQUILIBRIUM PRICES – COMPUTATION AND ASSUMPTIONS

The goal of feature enhancement is to improve profitability of the firm by introducing product with feature enhancement into an existing market. For this reason, we believe that the only sensible measure of the economic value of feature enhancement is the incremental profits that the feature enhancement will generate.

$$\Delta \pi = \pi \left(p^{eq}, m^{eq} | A^* \right) - \pi \left(p^{eq}, m^{eq} | A \right) \tag{11.4.1}$$

π is the profits associated with the industry equilibrium prices and shares given a particular set of competing products which is represented by the choice set defined by the attribute matrix. A^* denotes the set of products where one of the products has been enhanced by adding or improving a product feature. A represents the set of products without feature enhancement. The set of products in the market is defined via their vector of characteristics.

The equilibrium depends on the set of products offered in the market place. $\left(p^{eq}, m^{eq} \right)$ is the outcome of a price equilibrium with m denoting the vector of market shares. An equilibrium is defined as a set of prices and accompanying market shares which satisfy the conditions specified by a particular equilibrium concept. We use the standard Nash Equilibrium concept for differentiated products.

In the abstract, our definition of economic value of feature enhancement seems to be the only relevant measure for the firm that seeks to enhance a feature. All funds have an opportunity cost and the incremental profits calculation is fundamental to deploying product development resources optimally. In fairness, industry practitioners of conjoint analysis also appreciate some of the benefits of an incremental profits orientation. Often, marketing research firms construct "market simulators" that simulate market shares given a specific set of products in the market. Some even go further as to attempt to compute the "optimal" price by simulating different market shares corresponding to different "pricing scenarios." In these exercises, practitioners fix competing prices at a set of prices that may include their informal estimate of competitor response. This is not the same as computing a marketing equilibrium but moves in that direction.

11.4.1 Assumptions

Once the principle of incremental profits is adopted, the problem becomes one of defining the nature of competition, the competitive set and to choose an equilibrium concept. These assumptions must be added to the assumptions of a specific parametric demand system (we will use a heterogeneous logit demand system that is flexible but still parametric) as well as a linear utility function over attributes and the assumption (implicit in all conjoint analysis) that products can be well described by bundles of attributes. Added to these assumptions, our valuation method will also require cost information.

Specifically, we will assume

1. Demand Specification: A standard heterogenous logit demand that is linear in the attributes (including price)

2. Cost Specification: Constant marginal cost

3. Feature Exclusivity: The feature can only be added to one product

4. No Exit: Firms cannot exit or enter the market after product enhancement takes place

5. Static Nash Price Competition

Assumptions 2 and 3 can be easily relaxed. Assumption 1 can be replaced by any valid or integrable demand system. Assumptions 4 and 5 cannot be relaxed without imparting considerable complexity to the equilibrium computations.

11.4.2 A Standard Logit Model for Demand

Valuation of product features depends on a model for product demand. In most marketing and litigation contexts, a model of demand for differentiated products is appropriate as developed in 11.1.1. This model allows applies to a situations with a relatively small number of competitors, each one of which offers products differentiated by observable features.

11.4.3 Computing Equilibrium Prices

The standard static Nash equilibrium in a market for differentiated products is a set of prices such that simultaneously satisfy all firms profit-maximization conditions. Each firm chooses price to maximize firms profits, given the prices of all other firms. These conditional demand curves are sometimes called the "best response" of the firm to the prices of other firms. An equilibrium, if it exists,[7] is a set of prices that is simultaneously the best response or profit maximizing for each firm given the others. It should also be noted that in a Nash-equilibrium analysis with differentiated products, there is no formal supply curve. Instead, there is a locus of equilibrium outcomes which depend on both demand and supply inputs.

In a choice setting, the firm demand[8] is

$$\pi\left(p_j|p_{-j}\right) = M\mathbb{E}\left[Pr\left(j|p, A\right)\right]\left(p_j - c_j\right) \tag{11.4.2}$$

M is the size of the market, p is the vector of the prices of all J firms in the market, c_j is the marginal cost of producing the firms product. The expectation is taken with respect to the distribution of choice model parameters. In the logit case,[9]

$$\mathbb{E}\left[Pr\left(j|p, A\right)\right] = \int \frac{\exp\left(\beta'a_j - \beta_p p_j\right)}{\sum_j \exp\left(\beta'a_j - \beta_p p_j\right)} p\left(\beta, \beta_p\right) d\beta d\beta_p \tag{11.4.3}$$

The first order conditions of the firm are

$$\frac{\partial \pi}{\partial p_j} = \mathbb{E}\left[\frac{\partial}{\partial p_j} Pr\left(j|p, A\right)\right]\left(p_j - c_j\right) + \mathbb{E}\left[Pr\left(j|p, A\right)\right] \tag{11.4.4}$$

---

[7] There is no guarantee that a Nash equilibrium exists for heterogeneous logit demand.

[8] Again, we do not have an aggregate demand shock in the model. We think of the firm problem as setting prices given the observed characteristics and prices of all products in the marketplace. There is no sense in which firms are setting prices as a function of some unobserved characteristic as this is explicitly ruled out by the nature of the conjoint randomized experiment.

[9] We do not include a market wide shock to demand as we are not trying to build an empirical model of market shares. We are trying to approximate the firm problem. In a conjoint setting, we abstract from the problem of omitted characteristics as the products we use in our market simulators are defined only in terms of known and observable characteristics. Thus, the standard interpretation of the market wide shock is not applicable here. Another interpretation is that the market wide shock represents some sort of marketing action by the firms (e.g., advertising). Here we are directly solving the firm pricing problem holding fixed any other marketing actions. This means that the second interpretation of the market wide shock as stemming from some unobservable firm action is not applicable here.

The Nash equilibrium price vector is a root of the system of nonlinear equations which define the F.O.C. for all J firms. That is if we define

$$
h(p) = \begin{bmatrix} h_1(p) = \frac{\partial \pi}{\partial p_1} \\ h_2(p) = \frac{\partial \pi}{\partial p_2} \\ \vdots \\ h_J(p) = \frac{\partial \pi}{\partial p_J} \end{bmatrix}
\tag{11.4.5}
$$

then the equilibrium price vector, p^*, is a zero of the function $h(p)$.

There are two computational issues that arise in the calculation of Nash equilibrium prices. First, both the firm profit function (11.4.2) and the FOC conditions for the firm (11.4.4) require the computation of integrals to compute the expectation of the market share (market demand) and expectation of the derivative of market share in the FOC. Second, an algorithm must be devised for calculating the equilibrium price, given a method of approximating the integrals. The most straightforward method to approximate the requisite integrals is a simulation method. Given a distribution of demand parameters over consumers, we can approximate the expectations by simple average of draws from this distribution. Given that both the market share and the derivatives of market share are virtually costless to evaluate, an extremely large number of draws can be used to approximate the integrals (we routinely use in excess of 50,000 draws).

Given the method for approximating the integral, we must choose an iterative method for computing equilibrium prices. There are two methods available. The first is an iterative method where we start from some price vector, compute the optimal price for each firm given other prices, updating the price vector as we progress from the 1st to the Jth firm. After one cycle thru the J firms, we have updated the price vector to a second guess of the equilibrium. We continue this process until $\|p^r - p^{r-1}\| < tol$. The method of iterative firm profit maximization will work if there is a stable equilibrium. That is, if we perturb the price vector away from the equilibrium price, the iterative process will return to the equilibrium (at least in a neighborhood of the equilibrium). This is not guaranteed to occur even if there is exists a unique equilibrium.

The second method for computing equilibrium prices is to find the root of set of FOCs (11.4.5). The optimization problem

$$
\min_{p} \|h(p)\|
$$

can be solved via a quasi-Newton method that is equivalent to finding the roots directly using Newton's method with line search. This provides a more robust way of finding equilibria, if they exist, but does not provide a way of finding the set of equilibria if multiple equilibria exist. The existence of multiple equilibria would have to be demonstrated by construction via starting the optimizer/root finder from different starting points. In our experience with heterogeneous logit models, we have not found any instance of multiple equilibria; however we have found situations where we cannot find any equilibria (though only rarely and for extreme parameter values).

Thus, with additional supply side assumptions and information, we can compute equilibrium prices, quantities, and firm profits. This enables you to answer the questions of what can a firm charge for a feature enhancement and how will firm profits change if the feature is added. This should be the ultimate goal of conjoint analysis – to value a feature in terms of firm profits. This enables the firm to determine which features are the sources of the highest incremental profits. It is quite remarkable that this analysis is absent from the very substantial conjoint literature until (Allenby et al. [2014b]).

11.5 CAMERA EXAMPLE

In Case Study 1 (Chapter 10), we considered how conjoint panel data is stored for use by

R bayesm routines, explored prior specifications and explored the draws from the posterior distribution of respondent partworths as well as the posterior predictive distribution of partworths. In addition, we saw how to impose sign constraints. Sign constraints are absolutely vital for successful equilibrium calculations as we will see in this section.

11.5.1 WTP Computations

In Case Study 1, we computed what many term a "pseudo-WTP" and used this as an illustration as to how to compute the posterior predictive distribution of a relevant quantity. Recall that the mean of the posterior predictive distribution of $E\left[pWTP\right]$ was approximately \$45. In fact, the correct interpretation is that by adding the swivel screen feature to a digital camera product, this will increase WTP by approximately \$45. It is remarkable that this computation would suggest that adding the swivel screen to any product would increase utility by the same amount in dollar terms. Of course, this comes from the linearity of the utility function and it could well be that there should be an interaction between other variables such as brand and utility. It is a simple matter to add a brand-swivel screen interaction variable would achieve this greater flexibility.

However, there is also a fundamental defect in the pseudo-WTP measure in that this measure does not look at enhancement of the choice set that occurs when you add the swivel screen to one or more of the products in the choice. Any true WTP measure should reflect the utility afforded by the entire choice set. Clearly, this depends on the nature of the choices and also on the realization of random utility error terms. If we were to bestow the swivel screen on a very low total utility product, it is not clear that this would raise the utility of the choice set which depends on the highest level of utility that can be attained with the choice set. Adding the swivel screen to a product that is already highly regarded might well have a bigger impact on the maximum attainable utility. Working against this intuition or force are the utility errors. Even if a product has low average utility, the possibility that the errors might have a large positive realization also enhances the utility of the choice set and blunts some of the effect that make the enhancement of choice-set utility dependent on which product the swivel screen is added to.

Equation (11.3.8) provides a proper WTP calculation that should be interpreted the monetary value of an enhanced choice set some of the products have enhanced values of a feature. Let's add the swivel screen to the Sony digital camera choice alternative and compute the true WTP for the enhanced choice set of four cameras where only the Sony alternative has the swivel screen. Equation (11.3.8) provides the formula for WTP for a representative consumer with preferences given by the vector, $\left(\beta,\beta_p\right)$. Given that our model includes the distributional assumption, $\beta \sim N\left(\mu_\beta, V_\beta\right)$, we certainly can define the expected value of the WTP formula over the population distribution of preferences. Bayesian inference proceeds, as we have learned, by computing the posterior distribution of $E\left[WTP|\mu_\beta, V_\beta\right]$ via simulation (evaluating $E[WTP]$ for draws from the posterior distribution, $p(\mu_\beta, V_\beta|data)$).

Effectively we need a loop with a loop. The innermost loop will calculate $E\left[WTP|\mu_\beta, V_\beta\right]$ by drawing a large number of draws from the normal distribution of preferences for a specific value of μ_β, V_β. The outermost loop will loop over draws from the posterior. As a result, we will obtain draws from the posterior distribution of expected WTP.

First, we must set up the relevant matrices that describe the base and enhanced choice sets. Figure 11.1 provides the R code. The design matrix is a $p \times k$ matrix specifying the values of each attribute (there are k attributes) for each of the p alternatives (including the outside option). designBase provides the base specification with four cameras and the outside good. All "enhanced" values of digital camera features (including swivel screen) are turned off by setting their values to 0. designSony is exactly the same except that there is a 1 in the second row, 8th column to represent the addition of the swivel screen to the Sony camera. we also must set a vector prices for each of the alternatives. This provides a very simple and clear intuition as to why this WTP calculation is simply a demand-side computation and is based on an assumed choice set and preferences/utility estimates only. Even if we were to insert "market" prices into the WTP computation this would obviously not expand the calculation to include a valid supply analysis unless we were to include a different set of market prices for the world in which the choice set changes. Clearly, in most applications such market price information is unavailable since these exact choice sets are not observed in the marketplace. Finally, note that we have set the price of the outside good to zero which is simply a convenient normalization.

Now let's take a look at code (Figure 11.2) which does the work of simulating from the relevant posterior distribution. We create the vector to store each draw from the

```
prices=c(rep(175,4))/100   # base prices -- something like retail price
designBase <- rbind(c(1, 0, 0, 0, 0, 0, 0, 0, 0),
                    c(0, 1, 0, 0, 0, 0, 0, 0, 0),
                    c(0, 0, 1, 0, 0, 0, 0, 0, 0),
                    c(0, 0, 0, 1, 0, 0, 0, 0, 0),
                    c(0, 0, 0, 0, 0, 0, 0, 0, 0))

designSony <- rbind(c(1, 0, 0, 0, 0, 0, 0, 0, 0),
                    c(0, 1, 0, 0, 0, 0, 0, 1, 0),
                    c(0, 0, 1, 0, 0, 0, 0, 0, 0),
                    c(0, 0, 0, 1, 0, 0, 0, 0, 0),
                    c(0, 0, 0, 0, 0, 0, 0, 0, 0))


fullDesignBase <- cbind(designBase, c(prices, 0))
fullDesignSony <- cbind(designSony, c(prices, 0))
```

Figure 11.1 Specifying choice sets

```
select_iter = seq(from=(BurnIn+1),to=length(out$nmix$compdraw),by=4)
WTP <- double(length(select_iter))
ncnt=1
for(iter in select_iter) {
  betastar=rmixture(n = 4000, pvec = out$nmix$probdraw[iter],
                    comps = out$nmix$compdraw[[iter]])$x
                    # number of simulations to use to approximate E[WTP]
  beta=cbind(betastar[,1:9],-exp(betastar[,10]))
  xBetaBase <- beta%*%t(fullDesignBase) # XbetaBase is 4000 x 5
  xBetaSony <- beta%*%t(fullDesignSony)

  WTP[ncnt]=
    mean(log(rowSums(exp(xBetaSony))/rowSums(exp(xBetaBase)))/(-beta[,10]))
  cat("posterior sim iter = ",iter,fill=TRUE)
  ncnt=ncnt+1
  cat
}
```

Figure 11.2   Simulation from the posterior distribution of EWTP

R posterior. Recall that we ran *bayesm* routine for 50,000 MCMC iterations, discarding the first 10,000, and retaining every 10th iteration for a total of 4000 draws from the posterior of the normal distribution parameters. For each of these 4000 draws from the posterior distribution of μ_β, V_β, we compute EWTP. The object "out" holds the draws from the posterior and these draws are stored in the format for a general mixture of normals even though we are using only a one component normal mixture. The draws from this normal distribution much be transformed in order to enforce the parameterization used to impose the constraint that the price coefficient (which is the 10th partworth here) is negative. The *bayesm* routine, rmixture, makes these draws. "beta" is a 4000×10 matrix each row of which holds a draw from normal distribution of the "deep" parameters which is transformed via the standard exponential transformation. We then have to evaluate the WTP formulation for each of these 4000 draws and then average them in the last statement in the loop.

Figure 11.3 shows the posterior distribution of $E[WTP]$. The posterior mean of $E[WTP]$ is only \$7.31 with a 95% posterior interval from \$4.50 to \$11.00. The dramatically lower value of $E[WTP]$ as compared to pseudo WTP shows the importance of considering the choice alternatives in valuing any product enhancement. Consumers who have a premium for a brand other than Sony are likely to place a much lower value on the swivel screen feature as the maximum utility obtained from a choice set which bestows the feature on a less-preferred brand will be low *s*.

11.5.2 Equilibrium Price Calculations

Any form of WTP measure is constructed entirely from the demand side and does not reflect any supply-side analysis. If we are willing to specify marginal costs, a set of competitors and competitive offerings, and an equilibrium concept, we can complete the

E[*WTP*]

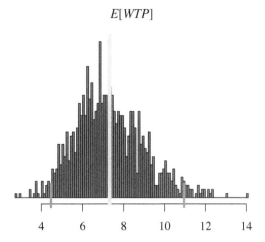

Figure 11.3   Posterior distribution of *E*[*WTP*]

analysis and compute the effect of a feature change on market prices, quantities, and firm profits as discussed in Section 11.4. With these additional assumptions, we can actually answer the question – what can a firm charge for a feature and what are the effects on firm profits?

In the case of the camera data, we will determine what can Sony charge for a "base" configuration with the swivel screen added to the product instead of simply the WTP of consumers for an enhanced choice set which includes this product as demonstrated above.

We will use several utility functions which have been written to compute Nash equilibria for a very general setting. These functions included in the source file, `nash_eq_functions.R`. The primary function is, `nash_eq`. `nash_eq` has a number of arguments which need to be specified:

1. `Ownership`: A nfirms × nbrands matrix which specifies which firm owns which brand or product.

2. `init_prices`: A nbrands dimensional vector of prices to start the iterative Nash equilibrium calculations from.[10]

3. `Design`: The design matrix specifying each product as a combination of attributes, specified in the same way as used in WTP computations.

4. `costs`: A nbrands dimensional vector with marginal costs for each product.[11]

5. `betadraws`: A nvars × nconsumers matrix of draws from the random coefficient distribution.

---

[10] Our experience is that this is not very critical but should be set in the neighborhood of market prices – certainly prices should be above marginal costs.

[11] Marginal costs should include the incremental cost of adding features, if any.

The ownership matrix allows for multi-product firms. By altering the ownership matrix, one can use the nash_eq function to compute merge or acquisition simulations where firm ownership of products change and this can be used to compute changes in prices or various diversion and other measures used in merger/acquisition analysis.

The betadraws matrix represents our discrete approximation to a continuum of consumers. That is, we are assuming a very large number of pseudo-consumers or "atoms" to approximate a market with an uncountably infinite number of consumers. For example, if we assume that preferences are normally distributed, $\beta \sim N\left(\mu_\beta, V_\beta\right)$, then we should take nconsumers draws from this normal distribution and use this to approximate demand represented by this continuous distribution. In general, if we use a very large number of consumers ($10{,}000$ or more), we can achieve a good approximation to market demand for any product j:

$$\int \frac{\exp\left(a'_j\beta - \beta_p p_j\right)}{1 + \sum_i \exp\left(a'_i\beta - \beta_p p_i\right)} p\left(\beta, \beta_p | \mu_\beta, V_\beta\right) d\beta$$

$$\approx \frac{1}{NC} \sum_{l=1}^{NC} \frac{\exp\left(a'_j\beta^l_j - \beta^l_p p_j\right)}{1 + \sum_i \exp\left(a'_i\beta^l - \beta^l_p p_i\right)} \tag{11.5.1}$$

where NC is the number of "atoms" or points in this discrete approximation and $\left(\beta^l, \beta^l_p\right)$ is the lth draw from the appropriate normal distribution.

There are two features of this discrete approximation that deserve comment. First and foremost, the investigator must impose a sign restriction on the price coefficient. If there is even a single atom with a positive price coefficient, then the profit-maximizing solution of the firm is to fire all of its consumers except one and raise price to a very high level. As we have discussed, we impose this sign restriction by re-parameterization and we assume that the "deep" parameters are multivariate normal. In practice, all this means is that we need to transform the normal draws of the price coefficient in a manner

R consistent with the reparameterization already performed by the *bayesm* routine.

Second, the use of a discrete approximation to the integral means that approximation to the demand-based market share is not differentiable and if we wish to use the iterative method for computing Nash equilibria then we must use a derivative free method. We have found that a simple Nelder Mean approach works very well.

Let's see all of the arguments (Figure 11.4) for nash_eq for our example of adding the swivel screen to the Sony product.

Here we set a single product firm ownership matrix with our four firms/four brands. In addition, we set marginal costs to $75 and assess an incremental cost for addition of the swivel screen of $5. We will use $10{,}000$ consumers to approximate the marketplace.

Posterior inference for equilibrium quantities proceeds using the same ideas illustrated for the case of WTP. We have draws from the posterior distribution of the parameters of the "deep" normal distribution and we must undertake whatever equilibrium calculations are desired for each of these draws as illustrated in Figure 11.5.

First, we must select from the available 4000 draws after the burn-in period. Since the Nash equilibrium calculations require one or two seconds per calculation, in the interest of expediting the process we only use every 10th draw in these 4000 thinned draws.

```
# define "ownership" matrix
nfirms = 4
nbrds= 4
firm_names = c("Canon", "Sony", "Nikon", "Panasonic")
Own = matrix(0,nrow=nfirms,ncol=nbrds)
diag(Own) = c(rep(1,nbrds))
rownames(Own) = firm_names
colnames(Own) = firm_names  # here we have single product firms

# costs are in hundreds and SS feature is added to only Sony
costs_wo_ss = c(75,75,75,75)/100
costs_w_ss = c(75,80,75,75)/100


nconsumers=10000
```

Figure 11.4 Arguments for nash_eq

```
select_iter = seq(from=(BurnIn+1),to=length(out$nmix$compdraw),by=10)
eq_prices_base_draws = matrix(double(length(select_iter)*nbrds),nrow=nbrds)
eq_prices_Sony_draws = matrix(double(length(select_iter)*nbrds),nrow=nbrds)
eq_profit_base_draws = matrix(double(length(select_iter)*nfirms),nrow=nfirms)
eq_profit_Sony_draws = matrix(double(length(select_iter)*nfirms),nrow=nfirms)


ncnt=1
for(iter in select_iter) {
  #simulate consumers
  R=backsolve(out$nmix$compdraw[[iter]][[1]]$rooti,diag(nvar))
  betadraws = crossprod(R,matrix(rnorm(nvar*nconsumers),nrow=nvar)) +
    out$nmix$compdraw[[iter]][[1]]$mu
  # transform price coef
  betadraws[nvar,] = - exp(betadraws[nvar,])
  # now compute equil prices for that draw of "market" atoms
  cat("posterior sim iter = ",iter,fill=TRUE)
  init_prices = c(rep(2,nbrds)) # initial prices in $100
  outbase = nash_eq(Own,init_prices,designBase,costs_wo_ss,betadraws)
  outSony = nash_eq(Own,init_prices,designSony,costs_w_ss,betadraws)
  eq_prices_base_draws[,ncnt] = outbase$prices
  eq_prices_Sony_draws[,ncnt] = outSony$prices
  eq_profit_base_draws[,ncnt] = outbase$profit
  eq_profit_Sony_draws[,ncnt] = outSony$profit
  ncnt=ncnt+1
}
```

Figure 11.5 Computing posterior distribution of equilibrium prices

"select_iter" is a vector of these selected MCMC iterations. Next we create room to store the resulting equilibrium calculations for each draw of the random coefficient or preference distribution parameters. We undertake two different equilibrium calculations: "base" references to the base configuration of all four products and "Sony" refers to situation in which the swivel screen is added only to the Sony product. The "for" loop sets the iterations over the MCMC draws. In the loop, we must extract the relevant parameters from the MCMC output, draw the "atoms" or pseudo-consumers, transform to enforce the appropriate reparameterization and then call nash_eq with the arguments set forth above. Figure 11.5 shows the appropriate R code.

This loop will take a few minutes (more than 2 and less than 10, depending on your computer) to complete and will leave the posterior draws in "*eq_prices_base_draws*" and "*eq_prices_Sony_draws*." We can now graph the difference between the Sony equilibrium price with the swivel screen and the Sony equilibrium price in the base configuration, see Figure 11.6. We can see that the enhanced demand created by the addition of swivel screen feature allows Sony to charge a higher equilibrium price. The posterior mean of that increment in price is about $25 but there is considerable posterior uncertainty as represented by the 95% posterior interval which starts at $17 and extends to $34. This is due to the relatively small number of respondents in the conjoint survey as noted in Rossi [2014a,b].

The addition of the swivel screen feature provides an important point of differentiation for the Sony camera as reflected not only by the fact that Sony is able to increase their equilibrium price (and the competitors are forced to reduce their prices) but by Sony's incremental profits. Figure 11.7 shows the posterior distribution of the percentage increase in Sony profits from the addition of the swivel screen. While there is considerable uncertainty in the posterior distribution, the addition of the feature increases Sony's variable cost-based profits by 58% as reflected by the posterior mean.

Of course, the large increase in Sony profits is predicated on only the Sony product having the differentiating feature. In reality, the addition of the swivel screen to the Sony product may induce not only a competitive price response but also the alteration of competing products. Clearly, if the industry transitions to a new set of products, all of which have the swivel screen, the rents Sony is earning will dissipate. In some cases, patents or technological barriers may prevent this kind of competitive response.

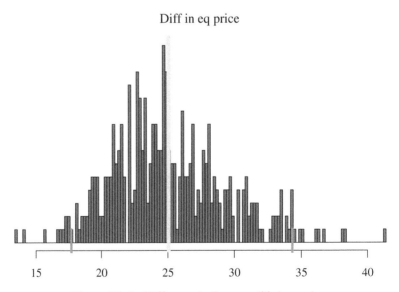

Figure 11.6 Difference in Sony equilibrium prices

%Change eq profits

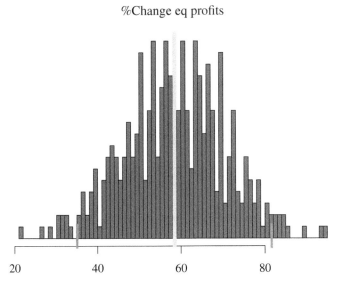

Figure 11.7   Percentage change in Sony equilibrium profits

These considerations may also be considered the basis of a distinction between long-run and short-run competitive response and are a subject for further research.

11.5.3 Lessons for Conjoint Design from WTP and Equilibrium Price Computations

As we have explained, the design of a conjoint survey that can enable a valid demand system imposes higher and different standards than in many conjoint survey applications. Not only must the survey sample be representative, the conjoint tasks designed in a manner that is understandable and clear to respondents, and free of bias, but the conjoint survey must include the outside option and the price coefficients or part-worths must be constrained to be negative.[12] It is a wise practice to inspect the conjoint data to insure that the outside option is not either completely ignored or the most frequent selected option. This is a form of pre-test or pilot study criteria that can indicate that the choice tasks and choice sets are poorly or unrealistically designed.

Even surveys that are designed to meet the criteria necessary for a valid demand systems can yield data that produce unreasonable equilibrium price estimates and WTP calculations. This is because these kinds of calculations impose a more strenuous stress test of the validity of the conjoint exercise than standard demand forecasting calculations. For example, one can obtain conjoint data which shows a reasonably high degree of both

---

[12] In the case of the "dummy" variable approach to conjoint designs with many different levels of price, not only must the price coefficients be negative, but monotonicity in the levels of price must be imposed. This can be achieved via a reparameterization to a "base" price coefficient and increments in price sensitivity which are constrained to be negative, i.e., the marginal dis-utility of price does not decrease as the level of price increases.

in-sample and out-of-sample goodness of fit as measured by log-likelihood or hit rates but equilibrium price or WTP calculations can yield unreasonable values such as extraordinarily high equilibrium prices or huge WTP values. This is because the price coefficients are unrealistically small. Low price sensitivity can yield unrealistic margins over cost and high WTP values. We have found that it is important to screen out respondents who do not exercise diligence in completing the choice tasks. This can be done by removing respondents whose likelihood values are close to those which would obtain by random guessing. Other possibilities include building in rigorous reliability tests in the portion of the survey that precedes the choice tasks.

In any event, proper WTP and equilibrium calculations impose a higher standard on conjoint data than is typically imposed in many conjoint survey applications. This represents the fundamental trade-off in the use of conjoint methods. Conjoint data is, by construction, free of many of the problems which plague observational data analyses such as endogeneity and errors-in-the-variables biases. However, poorly designed conjoint surveys and/or respondents who do not take the exercise seriously pose a different threat to the external validity of the conjoint survey method.

12
Case Study 3: Scale Usage Heterogeneity

Abstract

Questions that use a discrete ratings scale are commonplace in survey research. Examples in marketing include customer satisfaction measurement (CSM) and purchase intention. Survey research practitioners have long commented that respondents vary in their usage of the scale; common patterns include using only the middle of the scale or using the upper or lower end. These differences in scale usage can impart biases to correlation and regression analyses. In order to capture scale usage differences. We develop a model with individual scale and location effects and a discrete outcome variable. The joint distribution of all ratings scale responses is modeled rather than specific univariate conditional distributions as in the ordinal probit model. The model is applied to a customer satisfaction survey where it is shown that the correlation inferences are much different once proper adjustments are made for the discreteness of the data and scale usage. The adjusted or latent ratings scales is also more closely related to actual purchase behavior.

12.1 BACKGROUND

Customer satisfaction surveys, and survey research in general, often collect data on discrete rating scales. Figure 12.1 shows a sample questionnaire of this type from Maritz Marketing Research Inc, a leading CSM firm. In this sample questionnaire, a five point scale (excellent to poor) is used while in other cases 7 and 10 point scales are popular. Survey research practitioners have long commented that respondents vary in their usage of the scale; common patterns include using only the middle of the scale or using the upper or lower end. In addition, it has been observed that there are large cultural or cross-country differences in scale usage, making it difficult to combine data across cultural or international boundaries. These different usage patterns, which we term "scale usage heterogeneity," impart biases to many of the standard analyses

Bayesian Statistics and Marketing, Second Edition. Peter E. Rossi, Greg M. Allenby, and Sanjog Misra
© 2024 John Wiley & Sons Ltd. Published 2024 by John Wiley & Sons Ltd.

	Much Better Than	Better Than	Equal to	Less Than	Much Less Than	Not Applicable
Overall Performance	○	●	○	○	○	○
Service						
1. Efficiency of service call handling.	○	○	○	○	○	○
2. Professionalism of our service personnel.	○	○	○	○	○	○
3. Response time to service calls.	○	○	○	○	○	○
Contract Administration						
4. Timeliness of contract administration.	○	○	○	○	○	○
5. Accuracy of contract administration.	○	○	○	○	○	○

Figure 12.1   Example of customer satisfaction survey questionnaire

conducted with ratings data, including regression and clustering methods as well as the identification of individuals with extreme views.

The standard procedure for coping with scale usage heterogeneity is to center each respondent's data by subtracting the mean over all questions and dividing by the standard deviation of response. The use of respondent means and standard deviations assumes that the response data is continuously distributed and from an elliptically symmetric distribution. Furthermore, the estimates of the individual location and scale parameters obtained by computing the mean and standard deviation over a small number of questions are often imprecise.

In order to choose an appropriate modeling strategy for ratings data, we must consider the types of analyses that will be conducted with this data as well as the basic issues of what sort of scale information (ratio, interval or ordinal) is available in this data. To facilitate this discussion, assume that the data is generated by a customer satisfaction survey; however, the methods developed here apply equally well to any data in which a ratings scale is used (examples include purchase intentions and psychological attitude measurement). In the typical CSM survey, respondents are asked to give their overall satisfaction with a product as well as assessments of satisfactions with various dimensions of the product or service. Ratings are made on five, seven or ten point scales. We will focus on two major uses of CSM ratings data: 1. Measurement of the relationship between overall satisfaction and satisfaction with specific product attributes and 2. Identification of customers with extreme views. Scale usage heterogeneity can substantially bias analyses aimed at either use.

For example, if some respondents tend to use either the low or high end of the scale, this will tend to bias upward any measure of correlation between two response items.

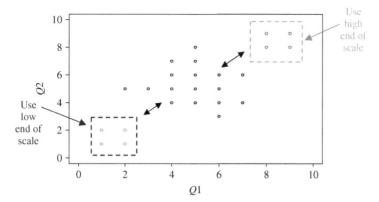

Figure 12.2   Scale usage heterogeneity and upward correlation bias

The middle group of points in Figure 12.2 represents a hypothetical situation in which all respondents have the same scale usage. If some respondents use the upper or lower end of the scale, this will move points outward from the middle grouping, creating a higher but spurious correlation. Thus, any covariance-based analysis of rating scale data such as regression or factor analysis can be substantially biased by scale usage heterogeneity, aside from the problems associated with using discrete data in methods based on the assumption of continuous elliptically symmetric distributions. In addition, any cluster analysis or filtering of the data for respondents with extreme views will tend to identify a group of nay or yeah sayers whose true preferences may not be extreme or even similar.

Practitioners have long been aware of scale usage patterns and often center the data. We observe N respondents answering M questions on a discrete rating scale consisting of the integers 1 to K; the data array is denoted $X = \{x_{ij}\}$, an $N \times M$ array of discrete responses, $x_{ij} = \{k\}$, $k = 1, \ldots, K$. Centering would transform the X array by subtracting row means and dividing by the row standard deviation.

$$X^* = \left[(x_{ij} - \bar{x}_i)\,/s_i\right]$$

After the data is centered, standard correlation and regression methods are used to examine the relationship between various questions. To identify extreme respondents or to cluster respondents, it is more typical to use the raw response data.

To select the appropriate analysis method, it is important to reflect on the nature of the scale information available in ratings data. Our perspective is that the discrete response data provides information on underlying continuous and latent preference/satisfaction. Clearly, the ratings provide ordinal information in the sense that a higher discrete rating value means higher true preference/satisfaction. It is our view that ratings data can also provide interval information, once the scale usage heterogeneity has been properly accounted for. However, we do not believe that even properly adjusted ratings data can provide ratio level information. For example, if a respondent gives only ratings at the top end of the scale, we cannot infer that he/she is extremely satisfied. We can only infer that the level of relative satisfaction is the same across all items for this respondent.

Centering acknowledges this fundamental identification problem, but introduces imprecise row mean and row standard deviation estimates which introduce considerable noise into the data. In most cases, fewer than 20 questions are used to form respondent means and standard deviation estimates. Furthermore, the use of centered data in correlation, regression or clustering analyses ignores the discrete aspect of this data (some transform the data prior to centering, but no standard transformation can change the discreteness of the data). In the next section, we develop a model which incorporates both the discrete aspects of the data and scale usage heterogeneity.

12.2 MODEL

Our model is motivated by the basic view that the data in the X response array is a discrete version of underlying continuous data. For $i = 1, \ldots, N$ and $j = 1, \ldots, M$, let y_{ij} denote the latent response of individual i to question j. Let $y'_i = [y_{i1}, \ldots, y_{iM}]$ denote the latent response of respondent i to the entire set of M questions. Assume there are $K + 1$ common and ordered cut-off points $\{c_k : c_{k-1} \leq c_k, k = 1, \ldots, K\}$ where $c_0 = -\infty, c_K = +\infty$, such that for all i, j and k

$$x_{i,j} = k \text{ if } c_{k-1} \leq y_{i,j} \leq c_k \tag{12.2.1}$$

and

$$y_i \sim N\left(\mu_i^*, \Sigma_i^*\right) \tag{12.2.2}$$

The interpretation of the model in (12.3.1) and (12.3.2) is that the observed responses are iid multinomial random variables, where the multinomial probabilities are derived from an underlying continuous multivariate normal distribution. The set of cutoffs $[c_0, \ldots, c_K]$ discretizes the normal variable $y_{i,j}$.

The probability that $x'_i = [x_{i,1}, \ldots, x_{i,M}]$ takes on any given realization (a vector of M integers between 1 and K) is given by the integral of the joint normal distribution of y over the appropriately defined region. For example, if $M = 2$, and $x'_i = [2, 8]$, then Figure 12.3 depicts this integral of a bivariate normal distribution over a rectangle defined by the appropriate cut-offs.

The above-proposed model is different in nature from some latent variable models used in Bayesian analyses of discrete data. Most of such common models deal with grouped data in form of contingency tables, where the prior distribution of multinomial probabilities is usually taken as Dirichlet over multidimensional arrays. There, the

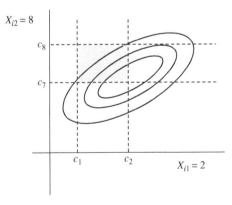

Figure 12.3 Computing the multinomial probabilities

problem of interest is usually of modeling probabilities forming certain patterns of statistical dependence. Here we are interested in modeling individual responses to make
individual measurements comparable for the sake of correlation and regression analysis.

It should be noted that the model in (12.3.1) is not a standard ordinal probit model.
We have postulated a model of the *joint* discrete distribution of the responses to all M
questions in the survey. Standard ordinal probit models would focus on the conditional
distribution of one discrete variable given the values of another set of variables. Obviously,
since our model is of the joint distribution, we can make inferences regarding various conditional distributions so that our model encompasses the standard conditional approach.
In analysis of ratings survey data, we are required to have the capability of making inferences about both the marginal distribution of specific variables as well as conditional
distributions so that a joint approach would seem natural.

The model in (12.3.1) and (12.3.2) is overparameterized since we have simply allowed
for an entirely different mean vector and covariance matrix for each respondent. In order
to allow for differences in respondent scale usage without over parameterization, the y
vector is written as a location/scale shift of common multivariate normal variable.

$$y_i = \mu + \tau_i \iota + \sigma_i z_i$$
$$z_i \sim N(0, \Sigma)$$
(12.2.3)

We have allowed for a respondent-specific location and scale shift to generate the mean
and covariance structure in (12.3.2) with $\mu_i^* = \mu + \tau_i \iota$ and $\Sigma_i^* = \sigma_i^2 \Sigma$. Analysis of the
joint distribution of questions as well as the identification of customers with extreme
views is based on the set of model parameters: $\{z_i\}$, μ, Σ.

The model (12.3.3) accommodates scale usage via the (τ_i, σ_i) parameters. For
example, a respondent who uses the top end of the scale would have a large value of τ and
a small value of σ. It is important to note that this model can easily accommodate lumps
of probability at specific discrete response values. For example, in many customer satisfaction surveys, there are a significant fraction of respondents who give only the top value
of the scale for all questions in the survey. In the model outlined in (12.3.1)–(12.3.3),
we would simply have a normal distribution centered far out (via a large value of τ) so
that there is a high probability that y will lie in the region corresponding to the top
rating. We can also set τ to zero and make σ_i large to create "piling-up" at both extremes.
However, our model cannot create lumps of probability at two different non-extreme
values (to accommodate a respondent who uses mostly 2 s and 9 s on a 10 point scale).

As a modeling strategy, we have chosen to keep the cuts-offs (c) common across
respondents and shift the distribution of the latent variable. Another strategy would be
to keep a common latent variable distribution for all respondents and shift the cut-offs
to induce different patterns of scale usage. The problem here would be choosing a flexible but not too overly-parameterized distribution for the cut-offs. Given that we often
have a small number of questions per respondent (M), a premium should be placed on
parsimony. There is also a sense in which the model in (12.3.3) can be interpreted as a
model with respondent specific cut-offs. If we define

$$c_i^* = \tau_i + \sigma_i c$$

where c is the vector of cut-offs, then we have the same model with respondent-specific
cut-offs and a common invariant distribution of latent preferences.

The model specified in (12.3.3) is not identified. The entire collection of τ_i parameters
can be shifted by a constant and with a compensating shift can be made to μ without

changing the distribution of the latent variables. Similarly, we can scale all σ_i and make a reciprocal change to Σ. As discussed below, we solve these identification problems by imposing restrictions on the hierarchical model. $(\tau_i, \ln \sigma_i)$ are assumed to be bivariate normal.

$$\begin{bmatrix} \tau_i \\ \ln \sigma_i \end{bmatrix} \sim N(\varphi, \Lambda) \tag{12.2.4}$$

The model in (12.3.4) allows for a correlation between the location and scale parameters. For example, if there is a sub-population that uses the high end of the scale, then we would expect an inverse relationship between τ and σ. In most applications of hierarchical models, the location and scale parameters are assumed to be independent.

We achieve identification of τ_i by imposing the restriction, $E[\tau_i] = 0$. Since the distribution of τ_i is symmetric and unimodal, we are also setting the median and mode of this distribution to zero. Greater care must be exercised in choosing the identification restriction for σ_i, as this distribution is skewed. One logical choice might be to set $E[\sigma_i^2] = 1$. This imposes the identification restriction, $\varphi_2 = -\lambda_{22}$. However, as the dispersion parameter (λ_{22}) is increased, this produces the distribution of σ_i becomes concentrated around a mode smaller than 1.0 with a fat right tail. Our view is that this is an undesirable family of prior distributions. The right panel of Figure 12.4 illustrates the prior on σ_i for a small and large value of λ_{22}.

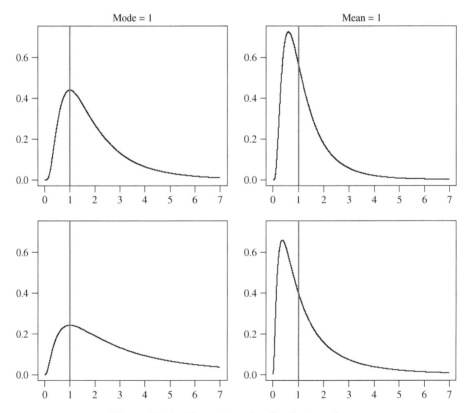

Figure 12.4 Alternative prior distributions for σ_i

A more reasonable approach is to restrict the mode of the prior on σ_i to be 1. This imposes the restriction, $\varphi_2 = \lambda_{22}$. The left panel of Figure 12.4 shows this family of distributions. As λ_{22} increases, these distributions retain the bulk of their mass around one, but achieve greater dispersion by thickening the right tail.[1] Thus, we employ two identification restrictions.

$$\varphi_1 = 0 \text{ and } \varphi_2 = \lambda_{22} \tag{12.2.5}$$

Even with the cut-offs $\{c_k\}$ assumed fixed and known, the model in (12.3.1) and (12.3.3) is a very flexible model which allows for many possible discrete outcome distributions. In particular, the model allows for "piling up" of probability mass at either or both endpoints of the ratings scale – a phenomenon frequently noted in CSM data. For further flexibility, we could introduce the cut-offs as free parameters to be estimated. However, we would have to recognize that identification restrictions must be imposed on the cut-off parameters since shifting all cut-offs by a constant or scaling all cut-offs is redundant with appropriate changes in μ and Σ. For this reason, we impose the following identification restrictions:

$$\begin{aligned} \sum_k c_k &= m_1 \\ \sum_k c_k^2 &= m_2 \end{aligned} \tag{12.2.6}$$

The model in (12.3.1)–(12.3.6) is fully identified but introduces $K - 2$ free cut-off parameters. In order to make the model more parsimonious, we will consider further restrictions on the c_k parameters. If we impose equal spacing of the cut-offs, then by (12.3.6) there would be no free c_k parameters. Once the identification restrictions in (12.3.6) are imposed, the only sort of flexibility left in the cut-offs is to introduce skewness or nonlinear spread in the values. In order to allow for nonlinear spread while keeping the number of parameters to a minimum, we impose the further restriction that the cut-off values lie on a quadratic equation.

$$c_k = a + bk + ek^2 \quad k = 1, \dots, K - 1 \tag{12.2.7}$$

For example, consider the case of a 10 point scale with $a = 5.5$, $b = 1$ and $e = 0$, then $c_1 = 1.5$, $c_2 = 2.5$, $\dots$, $c_9 = 9.5$. Johnson and Albert (1999) review univariate ordinal probit models in which the cut-offs are not parameterized and are estimated usually with a diffuse prior subject to different identification conditions.

Given the identification restrictions in (12.3.6) and the parameterization in (12.3.7), e is the only free parameter; that is, given m_1, m_2, and e, we can solve for a and b by substituting for c_k in (12.3.6) using (12.3.7). In our implementation, m_1 and m_2 are selected so that when $e = 0$, we obtain a standard equal spacing of cut-off values, with each centered around the corresponding scale value. This means that for a 10 point scale, $m_1 = \sum_{k=1}^{K-1}(k + 0.5) = 49.5$ and $m_2 = \sum_{k=1}^{K-1}(k + 0.5)^2 = 332.25$. The mapping from e to the quadratic coefficients in (12.3.7) is only defined for a bounded range of e values.

The role of e is to allow for a skewed spreading of the cut-off values as shown in Figure 12.5. A positive value of e spreads out the intervals which result in high scale ratings and compresses the intervals on the low end. This will result in massing of probability at the upper end.

---

[1] In Rossi et al. (2001), the strategy of setting the mean to 1 is used for identification.

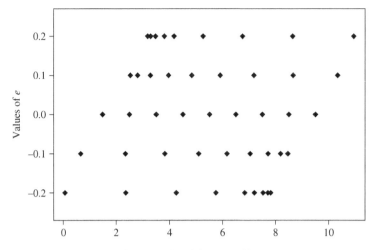

Figure 12.5   The role of the cut-off parameter, e

12.3 PRIORS AND MCMC ALGORITHM

To complete the model, we introduce priors on the common parameters.

$$\pi (\mu, \Sigma, \varphi, \Lambda, e) = \pi (\mu) \pi (\Sigma) \pi (\varphi) \pi (\Lambda) \pi (e) \qquad (12.3.1)$$

with

$$\begin{aligned} &\pi (\mu) \propto \text{constant} \\ &\pi (e) \propto unif\, [-0.2, 0.2] \\ &\Sigma \sim IW\left(v_\Sigma, V_\Sigma\right) \\ &\Lambda \sim IW\left(v_\Lambda, V_\Lambda\right) \end{aligned} \qquad (12.3.2)$$

That is, we are using flat priors on the means and the cut-off parameter and standard Wishart priors for the inverse of the two covariance matrices. We note that the identification restriction in (12.3.5) means that the prior on Λ induces a prior on φ. Figure 12.5 shows that the range of e in our uniform prior is more than sufficient to induce wide variation in the patterns of skewness in the cut-offs. We use diffuse but proper settings for the priors on Σ^{-1} and Λ^{-1}. These parameter values center the prior on Σ over the identity matrix

$$v_\Sigma = \dim (\Sigma) + 3 = K + 3, \quad V_\Sigma = v_\Sigma I \qquad (12.3.3)$$

The prior on Λ influences the degree of shrinkage in τ_i, σ_i estimates. Our hierarchical model adapts to the information in the data regarding the distribution of τ_i, σ_i, subject to the influence of the prior on the hyperparameter Λ. There will rarely be more than a small number of questions on which to base estimates of τ_i, σ_i. This means that the prior on Λ may be quite influential. In most hierarchical applications, there is a subset of units for which a good deal of information is available. This subset allows for determination of Λ via adaptive shrinkage. However, in our situation, this subset is not available and the prior on Λ has the potential to exercise more influence than normal. For these reasons, we will exercise some care in the choice of the prior on Λ. We will also consider and recommend prior settings somewhat tighter than typically used in hierarchical contexts.

The role of the prior on Λ is to induce a prior distribution on τ_i, σ_i. To examine the implications for choice of the prior hyperparameters, we will compute the marginal prior on τ_i, σ_i via simulation. The marginal prior is defined by

$$\pi(\tau,\sigma) = \int p(\tau,\sigma\,|\Lambda)\,\pi\left(\Lambda\,|v_\Lambda, V_\Lambda\right)\,d\Lambda \tag{12.3.4}$$

To assess v_Λ, V_Λ, we consider a generous range of possible values for τ_i, σ_i, but restrict prior variation to not much greater than this permissible range. For τ, we consider the range ± 5 to be quite generous in the sense that this encompasses much of a 10 point scale. For σ, we must consider the role of this parameter in restricting the range of possible values for the latent variable. Small values of σ correspond to respondents who only use a small portion of the scale, while large values would correspond to respondents who use the entire scale. Consider the ratio of the standard deviation of a "small" scale range user (e.g., someone who only uses the bottom or top 3 scale numbers) to standard deviation of a "mid range" user who employs a range of 5 points on the 10 pt scale. This might correspond to a "small" value of σ. The ratio for a respondent who uses end points and the "middle" of the scale (e.g., 1, 5, 10), a "large range" user, to the "mid range" user could define a "large value" of σ. These computations suggest that a generous range of σ values would be $(0.5, 2)$. We employ the prior settings corresponding to a relatively informative prior on τ_i, σ_i.

$$v_\Lambda = 20,\ \ V_\Lambda = \left(v_\Lambda - 2 - 1\right)\overline{\Lambda},\ \ \overline{\Lambda} = \begin{bmatrix} 4 & 0 \\ 0 & 0.5 \end{bmatrix} \tag{12.3.5}$$

The settings in (12.4.5) ensure that $E[\Lambda] = \overline{\Lambda}$. If v_Λ is set to 20, then the resulting marginal prior on τ_i, σ_i provides coverage of the relevant range without admitting absurdly large values as illustrated in Figure 12.6.

The model defined in equations (12.3.1)–(12.3.4) with priors given by (12.4.2) and identification restrictions (12.3.5)–(12.3.7) is a hierarchical model that can be analyzed with some modifications to the standard Gibbs sampler. Our interest centers not only on common parameters but on making inferences about the respondent specific scale usage and latent preference parameters, ruling out the use of classical statistical methods. Four problems must be overcome to construct the sampler.

1. Data augmentation requires a method for handling truncated multivariate normal random variables as in the MNP model (see Chapter 4).

2. One of the conditional distributions in the Gibbs sampler requires the evaluation of the integral of multivariate normal random variables over a rectangle. We use the GHK simulation method (see Chapter 2).

3. The random effects model and priors are not always conditionally conjugate.

4. We accelerate the Gibbs sampler by integrating out some of the latent variables to block e and $\{y_i\}$.

We refer the reader to the appendix of Rossi et al. (2001) for more details.

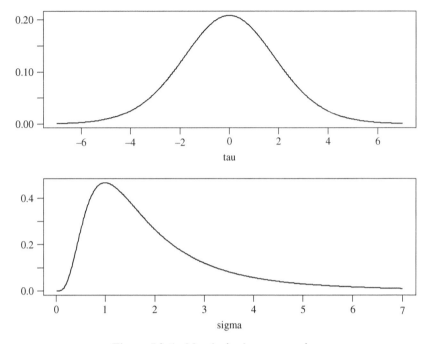

Figure 12.6 Marginal priors on τ and σ

12.4 DATA

To illustrate our method, we examine a customer satisfaction survey done in a business-to-business context with an advertising product. This dataset can be loaded once the package *bayesm* package has been installed using R command **R** `data(customerSat)`. 1811 customers were surveyed as to their views regarding satisfaction with overall product performance, aspects of price (three questions) and various dimensions of effectiveness (six questions). Figure 12.7 lists the specific questions asked. All responses are on a 10 point ratings scale.

12.4.1 Scale Usage Heterogeneity

Figure 12.8 plots the median over the ten questions vs. the range of responses for each of the 1811 respondents. Since all responses are integer, the points are "jittered" slightly so that the number of respondents at any given combination of range (0–9) and median (1–10) can be gauged. Figure 12.8 shows considerable evidence of scale usage heterogeneity. A number of respondents are using only the top end of the scale which is represented by points in the lower right hand corner of the figure. In fact, a reasonably large number give only the top rating response (10) to all questions. On the other hand, there are very few customers who use the lower end of the scale (lower left hand corner) and a large number who use a much of the scale.

On a scale from 1 to 10 where 10 means an "Excellent" performance and 1 means a "Poor" performance, please rate BRAND on the following items:

Q1. Overall value

Price:

Q2. Setting competitive prices.

Q3. Holding price increases to a reasonable minimum for the same ad as last year.

Q4. Being appropriately priced for the amount of customers attracted to your business.

Effectiveness:

Q5. Demonstrating to you the potential effectiveness of your advertising purchase.

Q6. Attracting customers to your business through your advertising.

Q7. Reaching a large number of customers.

Q8. Providing long-term exposure to customers throughout the year.

Q9. Providing distribution to the number of households and/or business your business needs to reach.

Q10. Proving distribution to the geographic areas your business needs to reach.

Figure 12.7 List of questions

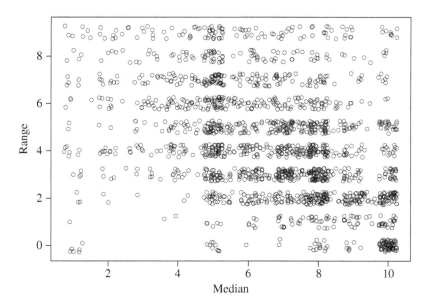

Figure 12.8 Respondent range vs. median (jittered values)

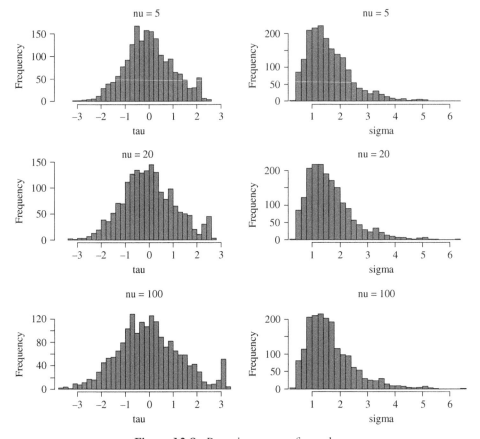

Figure 12.9   Posterior means of τ_i and σ_i

Our hierarchical model should capture the scale usage heterogeneity in the distribution of (τ_i, σ_i). Figure 12.9 provides histograms of the N posterior means of (τ_i, σ_i). Both τ_i and σ_i display a great deal of variation from respondent to respondent, indicating a high degree of scale usage heterogeneity. Figure 12.9 shows three prior settings, ranging from highly diffuse (top) to tight (bottom). For $v = 5$ and $v = 20$, the posteriors of both τ_i and σ_i are very similar indicating little sensitivity to the prior. The hierarchical prior adapts to the information in the data, centering on values of τ_i and σ_i that are shrunk quite a bit. When $v = 100$, the prior is reasonably tight about the mean value. This prior has a greater influence and reduces the extent of shrinkage. For example, when $v = 100$, there appears to be a small mode in the τ_i distribution around 3. This mode accommodates those respondents who always give the highest rating (10) in answering all questions. More diffuse priors induce more shrinkage from the information in the data and the τ_i distributions for those respondents that always respond with the highest rating are shrunk toward zero. With the higher prior, there is less of this shrinkage.

Figure 12.10 displays the posterior distribution of Λ. The top panel shows the posterior distribution of $\sqrt{\lambda_{11}}$, the standard deviation of τ_i. This distribution is reasonably tight around 1.55 or so and puts virtually no mass near 2 that corresponds to the location

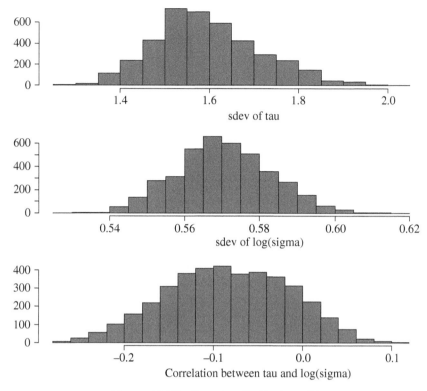

Figure 12.10   Posterior distribution of Λ

of the prior. Even more striking is the posterior distribution of $\sqrt{\lambda_{22}}$, the standard deviation of $\log\left(\sigma_i\right)$. This distribution is centered tightly about 0.57, indicating that the more than 1000 respondents are quite informative about this scale usage parameter. Finally, the posterior distribution of the correlation between τ_i and $\log\left(\sigma_i\right)$ puts most of its mass on small negative values, indicating that there is a tendency for those who give high ratings to use less of the scale.

Our model also differs from the standard centering and the standard normal approaches in that we explicitly recognize the discrete nature of the outcome data. Moreover, the discrete outcomes do not appear to be merely a simple filtering of the underlying latent data in which latent data is simply rounded to integers using equal size intervals. The posterior of the quadratic cut-off parameter, e, has a mean of 0.014 with a posterior standard deviation of 0.0039. This implies a skewed set of cutoffs.

12.4.2 Correlation Analysis

One of the major purposes of these surveys is to learn about the relationship between overall satisfaction and various dimensions of product performance. The presence of scale usage heterogeneity prevents meaningful use of the raw data for correlation purposes. Scale usage heterogeneity will bias the correlations upward. Table 12.1 provides

Table 12.1   Raw data means, together with covariance (lower triangle) and correlation (upper triangle) matrix

Q	Mean									
1	6.06	6.50	0.65	0.62	0.78	0.65	0.74	0.59	0.56	0.44	0.45
2	5.88	4.38	7.00	0.77	0.76	0.55	0.49	0.42	0.43	0.35	0.35
3	6.27	4.16	5.45	7.06	0.72	0.52	0.46	0.43	0.46	0.38	0.40
4	5.55	5.36	5.43	5.16	7.37	0.64	0.67	0.52	0.52	0.41	0.40
5	6.13	4.35	3.83	3.62	4.53	6.84	0.69	0.58	0.59	0.49	0.46
6	6.05	4.82	3.29	3.15	4.61	4.61	6.49	0.59	0.59	0.45	0.44
7	7.25	3.64	2.70	2.73	3.42	3.68	3.66	5.85	0.65	0.62	0.60
8	7.46	3.28	2.61	2.79	3.23	3.51	3.41	3.61	5.21	0.62	0.62
9	7.89	2.41	1.99	2.18	2.39	2.72	2.47	3.20	3.02	4.57	0.75
10	7.77	2.55	2.06	2.33	2.42	2.67	2.51	3.21	2.95	3.54	4.89

the means and correlations for the survey data. The correlations are uniformly positive and in the $(0.4, 0.7)$ range. Table 12.2 provides posterior means (standard deviations) and correlations of the standardized latent variable, z. Table 12.2 provides a dramatically different view of the correlation structure behind the data. In general, the correlations adjusted for scale usage are smaller than those computed on the raw data. This accords with the intuition developed in the discussion of Figure 12.2. In particular, the correlations between question 2 (price) and questions 9 and 10 (effectiveness in reach) are estimated to be one half the size of the raw data after adjusting for scale usage heterogeneity. Intuitively, it seems reasonable that those who are very satisfied with price (i.e., low price) should not also think that the advertising reach is also high. Table 12.3 provides the correlation matrix of the centered data. Centering the data changes the uniformly positive correlations to mostly negative correlations. It does not seem intuitively reasonable that questions probing very similar aspects of the product should have zero to small negative correlations.

12.5 DISCUSSION

There are two challenges facing an analyst of ratings scale data: 1. Dealing with the discrete/ordinal nature of the ratings scale and 2. Overcoming differences in scale usage across respondents. In the psychometric literature, various optimal scaling methods have been proposed for transforming discrete/ordinal data into data that can be treated in a more continuous fashion. However, it is difficult to adapt these methods to allow for respondent-specific scale usage. We adopt a model-based approach in which an underlying continuous preference variable is subjected to respondent-specific location and scale shifts. This model can easily produce data of the sort encountered in CSM where some respondents are observed to only use certain portions of the scale (in the extreme, only one scale value). Standard centering methods are shown to be inferior to our proposed procedure in terms of the estimation of relationships between variables.

   Our analysis demonstrates that scale usage heterogeneity can impart substantial biases to estimation of the covariance structure of the data. In particular, scale usage

Table 12.2 Posterior Inference for means, variances and correlations of latent response (z)

Q	Mean (μ)	Covariance/correlation (Σ)									
1	6.50 (0.08)	2.46 (0.29)	0.59	0.73	0.53	0.73	0.66	0.45	0.40	0.24	0.26
2	6.23 (0.08)		2.98 (0.31)	0.77	0.76	0.47	0.36	0.25	0.26	0.14	0.15
3	6.55 (0.08)			3.33 (0.32)	0.69	0.43	0.34	0.27	0.30	0.20	0.23
4	6.08 (0.08)				3.28 (0.35)	0.56	0.57	0.37	0.36	0.20	0.21
5	6.53 (0.08)					2.82 (0.30)	0.63	0.47	0.47	0.32	0.32
6	6.55 (0.08)						2.39 (0.30)	0.49	0.45	0.26	0.26
7	7.46 (0.08)							2.93 (0.28)	0.63	0.62	0.60
8	7.56 (0.08)								2.43 (0.23)	0.61	0.58
9	7.90 (0.08)									2.64 (0.25)	0.78
10	7.82 (0.08)										2.76 (0.26)

Posterior standard deviations are in parentheses.

Table 12.3 Means and correlation/covariances of standardized variables

Q	Mean										
1	-0.29	0.66	-0.07	-0.13	0.03	-0.14	0.06	-0.11	-0.16	-0.24	-0.21
2	-0.42	-0.05	0.82	0.35	0.20	-0.19	-0.36	-0.32	-0.25	-0.26	-0.27
3	-0.18	-0.10	0.31	0.93	0.14	-0.21	-0.33	-0.33	-0.24	-0.24	-0.22
4	-0.60	0.02	0.14	0.11	0.62	-0.23	-0.17	-0.24	-0.20	-0.26	-0.28
5	-0.28	-0.09	-0.15	-0.18	-0.16	0.76	0.04	-0.07	-0.01	-0.10	-0.11
6	-0.32	0.04	-0.28	-0.27	-0.12	0.03	0.74	0.03	0.03	-0.12	-0.14
7	0.33	-0.08	-0.23	-0.26	-0.16	-0.05	0.02	0.67	0.01	0.06	0.05
8	0.46	-0.09	-0.16	-0.17	-0.12	-0.01	0.02	0.01	0.56	0.01	-0.04
9	0.68	-0.14	-0.17	-0.18	-0.16	-0.07	-0.08	0.04	0.00	0.58	0.31
10	0.61	-0.14	-0.20	-0.18	-0.18	-0.08	-0.10	0.03	-0.02	0.19	0.67

heterogeneity causes upward bias in correlations and can be at the source of the colinearity problems observed by many in the analysis of survey data. The covariance structure is the key input to many different forms of analysis, including identification of segments via clustering and the identification of relationships through covariance-based structural modeling. Our procedures provide unbiased estimates of the covariance structure which can then be used as input to subsequent analysis.

While it is well documented that scale usage heterogeneity is prevalent in ratings scale data, not much is known about the determinants of this behavior. Important questions for future research include what sorts of respondents tend to exhibit a high degree of scale usage heterogeneity and how can questionnaires and items be designed to minimize scale usage heterogeneity. Our view is that it is desirable to reduce the magnitude of this phenomenon so as to increase the information content of the data. Whether it is possible to design survey instruments for customer satisfaction that are largely free of this problem is open to question. Ultimately, the answer must come from empirical applications of our model that will detect the extent of this problem.

12.6 R IMPLEMENTATION

R The R implementation of this method is embodied in the function, `rscaleUsage`, which is included in *bayesm*. We made a few changes to the basic algorithm in Rossi et al. (2001). First, we did not use an independence Metropolis step for drawing σ_i, instead we used a relatively fine grid over a relatively large interval. Second, we used a uniform prior for e on a grid on the interval $(-0.1, 0.1)$ and we used a random walk on this grid (moving to left and right grid points with equal probability except at the boundary where we propose next innermost point with probability one). We use 100 replications to compute the GHK estimates of normal probabilities. We use 500 as the default number of grid points for the griddy Gibbs parameter draws (σ_i, e, Λ). We use the default prior settings given in (12.4.2), (12.4.3), and (12.4.5) above.

13

Case Study 4: Volumetric Conjoint

Abstract

Consumer demand for products often result in the purchase of multiple goods at the same time. Corner solutions, or the nonpurchase of items, occur when consumers have strong preference for some goods that do not satiate and weak preference for other goods. However, if nonpurchase arises because a consumer finds particular brands and attributes unacceptable, leading to the formation of consideration sets, then estimates of preference will be too extreme and biased. In this case study, we extend the work on consideration sets and discrete choices to a wider class of models, and develop a model of multiple discreteness with conjunctive screening of the alternatives that remove offerings from consideration.

13.1 INTRODUCTION

Consumer choices among alternatives in any product class involve the screening of alternatives to reduce the cognitive demands of decision making. Some brands available for sale are potentially responsive to the needs of individuals and others are not. Resource conserving decision makers have long been known to rule out alternatives that are not candidates for purchase. Responsive attributes are those that can potentially serve as instruments in making changes in the state of the individual, and alternatives with these attributes are admitted into a consideration set for further evaluation and potential purchase. In this case study, we develop a general model for identifying brands and their characteristics, or attributes, used to screen out brands from consideration.

Models of a consideration set formation have been successfully developed for discrete choice models and in this case study we extend their application in models of multiple discreteness or horizontal variety (Kim et al. [2002]) where more than one alternative can be simultaneously chosen. When a choice alternative is included in the consideration set

Bayesian Statistics and Marketing, Second Edition. Peter E. Rossi, Greg M. Allenby, and Sanjog Misra.
© 2024 John Wiley & Sons Ltd. Published 2024 by John Wiley & Sons Ltd.

but not chosen, its ratio of marginal utility to price (i.e., its "bang for the buck") serves as a lower bound to the items that are chosen. When it is not part of the consideration set, its ratio does not play this role. This can result in different inferences about the importance of product attributes. We investigate the conjunctive screening occurring during consumers' decision. Conjunctive screening has been shown to be the most prevalent form of screening in the formation of consideration sets (Gilbride and Allenby [2004]).

13.2 MODEL DEVELOPMENT

Our model of demand is a generalization of multinomial models of choice where just one of the choice alternatives is selected. The standard discrete-choice setup can be formulated as a model with linear utility, $u(x) = \psi'x$ and constant marginal utility ψ. This results in a utility maximization problem where just one of the goods is selected – the one that yields the highest marginal utility divided by price, that is, ψ_k/p_k. We generalize this model by allowing for satiation, which results in indifference curves that are convex to the origin and the possibility that multiple units of demand are purchased.

Data for our generalized model of demand has a panel structure similar to that found for discrete choice models, except that more than one alternative can be chosen and purchase quantities can be present. Table 13.1 displays demand data from a survey in which respondents were asked how many days they would spend at a theme park during their vacation. The data are from a conjoint (i.e., stated preference) exercise where the price and features of the theme parks changed across the choice tasks. The data are sparse, with the most frequently observed entry being the number zero. The data take on only four different values $(0, 1, 2, 3)$, and it is difficult to imagine this short record of preferences providing a reliable estimate of preferences and price sensitivities without imposing additional structure on the analysis. The point of incorporating economic theory into the model is to allow us to parsimoniously explain demand.

Table 13.1   Example demand data – Florida theme parks

Choice	Magic kingdom	Epcot center	Animal kingdom	Hollywood studios	Universal studios	Islands of adventure	Busch gardens
1	0	0	2	2	1	0	0
2	1	1	2	0	1	0	0
3	1	2	0	2	0	0	0
4	0	2	2	1	0	0	0
5	1	0	2	2	0	0	0
6	1	2	2	0	0	0	0
7	2	0	0	0	1	0	1
8	2	0	0	0	2	0	0
9	1	1	1	0	0	0	1
10	1	0	0	0	3	0	0
11	0	1	0	0	2	0	1
12	0	1	1	1	0	0	1

Mathematically, we assume respondents:

$$\max u\,(x, z) \text{ subject to } p'x + z = E$$

where $u(x, z)$ is a utility function, x denotes a vector of quantities, z is the outside good that represents unspent money, p denotes a vector of prices and E is the budgetary allotment, or expenditure. E is sometimes called the "income constraint" and is interpreted as the maximum expenditure a consumer is willing to make in the product category.

The general solution to the problem of associating observed responses to a process of constrained maximization involves forming a function that combines the utility function and budget constraint with a parameter known as a Lagrangian multiplier:

$$L = u\,(x, z) + \lambda\left(E - p'x - z\right)$$

The purpose of the Lagrangian multiplier (λ) is to ensure that the derivative of $u\,(x, z)$ is proportional to the derivative of the constraint at the point of optimality. Differentiating with respect to x we obtain what are known as the Kuhn–Tucker first-order conditions:

$$u_j - \lambda p_j = 0 \quad \text{if } x_j > 0$$
$$u_j - \lambda p_j < 0 \quad \text{if } x_j = 0$$

where x is the vector of observed optimal demand, and u_j is the derivative of the utility function with respect to x_j. Rearranging terms, we can express the above as:

$$\text{if } x_j > 0 \text{ and } x_k > 0 \text{ then } \frac{u_j}{p_j} = \frac{u_k}{p_k} \text{ for all } j \text{ and } k$$

$$\text{if } x_j > 0 \text{ and } x_k = 0 \text{ then } \frac{u_j}{p_j} > \frac{u_k}{p_k} \text{ for all } j \text{ and } k$$

The Kuhn–Tucker conditions state a general principle of optimality associated with constrained maximization problems when the arguments are non-negative (i.e., either positive or zero). If we observe two choice options to be positively valued, then the bang-for-the-buck is equal when allocation is optimal. If a choice option is not chosen ($x_k = 0$), then the bang-for-the-buck for the "not" chosen good is less than that for the goods that are chosen. The bang-for-the-buck is the ratio of marginal utility to price, expressed as u_j/p_j. We will employ the Kuhn–Tucker conditions to relate our observed marketing data to model parameters.

We now expand this demand model to incorporate consideration sets. We take a strict view of consideration in that we do not introduce separate probability models for consideration and choice. That is, we develop our model assuming there is an abrupt change in the likelihood as a choice alternative enters the consideration set. Figure 13.1 illustrates the effect of alternative screening, or the presence of consideration sets, on demand for two alternatives, x_1 and x_2. The left side of Figure 13.1 shows utility maximization without screening, where point B is seen to yield a higher level of utility and is on a higher indifference curve. The right side of the Figure 13.1 has a reduced feasible set due to non-consideration of the second good, x_2, and the utility maximizing solution is at

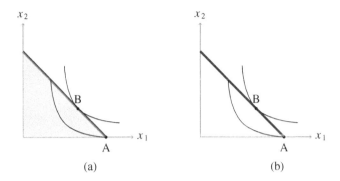

Figure 13.1   The effect of consideration sets on demand. (a) Both x_1 and x_2 considered. (b) Only x_1 considered.

point A. Not accounting for the effects of screening leads to estimated utility parameters that attempt to rationalize demand for A in terms of increased preference rather than a reduced choice set. The reduced feasible set also influences inference and predictions about price changes. Because the left side of the Figure 13.1 includes combinations of x_1 and x_2 in the feasible set, the price changes of x_2 can affect the consumers' final choice. However, the right side of the Figure 13.1 does not includes x_2 in the feasible set, and the price of x_2 does not affect consumer choice.

The consumer's set of considered brands is comprised of up to J goods and an outside good, z. We assume decreasing marginal utility for all goods, implying that consumers satiate in their preferences, and that utility is additively separable. Screening in the model is reflected in the indicator function I_j, which is equal to one if alternative j is included in the consideration set and is equal to zero otherwise.

We assume that consumers maximize their utility subject to constraints and consideration:

$$\text{Max } u(x, z) = \sum_{j=1}^{J} \frac{\psi_j}{\gamma} \ln\left(\gamma x_j + 1\right) I_j + \ln(z) \quad \text{subject to} \quad \sum_{j=1}^{J} p_j x_j I_j + z = E$$

where

$$I_j = \prod_{d=1}^{D} \left[1 - \tau_d \times I\left(A_{j,d} = 1\right)\right] \times I\left(p_j < \tau_p\right) \tag{13.2.1}$$

where the indicator function I_j accounts for screening behavior and enters both the utility function and budget constraint.

We have introduced the time subscript t to denote the choice task. Marginal utility is the derivative of the utility function with respect to x_t and z_t:

$$u_{jt} = \frac{\partial u\left(x_t, z_t\right)}{\partial x_{jt}} = \frac{\psi_{jt}}{\gamma x_{jt} + 1} = \frac{\exp\left(a_{jt}{}'\beta + \varepsilon_{jt}\right)}{\gamma x_{jt} + 1} \tag{13.2.2}$$

$$u_{z_t} = \frac{\partial u\left(x_t, z_t\right)}{\partial z_t} = \frac{1}{z_t} \tag{13.2.3}$$

and we see that marginal utility is diminishing in both. Here we assume that ψ_{jt}, the marginal utility associated with the first unit of consumption of alternative j, is specified as a regression model with $\psi_{jt} = \exp\left[a_{jt}'\beta + \varepsilon_{jt}\right]$. The term a_{jt} denotes the attributes of the jth alternative at time t and β are coefficients for their part worths. The error term is usually specified as Extreme Value, corresponding to a logit model, or Normal distribution, corresponding to a probit model in the case of discrete choice data. The error term allows the model to reflect data that is non-deterministic, possibly because of omitted variables. The coefficient γ is a parameter that governs the rate of satiation in the model, which is estimated from the data. When γ is small, then marginal utility declines slowly, and when it is larger the rate of satiation increases.

The matrix A specifies the presence of attributes for the choice alternatives, or products, with $A_{j,d} = 1$ indicating that alternative j has attribute d. The parameter $\tau_d \in \{0,1\}$ indicates whether the attribute d is used to screen alternatives. If $\tau_d = 1$, then consumers use attribute d to screen out choice alternatives and do not consider any product containing it. The conjunctive screening rule is represented as a product over the D attributes of indicator variables where if any screening attribute is present then the offering is deleted from the choice set. Additionally, we allow respondents to screen based on unit prices. Even though products may be within the budget E, consumers may reject offerings that are perceived as too expensive per unit. This is captured by the price threshold τ_p. Choice alternatives costing τ_p or more are excluded from the likelihood.

The Kuhn–Tucker conditions are derived as before by first forming the auxiliary function through the use of Lagrangian multipliers:

$$\text{Max } L = u(x_t, z_t, \{I_t\}) + \lambda \left\{ E - \sum_{j=1}^{J} p_{jt} x_{jt} I_{jt} - z_t \right\} \qquad (13.2.4)$$

which differs from standard models due to the presence of the screening indicators of $\{I_t\}$. Here we introduce the subscript t to reflect the panel structure of the data where there are t observations per respondent. Associating first-order conditions with observed demand yields:

$$u_{jt} = p_{jt} \cdot u_z \qquad \text{if} \qquad x_{jt} > 0 \qquad\qquad (13.2.5)$$

$$u_{jt} < p_{jt} \cdot u_z \qquad \text{if} \qquad x_{jt} = 0 \text{ and } I_{jt} = 1 \qquad\qquad (13.2.6)$$

Taking logarithms of equations (13.2.5) and (13.2.6) and substituting from equations (13.2.2) and (13.2.3) we have:

$$\varepsilon_{jt} = \mathcal{g}_{jt} \qquad \text{if} \qquad x_{jt} > 0 \qquad\qquad (13.2.7)$$

$$\varepsilon_{jt} < \mathcal{g}_{jt} \qquad \text{if} \qquad x_{jt} = 0 \text{ and } I_{jt} = 1 \qquad\qquad (13.2.8)$$

where

$$\mathcal{g}_{jt} = -a_j\beta + \ln\left(p_{jt}\right) + \ln\left(\gamma x_{jt} + 1\right) - \ln\left(z_t\right)$$

Assuming that the errors (ε_{jt}) are distributed i.i.d. $N(0, \sigma^2)$, the probability of observed demand where R_t goods are chosen out of C_t goods considered can be expressed as:

$$\Pr(x_t) = \Pr(x_{n_1,t} > 0, \quad x_{n_2,t} = 0, \quad n_{1,t} = 1, \ldots, R_t, \quad n_{2,t} = R_t + 1, \ldots, C_t)$$

$$= |J_{R_t}| \int_{-\infty}^{g_{C_t}} \cdots \int_{-\infty}^{g_{R_t+1}} f(g_1, \cdots, g_{R_t}, \varepsilon_{,R_t+1}, \cdots, \varepsilon_{C_t}) d\varepsilon_{R_t+1}, \cdots, d\varepsilon_{C_t}$$

$$= |J_{R_t}| \left\{ \prod_{i=1}^{R_t} \phi\left(g_{it}/\sigma\right)/\sigma \right\} \left\{ \prod_{j=R_t+1}^{C_t} \Phi\left(g_{jt}/\sigma\right) \right\} \tag{13.2.9}$$

where $f(\cdot)$ is the joint density distribution for ε, $|J_{R_t}|$ is the Jacobian, ϕ is the Normal PDF and Φ is the Normal CDF. Considered alternatives are picked according to the indicator function in equation (13.2.1).

For alternatives that are not chosen but considered, there is a mass contribution to the likelihood, whereas for chosen options there is a density contribution. If an alternative is not considered, there is no realization of the error term and consequently no mass or density contribution to the likelihood.

Transforming from random-utility error (ε) to the likelihood of the observed data (x), we need to consider the Jacobian $|J_{R_t}|$:

$$\left|J_{R_t}\right| = \prod_{i=1}^{R_t} \left(\frac{\gamma}{\gamma x_{it} + 1}\right) \left\{ \sum_{i=1}^{R_t} \frac{\gamma x_{it} + 1}{\gamma} \cdot \frac{p_{it}}{z_t} + 1 \right\} \tag{13.2.10}$$

Our approach to selecting brands for inclusion in the model is to introduce binary indicator variables I_j into the utility function and budget constraint to eliminate those brands not considered by the consumer. The indicator variables are deterministic functions of the Bernoulli draws τ_{dh}. The hyperparameters of the corresponding heterogeneity distribution, δ_d indicates the proportion of respondents for whom attribute d is used to screen out the alternative from consideration.

$$\tau_{dh} \sim \text{Bernoulli}(\delta_d) \tag{13.2.11}$$

where δ_d captures the aggregate propensity of attribute d being used in the screening process. The average log-price threshold is represented by δ_p.

$$\ln \tau_p \sim N(\delta_p, \sigma_p) \tag{13.2.12}$$

In the analysis reported below, we estimate the model as a hierarchical Bayes model with minimally informative prior distributions using Markov chain Monte Carlo (MCMC) methods. A multivariate normal distribution of heterogeneity is employed:

$$\theta_h \sim N(\bar{\theta}, V_\theta) \tag{13.2.13}$$

where $\theta_h = (\beta_h, \ln \sigma_h, \ln \gamma_h, \ln E_h)$ is a respondent-specific vector of the parameters governing the volumetric demand model.

13.3 ESTIMATION

Estimation follows a standard hierarchical Bayesian procedure of sampling from the full conditional distribution of the model parameters. This allows one to estimate a unique set of parameters for each respondent. The MCMC algorithm in this case involves seven steps:

1. $\{\theta_h\} \mid \bar{\theta}, V_\theta$
2. $\bar{\theta}, \Sigma_\theta \mid \{\theta_h\}, \bar{\theta}_0, A_0^{-1}$
3. $V_\theta \mid \{\theta_h\}, v_0, V_0$
4. $\tau_{dh} \mid \delta_d$
5. $\tau_p \mid \delta_p, \sigma_p$
6. $\delta_d \mid \{\tau_{dh}\}, \alpha_0, \beta_0$
7. $\delta_p, \sigma_p \mid \{\tau_p\}, \mu_{\delta_p}, \sigma_{\delta_p}, a_{\delta_p}, b_{\sigma_p}$

The Monte Carlo estimator proceeds by generating draws from the conditional distributions in steps 1–7, using previously generated draws as conditioning arguments. Details of these draws are as follows:

1. For each respondent, make a draw for the θ_h from a Multivariate Normal distribution. Here, m_h^2 is a tuning parameter that should be chosen to achieve acceptance rates of 15–45%. The superscript $\cdot^{(r)}$ refers to the rth draw.

$$\theta_h^{CAN} = \theta_h^{(r)} + \zeta \tag{13.3.1}$$

where

$$\zeta \sim (0, m_h^2 \cdot V_\theta^{(r)})$$

We use the random-walk Metropolis–Hastings algorithm so the probability of accepting the new draw is equal to:

$$\alpha_\theta = \min\left(1, \frac{l_h^{CAN} \exp\left[-\frac{1}{2}(\theta_h^{CAN} - \bar{\theta}^{(r)})'(V_\theta^{(r)})^{-1}(\theta_h^{CAN} - \bar{\theta}^{(r)})\right]}{l_h^{(r)} \exp\left[-\frac{1}{2}(\theta_h^r - \bar{\theta}^{(r)})'(V_\theta^{(r)})^{-1}(\theta_h^r - \bar{\theta}^{(r)})\right]}\right) \tag{13.3.2}$$

Here, l_h is likelihood from respondent h evaluated as the product of equations (13.2.9) and (13.2.10) across purchases.

2. $\bar{\theta}$ is drawn directly from its full conditional distribution using Gibbs sampling:

$$\bar{\theta}^{(r+1)} \sim N(\tilde{\theta}, \tilde{\Sigma}_\beta) \tag{13.3.3}$$

where

$$\tilde{\theta} = \left[A_0 + \left(\frac{V_\theta^{(r)}}{H} \right)^{-1} \right]^{-1} \left[\left(V_\theta^{(r)} \right)^{-1} \sum_{h=1}^{H} \theta_h^{(r+1)} + A_0 \bar{\theta}_0 \right]$$

$$\tilde{\Sigma}_\theta = \left[A_0 + \left(\frac{V_\theta^{(r)}}{H} \right)^{-1} \right]^{-1}$$

We assume $\bar{\theta}_0 = 0$ and $A_0 = 0.01I$ in our empirical application.

3. V_θ is drawn from its full conditional distribution:

$$V_\theta^{(r+1)} \sim \text{IW} \left(v_0 + H, V_0 + \sum_{h=1}^{H} \left(\theta_h^{(r+1)} - \bar{\theta}^{(r+1)} \right)^T \left(\theta_h^{(r+1)} - \bar{\theta}^{(r+1)} \right) \right) \quad (13.3.4)$$

We use $v_0 = 14$ and $V_0 = 14I$ in our empirical application.

4. τ_{dh} is drawn as follows:

$$I_{dh} = \begin{cases} 1 & \text{if } \sum_{t=1}^{T} \sum_{j=1}^{J} A_{j,d,t} \cdot x_{h,j,t} = 0 \\ 0 & \text{otherwise} \end{cases} \quad (13.3.5)$$

If I_{dh} is zero, it means that this respondent selected offering j at least once during the task period ($X_{h,j,t} \geq 1$) so that the respondent does not use this discrete variable as a screening criterion. That is, τ_{dh} cannot be equal to one, and therefore $\tau_{dh}^{(r+1)}$ is zero in this case. If I_{dh} is one, it means that respondent h does not select the alternatives when discrete variable d is present. In that case, τ_{dh} is 1 at the probability of α_τ, and 0 under the probability of $1 - \alpha_\tau$.

$$\alpha_\tau = \frac{l_h^{\tau_{dh}=1} \cdot \delta^{(r)}}{l_h^{\tau_{dh}=1} \cdot \delta^{(r)} + l_h^{\tau_{dh}=0} \cdot (1 - \delta^{(r)})} \quad (13.3.6)$$

Here, $l_h^{\tau_{dh}=1}$ is respondent h's likelihood value under $\tau_{dh} = 1$. $l_h^{\tau_{dh}=0}$ is respondent h's likelihood value under $\tau_{dh} = 0$. The likelihood is the product of equations (13.2.9) and (13.2.10) across purchases by the respondent.

5. For each respondent, a candidate for τ_p is drawn from a Normal distribution. Here, m_{ph}^2 is a tuning parameter that should be chosen to achieve acceptance rates of 15–45%. The superscript $\cdot^{(r)}$ refers to the rth draw.

$$\tau_{ph}^{CAN} = \tau_{ph}^{(r)} + \zeta \quad (13.3.7)$$

where

$$\zeta \sim (0, m_{ph}^2)$$

The draw is accepted with probability

$$\alpha_{\tau_p} = \begin{cases} \widehat{\alpha}_{\tau_p} & \text{if } \tau_p > p_{\max} \\ 0 & \text{otherwise} \end{cases} \tag{13.3.8}$$

where

$$p_{\max h} = \max \left\{ p_{hjt}, t = 1, \dots, T \right\} \forall j \in \{x_{hjt} > 0\} \tag{13.3.9}$$

is the highest price subject h has ever paid and

$$\widehat{\alpha}_{\tau_p} = \min \left(1, \frac{l_h^{CAN} \exp\left[-\frac{1}{2\sigma_p^{(r)}} (\tau_{ph}^{CAN} - \delta_p^{(r)})'(\tau_{ph}^{CAN} - \delta_p^{(r)}) \right]}{l_h^{(r)} \exp\left[-\frac{1}{2\sigma_p^{(r)}} (\tau_{ph}^{(r)} - \delta_p^{(r)})'(\tau_{ph}^{(r)} - \delta_p^{(r)}) \right]} \right) \tag{13.3.10}$$

6. δ_d draws as:

$$\delta_d \sim B\left(\sum_{h=1}^{H} \tau_{dh}^{(r+1)} + \alpha_0, H - \sum_{h=1}^{H} \tau_{dh}^{(r+1)} + \beta_0 \right) \tag{13.3.11}$$

Here, H is the total number of respondents, and B refers to the Beta distribution. We assumed $\alpha_0 = 1$, $\beta_0 = 1$ in our empirical application.

7. $\ln \delta_p$ and σ_p are drawn using Gibbs sampling:

$$\ln \delta_p \sim N\left(\frac{v_{\tau_p} \mu_{\tau_p} + H \bar{\tau}_p}{v_{\tau_p} + H}, \sqrt{v_{\tau_p} + \sigma_p^2 / H} \right) \tag{13.3.12}$$

where we assume $\mu_{\tau_p} = 0$ and $v_{\tau_p} = .01$ in our empirical application.

$$\sigma_p^2 \sim IG\left(\frac{a_0 + H}{2}, b_0 + \frac{\sum (\bar{\tau}_p - \tau_p)^2}{2} \right) \tag{13.3.13}$$

and we assume $a_0 = 3$ and $b_0 = 3$.

13.4 EMPIRICAL ANALYSIS

We apply the proposed model to two conjoint datasets that were collected from national panels. The first dataset is for purchases in the ice cream category and includes both discrete and volumetric responses, allowing us to compare the effect of screening on partworth estimates for the same set of individuals. The second dataset is for purchases in the frozen pizza category where we find that consumers purchase a greater variety of offerings. The frozen pizza purchases are for individuals who hold a store loyalty reward card and this data is more fully described in Allenby et al. [2019b], where partworth estimates from conjoint data are shown to agree with estimates based on actual store demand data.

13.4.1 Ice Cream

Individuals were qualified for inclusion in the survey if they routinely shopped for groceries or materially participated in making grocery decisions, including ice cream. Respondents were asked about their preference and demand for ice-cream varieties using a dual-response choice task in which they were first asked to indicate their preferred offering and then asked to indicate demand quantities for the array of offerings in the choice task. An example choice task is shown in Figure 13.2. Respondents were asked to provide responses to 12 choice tasks and the sample size is 490 respondents.

Table 13.2 displays the set of brands and flavors included in our analysis. Three package sizes are included – 4, 8, and 16 servings, which are included as product attributes in both

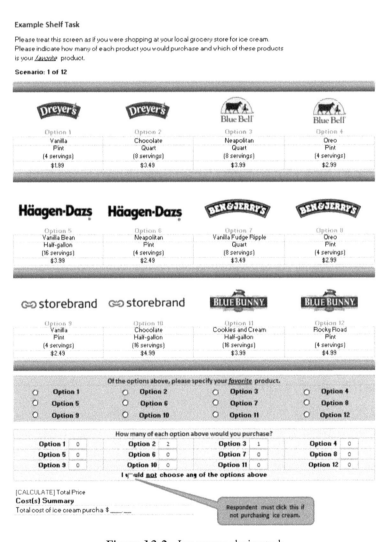

Figure 13.2 Ice-cream choice task

Table 13.2   Ice cream attributes

Attribute	Levels
Brand	Ben & Jerry's (BenNJerry), Blue Bell (BlueBell), Blue Bunny (BlueBunny), Breyers, Dryers, Haagen-Dazs, Store
Flavor	Chocolate, Chocolate Chip (ChocChip), Chocolate Dough (ChocDough), Cookies and Cream (CookieCream), Neapolitan, Oreo, Rocky Road (RockyRoad), Vanilla, Vanilla Bean (VanillaBean), Vanilla Fudge (VanillaFudge)
Size	4, 8, 16 servings

Table 13.3   Ice-cream model comparison

		Discrete choice		Volumetric	
		W/o screening	With screening	W/o screening	With screening
In sample	LMD	−6564	−5825	−17,135	−16,139
Holdout	MSE	0.049	0.049	1.259	1.180
	MAE	0.107	0.078	0.376	0.368
	Hit-probability	0.405	0.634		

the discrete choice and volumetric analysis. Demand quantities for the volumetric model, x_j, are measured in multiples of 4 servings. Prices vary between \$1.99 and \$4.99 per unit.

Table 13.3 compares model fit for the discrete and volumetric models, with and without screening. It is important to remember that the discrete choice data contains only corner solutions, while the volumetric response data contains for corner and interior solutions, where the data likelihood is a combination of mass (corners) and density (interior) contributions. Thus, it is not possible to directly compare the measure of model fit of the discrete choice model to the volumetric model, although a comparison within model (screening vs no screening) is appropriate. The measure of model fit indicates improvement for both models with screening, consistent with previous research on the effect of consideration sets on choice.

Parameter estimates for the discrete-choice (via standard hierarchical Logit model) and volumetric Ice-cream data are reported in Table 13.4. Reported are the means of the random-effects distribution $\bar{\theta}$ and δ_d and the posterior standard deviation of the mean.

Parameter estimates for the models without screening are plotted in Figure 13.3. The pattern of parameter estimates show agreement between the discrete and volumetric data, as expected. An exception is the package size attribute, which measures both the demand quantity and package size preference in the discrete choice model but only package size preference in the volumetric model. In addition, we find that in both models the magnitude of the coefficients is smaller for most attributes when screening is included.

The difference in the partworth estimates in models with and without screening is related to the probability that the product feature is used to screen to form the

Table 13.4   Ice cream – parameter estimates

	Discrete choice			Volumetric		
	W/o screening	With screening		W/o screening	With screening	
	$\bar{\theta}$	$\bar{\theta}$	δ	$\bar{\theta}$	$\bar{\theta}$	δ
β_0	0.03 (0.21)	**1.39** (0.25)		**−3.38** (0.10)	**−2.38** (0.12)	
Brand						
~BlueBell	**−1.91** (0.28)	**−1.54** (0.50)	0.16 (0.09)	**−0.76** (0.10)	**−0.40** (0.13)	0.34 (0.04)
~BlueBunny	**−1.43** (0.22)	**−1.21** (0.34)	0.17 (0.08)	**−0.72** (0.09)	**−0.49** (0.12)	0.29 (0.05)
~Breyers	**−0.26** (0.14)	−0.36 (0.20)	0.07 (0.04)	−0.02 (0.07)	−0.02 (0.10)	0.12 (0.04)
~Dryers	**−1.15** (0.18)	**−0.75** (0.26)	0.27 (0.06)	**−0.79** (0.09)	**−0.51** (0.11)	0.30 (0.04)
~HaagenDa	**−0.63** (0.15)	**−0.77** (0.19)	0.07 (0.04)	**−0.41** (0.07)	**−0.58** (0.11)	0.07 (0.04)
~Store	**−1.00** (0.18)	**−0.45** (0.24)	0.29 (0.05)	**−0.46** (0.08)	−0.12 (0.10)	0.30 (0.04)
~BenNJerry			0.19 (0.04)			0.20 (0.03)
Flavor						
~ChocChip	**−0.62** (0.16)	**−0.75** (0.25)	0.17 (0.07)	**−0.45** (0.08)	**−0.55** (0.11)	0.24 (0.06)
~ChocDough	**−0.47** (0.18)	0.35 (0.21)	0.40 (0.04)	**−0.47** (0.10)	**−0.21** (0.13)	0.37 (0.04)
~CookieCream	**−0.60** (0.16)	−0.26 (0.22)	0.33 (0.05)	**−0.39** (0.08)	**−0.50** (0.11)	0.29 (0.05)
~Neapolitan	**−1.21** (0.20)	**−0.69** (0.24)	0.45 (0.06)	**−0.64** (0.09)	**−0.47** (0.11)	0.47 (0.04)
~Oreo	**−0.55** (0.15)	−0.38 (0.21)	0.34 (0.05)	**−0.48** (0.08)	**−0.52** (0.10)	0.33 (0.05)
~RockyRoad	−0.22 (0.15)	**0.47** (0.14)	0.44 (0.03)	**−0.32** (0.08)	−0.04 (0.10)	0.42 (0.03)
~Vanilla	−0.24 (0.14)	−0.15 (0.18)	0.27 (0.06)	−0.14 (0.07)	**−0.39** (0.10)	0.21 (0.05)
~VanillaBean	**0.20** (0.11)	**0.29** (0.15)	0.31 (0.04)	0.03 (0.06)	**−0.29** (0.08)	0.21 (0.04)
~VanillaFudge	−0.04 (0.12)	−0.05 (0.16)	0.29 (0.05)	**−0.23** (0.07)	**−0.19** (0.08)	0.36 (0.04)
~Chocolate			0.40 (0.04)			0.45 (0.03)
Size						
~8	**0.38** (0.07)	**0.29** (0.08)	0.02 (0.03)	**−0.16** (0.04)	**−0.21** (0.05)	0.01 (0.02)
~16	**1.13** (0.09)	**1.07** (0.10)	0.01 (0.02)	**−0.19** (0.05)	**−0.16** (0.06)	0.01 (0.02)
~4			0.07 (0.03)			0.04 (0.03)
ln σ				−0.02 (0.04)	**0.19** (0.04)	
ln γ				**−2.24** (0.07)	**−1.94** (0.09)	
ln E				**2.32** (0.03)	**2.23** (0.03)	
ln β_p	**−1.27** (0.11)	**−1.26** (0.11)				
ln δ_p				4.97 (0.51)		3.07 (0.39)
σ_p				1.90 (0.38)		1.53 (0.23)

Boldfaced θ signify that the 95% posterior credible interval of the estimate does not include zero. Standard deviations printed in parentheses.

consideration set. Estimates of δ_d close to zero, indicating no screening on an attribute, is associated with more similar part-worth estimates than if the estimate of δ_d is large. Thus, inferences about the importance of the partworths are affected by whether or not the model allows for product screening.

We find that 49% of the choice options are screened out of each choice set using the discrete choice model and 54% of the choice options are screened out using the volumetric model. Figure 13.4 shows the distribution of average consideration set sizes across the 490 respondents. Volumetric data is more informative with respect to teasing apart consideration and preference, and therefore it allows for higher screening probabilities and smaller consideration set sizes for some respondents.

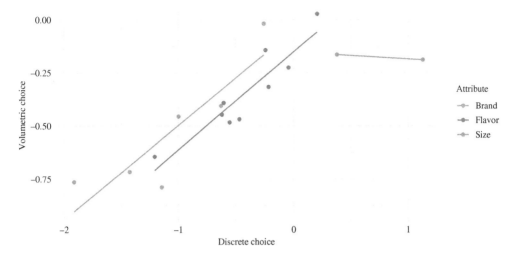

Figure 13.3   Ice cream – discrete choice vs volumetric choice "partworths"

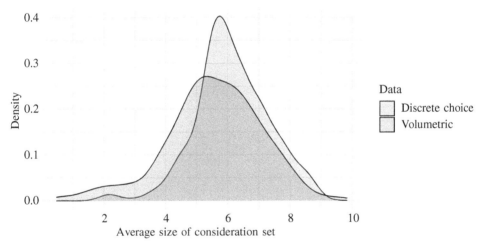

Figure 13.4   Distribution of average consideration set size (ice-cream data)

We find that ignoring screening results in an overprediction of demand as previously documented in the discrete choice literature (Pachali et al. [2020]), and that predicted differences in total volumetric demand between models with and without screening is minimal. Although screening eliminates people from the expected sales forecasts, the contribution of purchase volume from the remaining consumers offset this loss depending on the magnitude of their price sensitivities. The demand forecasts indicate that the effect of screening rules on a price promotion is (i) to constrain the incidence of purchase, while (ii) demand for respondents with positive demand increases more than predicted by models without screening. Thus, the effect of screening leads to an increase in the volumetric response among respondents with positive demand. Additional details are discussed in Kim et al. [2022].

13.4.2 Frozen Pizza

The Frozen Pizza dataset consists of responses from 181 consumer households who were randomly selected from card holders at a major grocery who purchased at least five frozen pizzas in the past two years. Respondents were asked to complete 12 choice tasks. One choice task was reserved for predictive testing, and the remaining 11 tasks were used for model calibration. In each task, respondents choose how many units of each of the six product alternatives they would purchase the next time they are looking to buy frozen pizza. Figure 13.5 displays an example choice task.

Table 13.5 displays the attributes and levels studied. Price levels were chosen in collaboration with the sponsoring grocer to mimic the typical price range used in stores. Three basic price levels ($1, $3, $4) were used and adjusted by brand, cheese quality and topping density premiums.

On average, respondents choose 1.7 different products with an average quantity of 2 each. The pizza data therefore exhibits greater horizontal demand than found in the

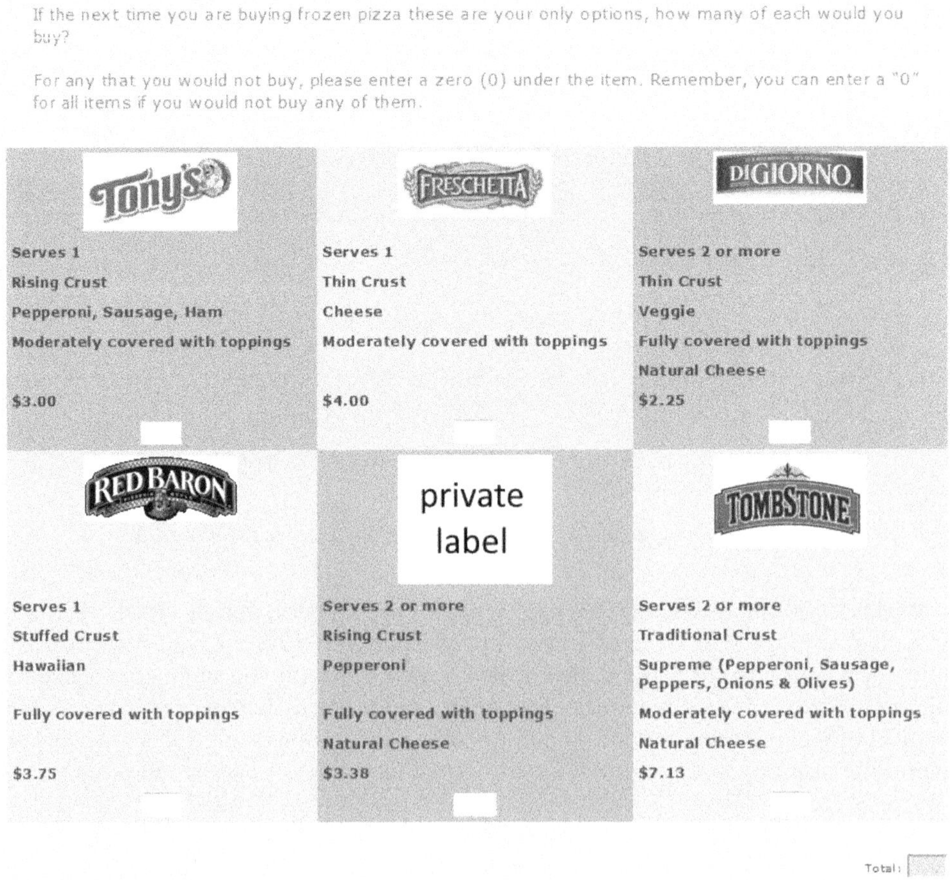

Figure 13.5   Frozen pizza choice task

Table 13.5 Frozen pizza attributes

Attribute	Levels
Brand	DiGiorno (DiGi), Frescetta (Fresc), Private Label (Priv)
	Red Baron (RedBa), Tombstone (Tomb), Tony's (Tony)
Size	Serves one (forOne), Serves two (ForTwo)
Crust	Thin, Rising (RisCr), Stuffed (StufCr), Traditional (TrCr)
Topping type	Pepperoni, Cheese, Hawaii (HI), Vegetarian (Veg),
	Pepperoni/Sausage/Ham (PepSauHam), Surpreme (Surp)
Topping density	Dense (densetop), Moderate (ModCover)
Cheese	No claim (NoInfo), Real cheese claim (realcheese)

Table 13.6 Frozen pizza volumetric model comparison

		Without screening	With screening
In sample	LMD	−8386	−7935
Holdout	MSE	1.40	1.27
	MAE	0.59	0.56

ice cream data where just one brand was selected on average. Overall, demand is characterized by more interior solutions compared with the ice cream data and a desire for variety and demand for multiple items.

Table 13.6 reports on the fit of the volumetric model with and without screening. The inclusion of a screening rule improves the fit of the model to the in-sample and holdout data.

Parameter estimates are reported in Table 13.7. We find that the intercept is smaller in the model with screening (−1.99 vs −2.61) because some of the nonchoices are screened out of the consideration set and not rationalized in terms of having lower utility. Brand preferences and other partworth estimates are also less negative and differ most from the model without screening when the screening parameter estimate for the attribute (δ) is larger. Satiation (γ) is understated when screening is ignored (−0.46 vs −0.06). We find that, on average, 39% of the choice options are screened out of each choice set.

Explicitly modeling consideration has implications for aggregate demand forecasts. The proposed model with the screening rule allows marketers to evaluate market demand by not including alternatives that would not be considered by consumers. To illustrate the effects of screening, we consider a small submarket of vegetarian pizza options across all brands and consider traditional crust vegetarian pizza with standard topping density and no "real cheese" attribute. The configuration of this case is displayed in Table 13.8.

Demand curves for the vegetarian pizzas are provided in Figure 13.6. The horizontal axis of the curves is the percentage discount from the regular price. The vertical axis is volume demanded. Predictions for the model with screening rules are given for three groups of respondents broken down by the probability of respondent screening. Indi-

Table 13.7 Parameter estimates – pizza data

	W/o screening $\bar{\theta}$		With screening $\bar{\theta}$		δ	
β_0	**−2.59**	(0.12)	**−1.95**	(0.13)		
Brand						
~Fresc	**−0.34**	(0.10)	−0.16	(0.11)	0.10	(0.03)
~Priv	**−0.69**	(0.10)	**−0.50**	(0.12)	0.15	(0.04)
~RedBa	**−0.60**	(0.10)	**−0.43**	(0.12)	0.13	(0.04)
~Tomb	**−0.67**	(0.10)	**−0.40**	(0.11)	0.17	(0.05)
~Tony	**−1.06**	(0.10)	**−0.73**	(0.12)	0.27	(0.06)
~DiGi					0.03	(0.02)
Cheese						
~realCheese	**0.12**	(0.05)	**0.14**	(0.06)	0.01	(0.02)
~NoInfo					0.01	(0.02)
Coverage						
~ModCover	−0.06	(0.05)	−0.09	(0.06)	0.01	(0.02)
~densetop					0.01	(0.02)
Crust						
~RisCr	0.03	(0.07)	0.04	(0.08)	0.02	(0.02)
~StufCr	−0.05	(0.07)	−0.02	(0.08)	0.05	(0.03)
~TrCr	0.10	(0.07)	0.11	(0.07)	0.02	(0.02)
~Thin					0.04	(0.03)
Size						
~ForTwo	**0.61**	(0.07)	**0.69**	(0.07)	0.01	(0.02)
~ForOne					0.04	(0.02)
Topping						
~Cheese	**−0.42**	(0.09)	**−0.42**	(0.09)	0.10	(0.03)
~HI	**−0.93**	(0.11)	**−0.28**	(0.12)	0.38	(0.06)
~PepSauHam	**−0.20**	(0.08)	−0.02	(0.09)	0.13	(0.04)
~Surp	**−0.31**	(0.09)	−0.03	(0.09)	0.20	(0.04)
~Veg	**−0.86**	(0.10)	**−0.60**	(0.11)	0.28	(0.06)
~Pepperoni					0.10	(0.03)
$\ln \sigma$	**−0.26**	(0.05)	**−0.10**	(0.06)		
$\ln \gamma$	**−0.46**	(0.07)	−0.06	(0.08)		
$\ln E$	**3.62**	(0.07)	**3.47**	(0.07)		
$\ln \delta_p$					2.41	(0.60)
σ_p					0.66	(0.31)

Table 13.8 Frozen pizza scenario for demand forecasts*

Alternative	Price	Brand	Size	Crust	Topping	Coverage	Cheese
1	3.50	DiGi	forOne	Thin	Veg	ModCover	NoInfo
2	3.00	Fresc	forOne	Thin	Veg	ModCover	NoInfo
3	2.00	Priv	forOne	Thin	Veg	ModCover	NoInfo
4	2.00	RedBa	forOne	Thin	Veg	ModCover	NoInfo
5	2.00	Tomb	forOne	Thin	Veg	ModCover	NoInfo
6	1.50	Tony	forOne	Thin	Veg	ModCover	NoInfo

*Abbreviations are described in Table 13.5.
Boldfaced θ signify that the 95% posterior credible interval of the estimate does not include zero. Standard deviations printed in parentheses.

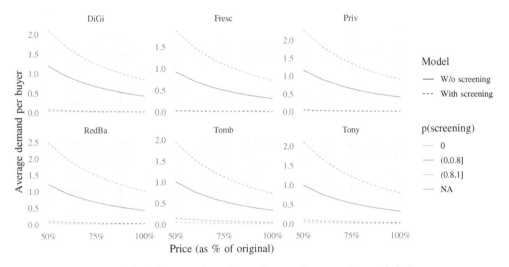

Figure 13.6 Frozen pizza demand curves by screening probability

viduals in the first group do not screen while those in second and third groups screen out the vegetarian pizza with probabilities $(0 < \tau < 0.8]$ and $(0.8 < \tau < 1.0]$, respectively. The curves show that demand predictions for individuals who do not screen out vegetarian toppings are higher, implying that models without screening rules undervalue these individuals in targeting and penetration analysis. The screening parameters are useful for identifying responsive individuals (i.e., prospects) willing to consider changes in the products they purchase.

13.5 DISCUSSION

Our model for consideration set formation identifies individuals with the interest and ability to make purchases in a product category. Some offerings are of no interest to some individuals and should be screened out of analysis and demand predictions. Our model provides a flexible format for exploring different counterfactual scenarios where products,

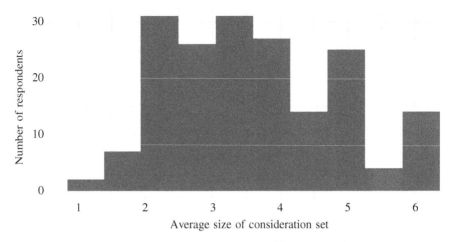

Figure 13.7   Frozen pizza consideration set size

or the object of analysis, are selected in a manner similar to models of variable selection and data partitioning.

Figure 13.7 illustrates the degree of heterogeneity in consideration set size in the frozen pizza data. A minority of respondents consider all six choice alternatives in every choice task, while most respondents only consider about half of the presented alternatives. Thus screening of some brands is commonly encountered. The decision to not purchase a product can be driven by a lack of interest in the product itself (i.e., lack of utility) or by the product not providing sufficient utility for its price. Consumer lack of interest is captured through the conjunctive screening rule and implies that there does not exist a price at which a product can become sufficiently attractive. Determining the reason for zero demand (i.e., low utility or screening) is important in positioning a brand and identifying its competition. It is also important for advertising and pricing decision that attempt to make the product attractive to individuals.

The proposed model allows for calculation of an inclusion probability due to the presence of specific attributes, distinguishing zero demand driven by non-consideration vs lack of preference. Figure 13.8 provides the summary for the two sources of zero demand for each alternative. Using the proposed conjunctive model, we compute the expected conditional probability of observing zero demand given that the product was considered (light grey) and the probability that the product was not even considered (dark grey). The corresponding bars are labeled "nopreference" and "screened." The share of corners based on screening varies between 37% for DiGiorno and 53% for Tony's pizza. Thus, a large proportion of observed zero demand is due to consumer screening and general lack of interest in the offering. The market size for each brand is therefore smaller than the number of respondents in the survey, with some brands and varieties competing more intensively depending on the patterns of screening co-occurrence.

While there are significant differences in purchase incidence and demand conditional on screening probability, we have shown that the effects on overall demand are similar. This is because consumer consideration and preference for an offering have similar, but not identical effects on demand. Respondents who have selected a vegetarian option at some point in their data history have no chance of excluding a vegetarian option from

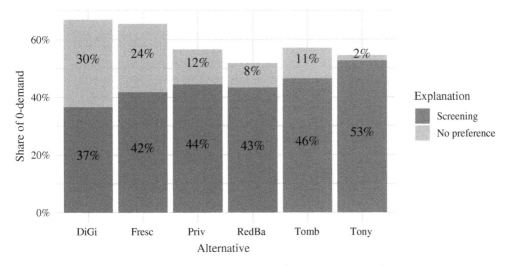

Figure 13.8   Predicting zero demand – screening vs preference

consideration. However, for those who have never selected a vegetarian option there are two possibilities: the vegetarian option may have been screened out, or it may considered but with low probability.

Figure 13.9 shows the relationship between consideration (τ) and preference (β) for our model. On vertical axis we plot the individual-level estimates of τ_{veg} based on the

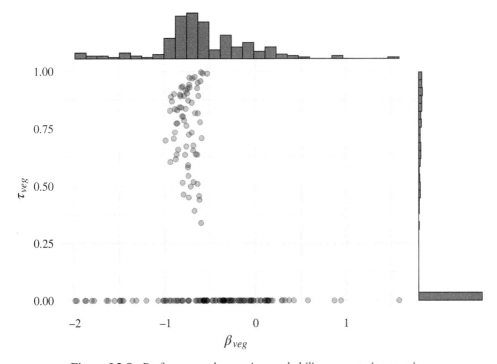

Figure 13.9   Preference and screening probability – vegetarian topping

draws of 1s and 0s over the MCMC iterations. Individuals with no screening ($\tau_{veg} = 0$) are plotted at the bottom and those with higher levels of screening appear toward the top of the plot. The horizontal axis is the mean of the individual-level partworth estimate for the vegetarian topping (β_{veg}). We find that respondents with high screening probabilities tend to have β_{veg} close to the mode, indicating that individual-level information about preferences is not available when screening is present. More extreme estimates of β_{veg} occur when screening is not present. Thus, nonpurchase is not rationalized as screening when there exists sufficient individual-level information to estimate preference.

13.6 *USING THE CODE*

The R package "echoice2" has been developed by Nino Hardt for estimating volumetric demand models including conjunctive screening rules. The package includes a vignette for the analysis of the Pizza data described above.[1]

13.7 *CONCLUDING REMARKS*

The importance of choice sets and screening rules has long been recognized in the literature. Previous models of screening and consideration zero-out the choice probability for alternatives not considered, whereas our model assumes that neither a consumer's utility nor their budget is affected by goods not considered. Our model nests the discrete choice model as a special case, and is shown to fit our packaged goods data with corner and interior solutions better than a model without screening. Heterogeneity in the screening parameters also indicates that the consideration sets are different across consumers. Distinguishing true corner solutions from non-consideration reveals that the rate of satiation is higher than what can be found in models without screening. We find that models of volumetric demand indicate higher rates of screening, smaller considerations sets and heightened demand among respondents not screening when compared to models of discrete choice.

---

[1] It is available on CRAN: https://cran.r-project.org/package=echoice2.

14

Case Study 5: Approximate Bayes and Personalized Pricing

Abstract

Approximate Bayes methods have become increasingly popular in recent years given the need for scalability and speed in business applications. This case study demonstrates the use of such methods in the context of personalized pricing.

With the wide availability of granular customer data there has been a push toward the scaling of decisions to the individual level so as to provide personalized marketing offerings. The core idea in this endeavor is to estimate heterogeneous effects using cross-sectional data and then use these estimates to customize marketing mix elements. Typically, the challenge lies in the high-dimensionality of the covariate set as well as the number of individuals which can be addressed using ML tools. In this case study, we examine the issue of personalization of pricing decisions using a approximate Bayes implementation of binary Logit model with L1 penalization.

14.1 HETEROGENEITY AND HETEROGENEOUS TREATMENT EFFECTS

The building block of any personalization scheme is heterogeneity. Typically, we think of heterogeneity in terms of the differences in how customers react to a particular marketing treatment. For example, different customers react differently to the same price or to the same advertising message. Effective personalization requires not only that customers exhibit such heterogeneity in responsiveness but also that the firm be able to offer differentiated treatments targeting this heterogeneity. To do so, the firm needs to have a predictive model via which a prospective or returning customer's responsiveness can be ascertained for a given treatment. These effects are commonly referred to as heterogenous treatment effects (HTE) and involve the projection of an individual's preference parameters onto a (often high-dimensional) set of observable characteristics. A key point of distinction between the HTE approach and random coefficients is that we ignore the possibility of unobservables in the HTE approach. The argument is that given the vast array of descriptors available (past purchases and interactions, demographics, device data, third part behavioral data, etc.) at the individual level the need for unobservables is obviated. This of course is an empirical question that we will not debate here. In what follows, we take a stylized look at the approach outlined by Dubé and Misra [2023].

14.2 THE FRAMEWORK

Much of what is presented here will follow Dubé and Misra [2023] and the discussion of their work in Chapter 8. Assume that a prospective customer i with observable features x_i obtains the following incremental utility from purchasing vs not purchasing

$$\Delta U_i = \alpha_i + \beta_i p_i + \varepsilon_i$$
$$= \alpha\left(x_i; \theta_\alpha\right) + \beta\left(x_i; \theta_\beta\right) p_i + \varepsilon_i$$

where $\alpha\left(x_i; \theta_\alpha\right)$ is an intercept and $\beta\left(x_i; \theta_\beta\right)$ is a slope associated with the price, p_i. In our implementation we will assume[1] that

$$\alpha\left(x_i; \theta_\alpha\right) = x_i' \theta_\alpha$$
$$\beta\left(x_i; \theta_\beta\right) = x_i' \theta_\beta. \qquad (14.2.1)$$

To complete their model, the authors make the usual assumption that the random utility error ε_i is distributed i.i.d. Logistic and obtain:

$$\mathbb{P}\left(y_i = 1 | p_i, x_i; \Theta\right) = \frac{\exp\left(\alpha\left(x_i; \theta_\alpha\right) + \beta\left(x_i; \theta_\beta\right) p_i\right)}{1 + \exp\left(\alpha\left(x_i; \theta_\alpha\right) + \beta\left(x_i; \theta_\beta\right) p_i\right)}, \qquad (14.2.2)$$

where we use the notation Θ to denote all parameter and Θ_j to denote a particular element contained in it.

---

[1] Note that changing the specification to deep neural nets or other flexible ML frameworks would follow very similar logic.

The log-likelihood follows immediately and has the usual structure with

$$\ell\left(\mathbf{D}_i|\Theta\right) = \left[y_i \ln \mathbb{P}\left(y_i = 1|p_i, x_i; \Theta\right) + \left(1 - y_i\right) \ln \left(1 - \mathbb{P}\left(y_i = 1|p_i, x_i; \Theta\right)\right)\right] \quad (14.2.3)$$

and

$$\ell = \sum_{i=1}^{N} \ell\left(\mathbf{D}_i|\Theta\right) \quad (14.2.4)$$

14.2.1 Introducing the ML Element

Recall that the maximum a posteriori (MAP) estimator (or the penalized objective) can be written as

$$\sum_{i=1}^{N} \ell\left(\mathbf{D}_i|\Theta\right) - \lambda \sum_{j=1}^{J} \left|\Theta_j\right| \quad (14.2.4\text{a})$$

where the (L1) penalty term is $\lambda \sum_{j=1}^{J} \left|\Theta_j\right|$. As we discussed elsewhere in this text, this can be interpreted as a Laplace prior on the parameters. Here, however, we treat it as part of the model and to complete this model, we need to specify λ. We adopt a Bayesian approach in the spirit of Newton et al. [2021] and allow for a hyper prior on $\lambda \sim F_\lambda$ where the prior distribution F_λ is specified by the researcher. Consequently, our Bayesian Bootstrap (BB) estimator[2] is then described as

$$\hat{\Theta}^b = \underset{\Theta \in \mathbb{R}^J}{\arg\max} \left\{ \sum_{i=1}^{N} w_i^{(b)} \ell\left(D_i|\Theta\right) - \lambda^{(b)} \sum_{j=1}^{J} |\Theta_j| \right\}. \quad (14.2.5)$$

The draws $\hat{\Theta}^b$ obtained from the BB procedure can be treated as if they were draws from the posterior of Θ. As such, they can be used to compute posterior quantities and to ascertain optimal marketing quantities and decisions in the usual way.

14.3 CONTEXT AND DATA

The context of the application is that of personalized pricing. In their paper, Dubé and Misra [2023] work with a digital B2B that offers a subscription service. The customers are mostly small to medium-sized businesses looking for candidates for a job that they have in mind. The firm provides customers matched resumes for the job opening described

---

[2] Our estimator is more accurately described as a weighted likelihood bootstrap (WLB) as coined by Newton and Raftery [1994]. We use the Rubin's terminology "Bayesian Bootstrap" (BB) here since it is a more general term that encompasses the WLB as well.

```
> table(datz$P)

0.019 0.039 0.059 0.079 0.099 0.119 0.139 0.159 0.179 0.199
  301   294   293   319   354   301   295   352   331   323
0.219 0.239 0.259 0.279 0.299 0.319 0.339 0.359 0.379 0.399
  330   328   316   314   306   305   330   308   304   358
0.419 0.439 0.459 0.479 0.499
  322   355   322   332   307
```

Figure 14.1 Range and distribution of prices

based on proprietary algorithms. Upon arriving at the website, customer input some data about themselves (demographics) and upload a job description which is then quantized into a vector of attributes. These elements together make up the characteristics of the customer. The authors also obtain data by running a randomized experiment where each incoming customer is randomly shown a price and their purchase decision recorded. The goal of the exercise is to use such data to compute heterogeneous treatment effects for price and then construct a personalized pricing scheme for the firm.

The dataset we use below is a simulated analog of the data used in Dubé and Misra [2023]. The original data is not available to us for privacy reasons but the authors make available code to simulate data which is close to the true dataset. We have modified that code a bit but the main features including the number of covariates, the range of prices and parameters are in line with their original code and data. The code to simulate the data is available in the script file mksimdata.R but we will not go into details of that here for the sake of brevity and because the code is self-explanatory.

There are two files that are created by mksimdata.R. The first is a file called estim-data.Rdata and it contains the data we will use for estimation. Loading this file into **R** reveals a dataframe object called datz that contains three components: a binary outcome (Y), a price variable (P), and a set of 133 customer characteristics $(\mathbf{X} = X_1, \ldots, X_{133})$. The other file, trueprefs.Rdata, contains the primitives of the data generating process including the true individual $\{\alpha(x_i), \beta(x_i)\}_{i=1}^{N}$ (atrue,btrue), the parameters $\{\theta_\alpha, \theta_\beta\}$ used to construct them named (thetatrue), and price elasticities at the individual level evaluated at prices in the data named (elastrue). We will use these quantities to verify the validity and accuracy of our estimation procedures.

The prices (similar to the actual application) range between \$19 and \$499. We have scaled them to be in thousands for convenience.

As we can see (Figure 14.1) there are approximately 300–350 observations in each price cell. We can also visualize the demand curve by computing the average subscription rate at each price level (Figure 14.2). The data seem to suggest a relatively inelastic demand curve.

14.4 DOES THE BAYESIAN BOOTSTRAP WORK?

Before moving on to the main application we conduct some simple tests to verify that the Bayesian Bootstrap (BB) gives us reasonable approximations to the posterior quantities we care about. To do so we will run a simple (nonheterogeneous) logistic

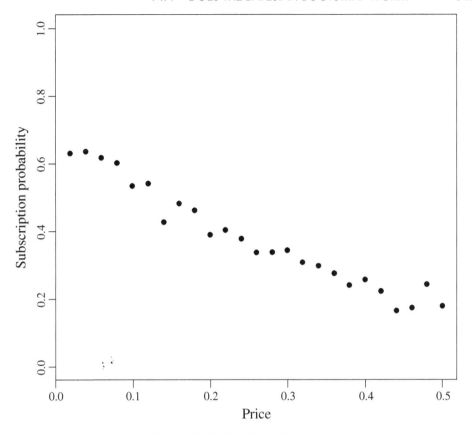

Figure 14.2 The demand curve

regression using MCMC methods as well as BB ideas. To start, we simplify our model so that,

$$\alpha\left(x_i; \theta_\alpha\right) = \alpha$$
$$\beta\left(x_i; \theta_\beta\right) = \beta.$$

Then the model is the usual homogenous binary logit model,

$$\mathbb{P}\left(y_i = 1 | p_i, x_i; \Theta\right) = \frac{\exp\left(\alpha + \beta p_i\right)}{1 + \exp\left(\alpha + \beta p_i\right)}. \qquad (14.4.1)$$

R   We use the *bayesm* package `rmnlIndepMetrop` to run an MCMC sampler obtaining 1000 draws using default priors (Figure 14.3).

The equivalent Bayesian Bootstrap procedure is relatively straightforward. Recall from earlier that

$$\ell_i = \left[y_i \ln \mathbb{P}\left(y_i = 1 | p_i, x_i; \Theta\right) + \left(1 - y_i\right) \ln\left(1 - \mathbb{P}\left(y_i = 1 | p_i, x_i; \Theta\right)\right)\right]. \qquad (14.4.2)$$

```
# MCMC for Homogenous Logit
library(bayesm)
xx = createX(p=2, 1, 0, cbind(0,as.matrix(datz$P)), NULL,INT = TRUE, base=1)
yy = as.numeric(datz$Y+1)
bdat = list(y=yy,X=xx,p=2); mcmc = list(R=1000,keep=1)
out = rmnlIndepMetrop(Data=bdat, Mcmc=mcmc)
summary(out$betadraw)
```

Figure 14.3   Homogenous logit Bayesian MCMC code

We adjust this to accommodate weights so that the weighted likelihood is,

$$\ell^{(b)} = \sum_i w_i^{(b)} \ell_i \qquad (14.4.3)$$

with weights defined by

$$w_i^{(b)} \sim \text{Exp}(1). \qquad (14.4.3a)$$

Note that when we obtain weights from the standard Exponential distribution they are proportional to the standard Dirichlet. As such, we can normalize the weights if we wish, but this is not required. The implementation of the above BB procedure can be done using the native $\texttt{glm}$ function in **R** since it allows us to pass through the case weights as arguments. The code in Figure 14.4 demonstrates this.

Both the MCMC and BB procedures run reasonably fast and provide us with draws from the posterior distribution of $\Theta = \{\alpha, \beta\}$. Even with the small number of bootstrap draws $(B = 100)$ the posteriors from the two procedures are quite similar (see Figure 14.5). While this comparison is for the simple, homogenous logit model it gives us some assurance that the BB as an approximate procedure is not altogether unreasonable. We now move on to the main application.

```
# Bayesian Bootstrap for Homogeneous Logit
N = nrow(datz)
set.seed(123)
cf=NULL
B=100
for(b in 1:B){
  wts =rexp(N)
  # Suppress warnings since glm warns about
  # non-integer outcomes because of weights
  dmo = suppressWarnings(glm(family=binomial,data=datz,Y~P,weights = wts))
  cf = rbind(cf,coef(dmo))
  cat(b, "\r")
}
```

Figure 14.4   Homogenous logit Bayesian bootstrap code

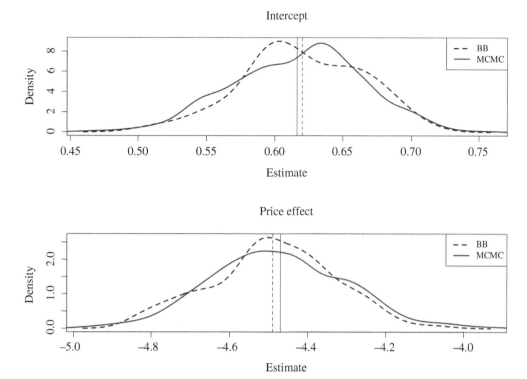

Figure 14.5   Homogenous logit MCMC and Bayesian bootstrap results

14.5 A BAYESIAN BOOTSTRAP PROCEDURE FOR THE HTE LOGIT

14.5.1 The Estimator

Our goal is to obtain HTE estimates and use those to build a personalization algorithm. To obtain those, we have to construct and implement the Bayesian Bootstrap version of the binary Logit model with (L1) regularization as described earlier. First, we sample and store the weights needed for the Bayesian Bootstrap. Recall that the weights, as before, are $w_i^{(b)} \sim \text{Exp}(1)$ and we store B=100 sets of these weights. We also normalize the weights so we have draws from the standard Dirichlet (See Figure 14.6).

```
# Draw weights and store
wts=list()
for(b in 1:B){
  wts[[b]] = rexp(N)
  wts[[b]] = wts[[b]]/sum(wts[[b]]) # This is optional
}
```

Figure 14.6   Sampling BB weights

```
doBB = function(wts){
  # Apply Model
  # Suppress warnings as before
  dmo = suppressWarnings(glmnet(penalty.factor=pf,family="binomial",x = mx,
                         y=my,alpha=1,weights=wts,lambda = rexp(1)/200))
  # We draw lambda ~ E(1/200)
  # but alternatively could do cross-validation (warning: much slower)

  # Extract Coefficients and store
  cfj = coef(dmo,'lambda.min')
  cf = rbind(cf,cfj)

  # Split cfj into relvant sets for a,b
  cfz=(matrix(cfj[3:(3+NZ-1)],nrow=NZ))
  cfz2=(matrix(cfj[(3+NZ):length(cfj)],nrow=NZ))

  # Construct relevant model primitives
  ahat = cfj[1]+zdat%*%cfz
  bhat = cfj[2]+zdat%*%cfz2

  # Return as list (could add dmo if one needs other elements)
  list(ahat=ahat,bhat=bhat,cf=cf)
}
```

Figure 14.7   LASSO Bayesian bootstrap code

Each set of weights $w^{(b)}$ are then sent to a procedure that maximizes the weighted penalized likelihood and then returns the relevant parameters. The code to do this is collected in a function called doBB (Figure 14.7).

The function doBB follows the logic of the earlier (homogeneous) binary Logit Bayesian bootstrap. We use the glmnet package which allows us to pass in case weights just like glm. The penalty factor option allows us to choose which covariates we wish to exclude from penalization. For our application, the intercept and the main price effect are not penalized. This is accomplished by setting the variable pf in the code to be a vector of ones except for the location of the price effect which is set to zero. The alpha=1 option is to choose the L1 penalty rather than the elastic-net penalty which allows for a convex combination of L1 and L2 penalties (governed by the value of alpha option).

An issue we have to deal with is choice of the penalization parameter and there are a number of options available. For one, we could do cross-validation as in the usual frequentist LASSO procedure or we could simply hold the parameter fixed. As we discussed earlier, we chose a different route in the spirit of Newton et al. [2021] and in particular impose the prior that $\lambda_i^{(b)} \sim \mathrm{Exp}\left(\frac{1}{200}\right)$. That completes the specification of our estimator and the output from each run is stored in the object dmo. For each run, we extract the parameter vector $\{\theta_\alpha, \theta_\beta\}$,[3] store those in cf, we then use those estimates to construct $\{\alpha\left(x_i; \theta_\alpha\right), \beta\left(x_i; \theta_\beta\right)\}$ and return all relevant information.

---

[3] We use "lambda.min" option even though we have supplied the parameter just in case the user changes the procedure to a cross-validated version. This can be done by using the cv.glmnet function. The user could also use the lambda.1se if they prefer that.

The function doBB is applied to each set of weights that we have sampled and stored. Since this procedure, unlike MCMC, is embarrassingly parallelizable we can avail of the various tools available in **R**. We use the pblapply function from the package with the same name to speed up the computations. In our application (on an 2.3 GHz 18-Core Intel Xeon iMac) the code takes about 5–6 seconds to run. Note that the computational burden of running a SVSS sampler using MCMC is significantly higher.

14.5.2 Results

The estimates obtained from the Bayesian Bootstrap can be treated exactly as we would draws from an MCMC procedure. They represent draws from the (approximate) posterior of interest. As such we can approximate quantities such as individual-level posterior means simply as

$$\mu_i = \sum_{b=1}^{B} h\left(z_i | \hat{\Theta}^{(b)}, x_i\right) \tag{14.5.1}$$

In the application (as for most individual level demand estimation problems) we have to deal with the possibility of draws of the price effect that are positive or even just very small. Ignoring this can have dramatic effects on prices and other economic calculations. While there is no established procedure we simply replace the "problem" draws with the mean on the negative draws for that individual. While this is a plugin-in it only effects about 1% of the draws so we are not too worried about it.

In Figure 14.8, we present the distributions of posterior means for $\{\alpha(x_i; \theta_\alpha), \beta(x_i; \theta_\beta)\}$ across individuals along with the true parameters.

The distributions seem to be reasonable approximations but exhibit some slight shifts from the true densities. To examine this further, we compute the posterior means of the individual level elasticities and compare those to the truth in Figure 14.9.

The similarity of the elasticity densities suggests that the procedure is doing its job not just statistically but also in recovering the economic primitives of interest. We now are set to use our estimates to conduct some policy counterfactuals such as creating a personalized pricing scheme.

14.6 PERSONALIZED PRICING

The optimal personalized pricing scheme requires us to obtain optimal prices for individual customers based on their observable characteristics (x_i). If we had access to the true posterior $f(\Theta|data)$ of the parameters, we would represent optimal individual level prices as,

$$p^*(x_i) = \arg\max_p \int_\Theta (p - c)\, \mathbb{P}\left(p; \alpha(x_i; \theta_\alpha), \beta(x_i; \theta_\beta)\right) f(\Theta|data)\, d\Theta \tag{14.6.1}$$

where, as before,

$$\mathbb{P}\left(p; \alpha(x_i; \theta_\alpha), \beta(x_i; \theta_\beta)\right) = \frac{\exp\left(\alpha(x_i; \theta_\alpha) + \beta(x_i; \theta_\beta)\, p\right)}{1 + \exp\left(\alpha(x_i; \theta_\alpha) + \beta(x_i; \theta_\beta)\, p\right)}. \tag{14.6.2}$$

Intercept

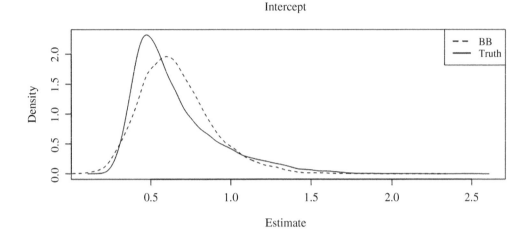

Price effect

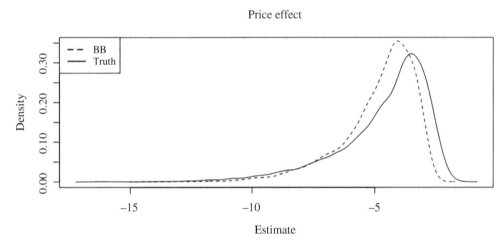

Figure 14.8   BB posterior means of $\left\{ \alpha\left(x_i; \theta_\alpha\right), \beta\left(x_i; \theta_\beta\right) \right\}$

Given our estimates, we can approximate the problem using our BB estimates as follows,

$$p^*_{BB}\left(x_i\right) = \arg\max_p \sum_{b=1}^{B} \left(p - c\right) \mathbb{P}\left(p; \alpha\left(x_i; \hat\theta_\alpha^{(b)}\right), \beta\left(x_i; \hat\theta_\beta^{(b)}\right)\right). \qquad (14.6.3)$$

Note, again, that this is completely analogous to how we would proceed with MCMC draws. The code to implement (Figure 14.10) this is simple and is simplified further by the context of the application where the marginal costs are zero.

We implement the procedure for both the BB draws as well as for the true individual parameters which we have access to. This allows us to see how well the Bayesian bootstrap does in obtaining the optimal prices and maximizing profits. We plot the prices based on the BB as well as true parameters in Figure 14.11.

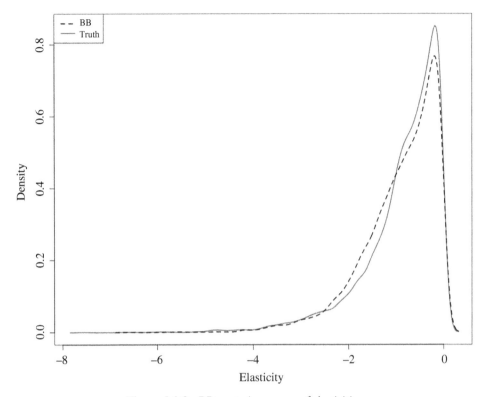

Figure 14.9 BB posterior means of elasticities

```
#  Prices and Profits
pistar = function(di){
  ahi = di$ahi
  bhi = di$bhi
  pi.i = function(price){
    # Apply
    u = ahi+bhi*price
    pr = 1/(1+exp(-u))
    # Profit
    mean(pr*price)
  }
  pstar = (optimize(pi.i,c(0,2),maximum = TRUE))
  c(pstar$max,pstar$obj)
}
```

Figure 14.10 Code to compute optimal prices

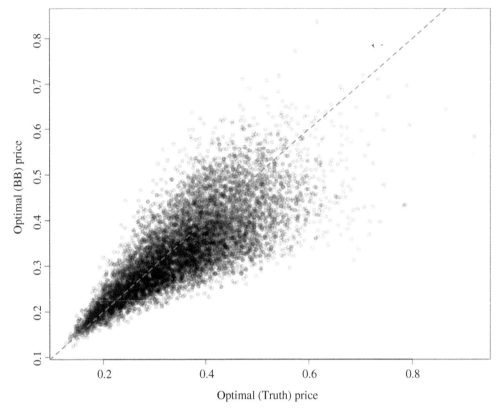

Figure 14.11   Optimal prices – true parameters vs BB

As is clear, the BB prices correlate extremely well with those based on the true parameters, although there is some fraying at the higher prices. A similar story obtains for profits. We compute profits for the personalized prices using only the *true* parameters (Figure 14.12). This is to see how the prices would perform in under the true DGP.

Evaluating the profits shows that the difference between the two personalization policies is only about −2.16% which is quite remarkable (Figure 14.12). We do remind the reader that this is using synthetic data and results may be quite different with real data.

```
# Eval Profits
Pstar = res[,1] # Prices
PRstar = 1/(1+exp(-(atrue+btrue*Pstar))) # Evaluation is always at true parameters
profit = (PRstar*Pstar) # Since MC=0 by assumption

# Profits based on true parameters
truPstar = trures[,1] # For true optimal
truPRstar = 1/(1+exp(-(atrue+btrue*truPstar)))
truprofit = (truPRstar*truPstar)
```

Figure 14.12   Code to compute profits under personalization

```
> # Compare Profits
> c(sum(profit),sum(truprofit))
. [1] 832.0406 852.2014
> # Percentage difference
> 100*mean((profit-truprofit)/truprofit)
[1] -2.162816
```

Figure 14.13 Comparison of aggregate profits

We can also examine the distribution of profits across customers. As we see in Figure 14.14 on the left panel, the density of profits across individuals is quite similar. The right panel of the same figure also shows that the two profits are very much in sync with the BB profits being bounded above by those obtained under the true parameters.

The above discussion provides all relevant elements needed to conduct various other analysis as pursued by Dubé and Misra [2023]. These include uniform prices, welfare calculations, and other counterfactuals. More generally, the framework provided above can easily be adapted to other marketing contexts such as catalog or advertising targeting and personalization.

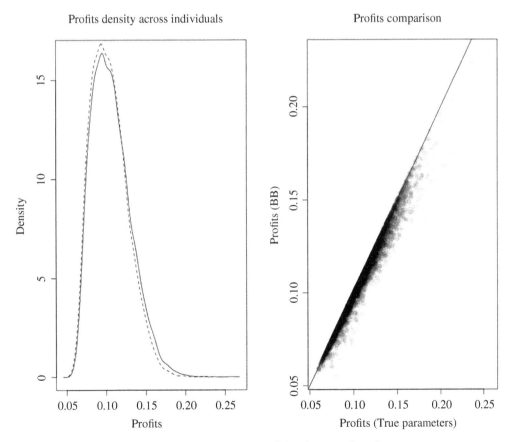

Figure 14.14 Comparison of distribution of profits

Appendix A
An Introduction to R and bayesm

In order to facilitate computation of the models in this book, we created a set of programs written in R and C++. R is a general purpose programming and statistical analysis system. R is free and available on the web. We have made our suite of programs into what is called an R "package." Our package is named bayesm. This package is easy to download and install from within R and is thoroughly documented, including test examples and illustrative datasets.

This appendix provides an introduction to the R environment and bayesm. In addition, there are many excellent books and video tutorials available on the R environment and RStudio enhancements.

A.1 SETTING UP THE R ENVIRONMENT AND BAYESM

Virtually, all users of the R statistical language use the RStudio Integrated Development environment (IDE). RStudio works on top of R and provides a convenient visual and programming interface. Of particular interest is that RStudio provides a graphical interface that makes R programming and package installation very convenient. RStudio provides a very nice code development environment, which includes dynamic syntax checking of R code and integration of the interface with C++. RStudio also supports a very wide variety of enhancements to the basic R interface including dynamic report generation using Rmarkdown.

A.1.1 Obtaining R

Visit http://cran.r-project.org/ or google "R language." CRAN is a network of mirror sites that allow you to download precompiled binary versions of R or source. There are pre-compiled versions of R for both Linux, Windows and MacOS. Download the appropriate version and install on your computer. We recommend a minimum of 16GB of memory for small to moderate sized datasets and MCMC inference.

Bayesian Statistics and Marketing, Second Edition. Peter E. Rossi, Greg M. Allenby, and Sanjog Misra
© 2024 John Wiley & Sons Ltd. Published 2024 by John Wiley & Sons Ltd.

A.1.2 Getting Started in RStudio

After installing R, google "RStudio download," and install the software. Invoke RStudio and a window will open on your desktop that is organized (by default) into three or four areas. Note that RStudio is best viewed on a large screen given the size of each of the four areas or panes.

Below is the MacOS RStudio environment that will appear after installation of RStudio for the first time.

The left side provides a standard R command console window. You may type any R command into this window. If the command results in some output that is nongraphics, it will be shown below the command you typed in. If your R commands produce a graphical result, this will be shown in the lower right pane under the "plots" tab.

In the example, below I used the `rnorm()` function to create the object "x" which is a vector that holds 1000 pseudo-normal random draws. The upper right hand pane of RStudio contains a number of panes, one of which shows the "environment," which provides information about R objects that are currently in the portion of memory associated with the current R session. Please note that R only stores objects in memory and anything created in R must be saved prior to exiting R or it will be erased on exit. You can store the current workspace as one R file on exit and this environment will be restored when you start up RStudio again.

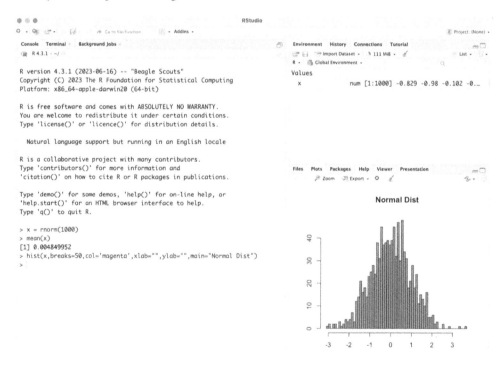

In the example, I also used the `mean()` function to find the sample average and used the hist() function to plot a histogram. The "output" of the mean function is displayed in the console window, while the graphical output of the histogram function is shown in the lower right "plots" tab of that pane. The "export" pull-down menu can be used to save the plot or "export" it as a file or copy to the clipboard.

It is a far better practice to create an R script file and type R commands in this file. This simplifies debugging of code and preserves your work for future use. To create an R script file use the green "+" pull-down menu and select the option to create a new R script. When you are done, save this file into the appropriate part of your directory tree.

Below we show the upper left corner of the RStudio environment that illustrates the creation of an R script as well as the dynamic facilities that show the arguments for functions as they are typed in. Here we see the standard arguments of the rnorm() function shown. RStudio also supports automatic command and object completion. If you type in the first new letters of an object, RStudio will bring up a window with possible completion possibilities.

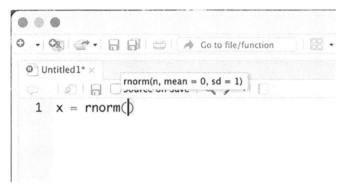

A.1.3 Obtaining Help in RStudio

There are a number of different ways to obtain help in RStudo. The "Help" tab of the lower right window allows you to access the manuals in PDF or HTML form as well as to search for help for specific packages and functions and datasets included in those packages.

You can also use the help and help.search commands in the R console windows. For example, we just used two commands, hist and rnorm.

Each R help window has the same sections:

Description

Usage

Arguments

Details

Value

References

See Also

Examples (note this is "cut-off" in the screen shot above)

Usage/Arguments/Examples are the most useful.

In particular, all help files include an examples section (including all functions in bayesm), these example files can also serve as a template for function usage which can be useful for functions with a large number or complicated arguments.

If you are not sure what command you need, `help.search("key word")` can be very useful.

A.1.4 Installing `bayesm`

`bayesm` can be installed from the RStudio mirror site by selecting the "Packages" tab in the lower right RStudio pane. Simply click on the "install" button and start typing the name of any package you desire to install. You can update any package by using the same pull-down menu to select packages to update.

In order to use bayesm efficiently, it is important to have a grasp of the basics of the R language and the use of R functions. Case Study 1 in this text illustrates using `bayesm` to analyze the built-in "camera" dataset using a standard hierarchical multinomial logit model. For newcomers to the R language, we recommend that you read the introduction to the R language below and follow the R code the accompanies Case Study 1.

A.2 THE R LANGUAGE

R is a functionally-oriented language. All commands are functions that act upon objects of various types. All commands produce objects as well. The basic R command is of the form: object = function(object). Functions can be composed to produce powerful (but sometimes hard-to-read) expressions. Users can define their own functions. Writing these user functions constitutes R-programming.

Let's start by reading in some data. Suppose we have a file in a spreadsheet that with some regression data on several different units. The file has a UNIT variable to identify which unit the data comes from, a dependent variable Y, and two independent variables X1 and X2.

```
UNIT   Y        X1          X2
  A    1      0.23815      0.4373
  A    2      0.55508      0.47938
  A    3      3.03399     -2.17571
  A    4     -1.49488      1.66929
  B   10     -1.74019      0.35368
  B    9      1.40533     -1.2612
  B    8      0.15628     -0.27751
  B    7     -0.93869     -0.0441
  B    6     -3.06566      0.14486
```

We write this data out of Excel by saving it as a text (tab-delimited file), data.txt (use the **save as** option on the **file** menu and choose text file in the file type box).

We can read this file into R using the `read.table()` command. In addition, RStudio supports import functions from the "import" sub-menu of the file menu. Here, we use basic R commands in order to explain the basics of the R language, not for efficiency.

```
> df=read.table("data.txt",header=TRUE)
> df
  UNIT  Y       X1        X2
1   A   1   0.23815   0.43730
2   A   2   0.55508   0.47938
3   A   3   3.03399  -2.17571
4   A   4  -1.49488   1.66929
5   B  10  -1.74019   0.35368
6   B   9   1.40533  -1.26120
7   B   8   0.15628  -0.27751
8   B   7  -0.93869  -0.04410
9   B   6  -3.06566   0.14486
```

The `read.table` function has two arguments: the name of the file and the argument "header." There are many other arguments but they are optional and often have defaults. The default for the header argument is the value FALSE. Using the argument, `header=TRUE`, tells the `read.table` function to expect that the first line of the file will contain (delimited by spaces or tabs) the names of each variable. TRUE and FALSE are examples of reserved values in R indicating a logical switch for true or false. Another useful reserved value is NULL that is often used to create an object with nothing in it.

The command `df=read.table(...)` assigns the output of the read.table function to the R object named "df." The object "df" is a member of a class or type of object called a data frame. A "data frame" is preferred by R as the format for datasets. A data frame contains a set of observations on variables with the same number of observations in each variable. In this example, each of the variables, Y, X1, and X2, is of type numeric (R does not distinguish between integers and floating point numbers), while the variable UNIT is character.

There are two reasons to store your data as a data frame: (1) Most R statistical functions require a data frame and (2) the data frame object allows the user to access the data either via the variables names or by viewing the dataframe as a two-dimensional array.

```
> df$Y
[1]  1  2  3  4 10  9  8  7  6
> mode(df$Y)
[1] "numeric"
> df[,2]
[1]  1  2  3  4 10  9  8  7  6
```

We can refer to the Y variable in df by using the df$XXX notation (where XXX is the name of the variable). The "mode" command confirms that this variable is, indeed, numeric. We can also access the Y variable by using notation in R for subsetting a portion of an array. The notation `df[,2]` means the values of the 2nd column of df. Below we will explore the many ways we can subset an array or matrix.

A.2.1 Using Built-In Functions: Running a Regression

Let's now use the built-in linear model function in R to run a regression of Y on X1 and X2, pooled across both units A and B.

```
> lmout=lm(Y ~ X1 + X2, data=df)
> names(lmout)
 [1] "coefficients"   "residuals"       "effects"         "rank"
 [5] "fitted.values" "assign"          "qr"              "df.residual"
 [9] "xlevels"        "call"            "terms"           "model"
> print(lmout)

Call:
lm(formula = Y ~ X1 + X2, data = df)

Coefficients:
(Intercept)              X1               X2
      5.084          -1.485           -2.221
> summary(lmout)

Call:
lm(formula = Y ~ X1 + X2, data = df)

Residuals:
Min           1Q  Median        3Q      Max
-3.3149 -2.4101  0.4034   2.5319   3.2022

Coefficients:
            Estimate Std. Error t value Pr(>|t|)
(Intercept)   5.0839     1.0194   4.987  0.00248 **
X1           -1.4851     0.8328  -1.783  0.12481
X2           -2.2209     1.3820  -1.607  0.15919
---
Signif. codes:  0 '***' 0.001 '**' 0.01 '*' 0.05 '.' 0.1 ' ' 1

Residual standard error: 2.96 on 6 degrees of freedom
Multiple R-Squared: 0.3607,     Adjusted R-squared: 0.1476
F-statistic: 1.693 on 2 and 6 DF,   p-value: 0.2612
```

lm is the function in the package (stats) that fits linear models. Note that the regression is specified via a "formula" that tells lm, which is the dependent and independent variables. We assign the output from the lm function to the object, lmout. lmout is a special type of object called a "list." A list is simply an ordered collection of objects of any type. The names command will list the names of the elements of the list. We can access any element of the list by using the $ notation.

```
> lmout$coef
(Intercept)              X1               X2
   5.083871      -1.485084      -2.220859
```

Note that we only need to specify enough of the name of the list component to uniquely identify it, for example, lmout$coef is the same as lmout$coefficients.

We can "print" the object lmout and get a brief summary of its contents. Print is a generic command that uses a different "print method" for each type of object. Print recognizes that lmout is a list of type lm and uses a specific routine to printout the contents of the list. A more useful summary of contents of lmout can be obtained with the summary command.

A.2.2 Inspecting Objects and the R Workspace

When you start up R, R looks for a file .Rdata in the directory in which R is started from (you can also double-click the file to start R). This file contains a copy of the R "workspace," which is a list of R objects created by the user. For example, we just created two R objects in the example above: df (the data frame) and lmout, the lm output object.

To list all objects in the current workspace, use the command ls(). The RStudio "environment" pane also lists all objects and some information on the type of object and contents.

```
> ls()
[1] "df"      "lmout"
```

This doesn't tell us too much about the objects. If you just type the object name at the command prompt and return, then you will invoke the default print method for this type of object as we saw above in the data frame example.

As useful command is the structure (str for short) command.

```
> str(df)
'data.frame':    9 obs. of   4 variables:
 $ UNIT: Factor w/ 2 levels "A","B": 1 1 1 1 2 2 2 2 2
 $ Y   : int   1 2 3 4 10 9 8 7 6
 $ X1  : num    0.238   0.555   3.034 -1.495 -1.740 ...
 $ X2  : num    0.437   0.479 -2.176  1.669  0.354 ...
```

Note that the str command tells us a bit about the variables in the data frame. The UNIT variable is of type "factor" with two levels. Type "factor" is used by many of the built-in R functions and is way to store qualitative variables.

The R workspace exists only in memory. You must either save the workspace when you exist (you will be prompted for this) or you must recreate the objects again.

A.2.3 Vectors, Matrices, and Lists

From our point of view, the power of R comes from statistical programming at a relatively high level. To do so, we will need to organize data as vectors, arrays, and lists. Vectors are ordered collections of the same type of object. If we access one variable from our data frame above, it will be a vector.

```
> df$X1
[1]   0.23815   0.55508   3.03399 -1.49488 -1.74019   1.40533   0.15628
    -0.93869 -3.06566
> length(df$X1)
[1] 9
> is.vector(df$X1)
[1] TRUE
```

The function is.vector returns a logical flag as to whether or not the input argument is a vector. We can also create a vector with the c() command.

```
> vec=c(1,2,3,4,5,6)
> vec
```

```
[1] 1 2 3 4 5 6
> is.vector(vec)
[1] TRUE
```

A matrix is a two-dimensional array. Let's create a matrix from a vector.

```
> mat=matrix(c(1,2,3,4,5,6),ncol=2)
> mat
     [,1] [,2]
[1,]    1    4
[2,]    2    5
[3,]    3    6
```

matrix() is a command to create a matrix from a vector. The option "ncol" is used to create the matrix with a specified number of columns (see also nrow). Note that the matrix is created column by column from the input vector (first subscripts varies the fastest). We can create a matrix row by row with the following command:

```
> mat=matrix(c(1,2,3,4,5,6),byrow=TRUE,ncol=2)
> mat
     [,1] [,2]
[1,]    1    2
[2,]    3    4
[3,]    5    6
```

We can also convert a data frame into a matrix.

```
> dfmat=as.matrix(df)
> dfmat
  UNIT Y    X1          X2
1 "A"  " 1" " 0.23815" " 0.43730"
2 "A"  " 2" " 0.55508" " 0.47938"
3 "A"  " 3" " 3.03399" "-2.17571"
4 "A"  " 4" "-1.49488" " 1.66929"
5 "B"  "10" "-1.74019" " 0.35368"
6 "B"  " 9" " 1.40533" "-1.26120"
7 "B"  " 8" " 0.15628" "-0.27751"
8 "B"  " 7" "-0.93869" "-0.04410"
9 "B"  " 6" "-3.06566" " 0.14486"
> dim(dfmat)
[1] 9 4
```

Note that all of the values of the results matrix are character as one of the variables in the data frame (UNIT) is character-valued. Finally, matrices can be created from other matrices and vectors using the cbind (column bind) and rbind (row bind) commands.

```
> mat1
     [,1] [,2]
[1,]    1    2
[2,]    3    4
[3,]    5    6
> mat2
     [,1] [,2]
[1,]    7   10
[2,]    8   11
```

```
[3,]    9    12
> cbind(mat1,mat2)
      [,1] [,2] [,3] [,4]
[1,]    1    2    7    10
[2,]    3    4    8    11
[3,]    5    6    9    12
> rbind(mat1,mat2)
      [,1] [,2]
[1,]    1    2
[2,]    3    4
[3,]    5    6
[4,]    7   10
[5,]    8   11
[6,]    9   12
> rbind(mat1,c(99,99))
      [,1] [,2]
[1,]    1 .  2
[2,]    3    4
[3,]    5    6
[4,]   99   99
```

R supports multi-dimensional arrays as well. Below is an example of creating a three-dimensional array from a vector.

```
> ar=array(c(1,2,3,4,5,6),dim=c(3,2,2))
> ar
, , 1

      [,1] [,2]
[1,]    1    4
[2,]    2    5
[3,]    3    6

, , 2

      [,1] [,2]
[1,]    1    4
[2,]    2    5
[3,]    3    6
```

Again, the array is created by using vector for the first dimension, then the second, and then third. A $3 \times 2 \times 2$ array as 12 elements not the six provided as an argument. R will repeat the input vector as necessary until the required number of elements are obtained.

A list is an ordered collection of objects of any type. It is the most flexible object in R that can be indexed. As we have seen in the lm function output, lists can also have names.

```
> l=list(1,"a",c(4,4),list(FALSE,2))
> l
[[1]]
[1] 1

[[2]]
[1] "a"
```

```
[[3]]
[1]  4  4

[[4]]
[[4]][[1]]
[1]  FALSE

[[4]][[2]]
[1]  2

>  l=list(num=1,char="a",vec=c(4,4),list=list(FALSE,2))
>  l$num
[1]  1
>  l$list
[[1]]
[1]  FALSE

[[2]]
[1]  2
```

In the example, we created a list of a numeric value, character, vector, and another list. We also can name each component and access them with the $ notation.

A.2.4 Accessing Elements and Subsetting Vectors, Arrays, and Lists

To access an element of a vector, simply enclose index of that element in square brackets.

```
>  vec=c(1,2,3,2,5)
>  vec[3]
[1]  3
```

To access a sub-set of elements, there are two approaches: (1) specify a vector of integers of the required indices, (2) specify a logical variable that is TRUE for the desired indices.

```
>  index=c(3:5)
>  index
[1]  3  4  5
>  vec[index]
[1]  3  2  5
>  index=vec==2
>  index
[1]  FALSE   TRUE FALSE   TRUE FALSE
>  vec[index]
[1]  2  2
>  vec[vec!=2]
[1]  1  3  5
```

c(3:5) creates a vector from the "pattern" or sequence from 3 to 5. The seq command can create a wide variety of different patterns.

To properly understand the example of the logical index, it should be noted that "=" is an assignment operator while "==" is a comparison operator. Vec==2 creates a logical vector with flags for if the elements of vec are 2. The last example uses the "not equal" comparison operator !=. We can also access the elements not in a specified index vector.

```
> vec[-c(3:5)]
[1] 1 2
```

To access elements of arrays, we can use the same ideas for vectors but we must specify a set of row and column indices. If no indices are specified, we get all of the elements on that dimension. For example, earlier we used the notation df[,2] to access the second column of the data frame df.

We can pull off the observations corresponding to unit "A" from the matrix version of dfmat using the commands:

```
> dfmat
  UNIT Y     X1              X2
1 "A"  " 1"  " 0.23815"  " 0.43730"
2 "A"  " 2"  " 0.55508"  " 0.47938"
3 "A"  " 3"  " 3.03399"  "-2.17571"
4 "A"  " 4"  "-1.49488"  " 1.66929"
5 "B"  "10"  "-1.74019"  " 0.35368"
6 "B"  " 9"  " 1.40533"  "-1.26120"
7 "B"  " 8"  " 0.15628"  "-0.27751"
8 "B"  " 7"  "-0.93869"  "-0.04410"
9 "B"  " 6"  "-3.06566"  " 0.14486"
> dfmat[dfmat[,1]=="A",2:4]
  Y     X1              X2
1 " 1"  " 0.23815"  " 0.43730"
2 " 2"  " 0.55508"  " 0.47938"
3 " 3"  " 3.03399"  "-2.17571"
4 " 4"  "-1.49488"  " 1.66929"
```

The result is a 4 × 3 matrix. Note that we are using the values of the dfmat to index into itself. This means that R evaluates the expression dfmat[,1] == "A" and passes the result into the matrix subsetting operator [ ], which is a function that processes dfmat.

To access elements of lists, we can use the $ notation if the element has a name or we can use a special operator [[ ]]. To see how this works, let's make a list with two elements, each corresponding to the observations for unit A and B. Note that the size of the matrices corresponding to each unit is different – unit A has four obs and unit B has five! This means that we can't use a three-dimensional array to store this data (we would need a "ragged" array).

```
> ldata=list(A=dfmat[dfmat[,1]=="A",2:4],B=dfmat[dfmat[,1]=="B",2:4])
> ldata
$A
  Y     X1              X2
1 " 1"  " 0.23815"  " 0.43730"
2 " 2"  " 0.55508"  " 0.47938"
3 " 3"  " 3.03399"  "-2.17571"
4 " 4"  "-1.49488"  " 1.66929"

$B
  Y     X1              X2
```

```
5 "10" "-1.74019" " 0.35368"
6 " 9" " 1.40533" "-1.26120"
7 " 8" " 0.15628" "-0.27751"
8 " 7" "-0.93869" "-0.04410"
9 " 6" "-3.06566" " 0.14486"

> ldata[1]
$A
   Y    X1          X2
1 " 1" " 0.23815" " 0.43730"
2 " 2" " 0.55508" " 0.47938"
3 " 3" " 3.03399" "-2.17571"
4 " 4" "-1.49488" " 1.66929"

> is.matrix(ldata[1])
[1] FALSE
> is.list(ldata[1])
[1] TRUE
> ldata$A
   Y    X1          X2
1 " 1" " 0.23815" " 0.43730"
2 " 2" " 0.55508" " 0.47938"
3 " 3" " 3.03399" "-2.17571"
4 " 4" "-1.49488" " 1.66929"
> is.matrix(ldata$A)
[1] TRUE
```

If we specify ldata[1], we don't get the contents of the list element (which is a matrix) but we get a list! If we specify ldata$A, we obtain the matrix. If we have a long list or we don't wish to name each element, we can use the [[ ]] operator to access elements in the list.

```
> is.matrix(ldata[[1]])
[1] TRUE
> ldata[[1]]
   Y    X1          X2
1 " 1" " 0.23815" " 0.43730"
2 " 2" " 0.55508" " 0.47938"
3 " 3" " 3.03399" "-2.17571"
4 " 4" "-1.49488" " 1.66929"
```

A.2.5 Loops

As with all interpreted languages, loops in R are slow. That is, they typically take more time than if implemented in a compiled language. On the other hand, matrix/vector operations are typically faster in R than in compiled language such as C and Fortran unless the optimized BLAS is called. Thus, wherever possible, "vectorization" or writing expressions as only involving matrix/vector arithmetic is desirable. This is more of an art than a science, however.

If a computation is fundamentally iterative (such as maximization or MCMC simulation), a loop will be required. A simple loop can be accomplished with the for structure. The syntax is of the form

```
for (var in range) { }
```

var is a numeric loop index. Range is a range of values of var. Enclosed in the braces is any valid R expression. There can be more than one R statement in the R expression. The simplest example is a loop over a set of *i* values from 1 to *N*.

```
x=0
for (i in 1:10)
    {
    x=x+1
    }
```

Let's loop over both units and create a list of lists of the regression output from each.

```
> ldatadf=list(A=df[df[,1]=="A",2:4],B=df[df[,1]=="B",2:4])
> lmout=NULL
> for (i in 1:2) {
+     lmout[[i]]=lm(Y ~ X1+X2,data=ldatadf[[i]])
+     print(lmout[[i]])
+     }

Call:
lm(formula = Y ~ X1 + X2, data = ldatadf[[i]])

Coefficients:
(Intercept)           X1              X2
      4.494        -2.860          -3.180


Call:
lm(formula = Y ~ X1 + X2, data = ldatadf[[i]])

Coefficients:
(Intercept)           X1              X2
      9.309         1.051           1.981
```

Here, we subset the data frame directly rather than the matrix created from the data frame to avoid the extra-step of converting character to numeric values and so that we can use the lm function, which requires data frame input. We can see that the same sub-setting command that works on arrays will also work on data frames.

A.2.6 Implicit Loops

In many contexts, a loop is used to compute the results of applying a function to either the row or column dimensions of an array. For example, if we wish to find the mean of each variable in a data frame, we want to apply the function "mean" to each column. This can be done with the `apply()` function.

```
> apply(df[,2:4],2,mean)
         Y         X1         X2
 5.5555556 -0.2056211 -0.0748900
```

The first argument specifies the array, the second the dimension (1 = row, 2 = col), and the third the function to be applied. In R, the apply function is simply an elegant loop so don't expect to speed things up with this. Of course, we could write this as a matrix operation that would be much faster.

A.2.7 Matrix Operations

One of the primary advantages of R is that we can write matrix/vector expressions directly in R code. Let's review some of these operators by computing a pooled regression using matrix statements.

The basic functions needed are:

`%*%`	Matrix multiplication e.g. (X %*% Y)
	note: X or Y or both can be vectors
`chol(X)`	Compute "square" or Cholesky root of square, pd matrix
	X=U'U where `U=chol(X)`; U is upper triangular
`chol2inv(chol(X))`	Compute inverse of square pd matrix using its Cholesky root
`crossprod(X,Y)`	t(X) %*% Y -- very efficient
`diag`	Extract diagonal of matrix or create diagonal matrix from a vector

Less frequently used are:

`%x%`	Kronecker product (to be used carefully as Kronecker products can create very large arrays)
`backsolve()`	Used to compute inverse of a triangular array

The R statements are:

```
y=as.numeric(dfmat[,2])
X=matrix(as.numeric(dfmat[,3:4]),ncol=2)
X=cbind(rep(1,nrow(X)),X)
XpXinv=chol2inv(chol(crossprod(X)))
bhat=XpXinv%*%crossprod(X,y)
res=as.vector(y-X%*%bhat)
ssq=as.numeric(res%*%res/(nrow(X)-ncol(X)))
se=sqrt(diag(ssq*XpXinv))
```

The first two statements create y and X. Then we add a column of ones using the `rep()` function for the intercept and compute the regression using Cholesky roots. Note that we must convert res to a vector to use the statement res %*% res. We also must convert ssq to a scalar from a 1 × 1 matrix to compute the standard errors in the last statement. We note that the method above is very stable numerically, but some users would prefer the QR decomposition. This would be simpler, but our experience has shown that the method above is actually faster in R.

A.2.8 Other Useful Built-In R Functions

R has thousands of built-in function and thousands more than can be added from contributed packages. Some functions used in the book include:

rnorm	Draw univariate normal random variates
runif	Draw uniform random variates
rchisq	Draw chi-sq random variates
mean	Compute mean of a vector
var	Compute Covariance matrix given matrix input
quantile	Computes quantiles of a vector
optim	General purpose optimizer
sort	Sort a vector
if	Standard if statement (includes else clause)
while	While loop
scan	Read from a file to a vector
write	Write a matrix to a file
sqrt	Square root
log	Natural log
%%	Modulo (e.g. $100\%\%10 = 0$)
round	Round to a specified number of sign digits
floor	Greatest integer < argument
.C	Interface to C and C++ code (more later)

A.2.9 User-defined Functions

The regression example above is a situation for which a user-defined function would be useful. To create a function object in R, simply enclose the R statements in braces as assign this to a function variable.

```
myreg=function(y,X){
#
# purpose: compute lsq regression
#
# arguments:
#    y -- vector of dep var
#    X -- array of indep vars
#
# output:
#    list containing lsq coef and std errors
```

```
#
XpXinv=chol2inv(chol(crossprod(X)))
bhat=XpXinv%*%crossprod(X,y)
res=as.vector(y-X%*%bhat)
ssq=as.numeric(res%*%res/(nrow(X)-ncol(X)))
se=sqrt(diag(ssq*XpXinv))
list(b=bhat,std_errors=se)
}
```

The code above should be executed either by cutting and pasting into R or by sourcing a file containing this code. This will define an object called "myreg."

```
ls()
 [1] "ar"    "bhat"          "df"      "dfmat"      "i"          "index"
 [7] "l"     "last.warning"  "ldata"   "ldatadf"    "ldataidf"   "lmout"
[13] "mat"   "mat1"          "mat2"    "myreg"      "names"      "res"
[19] "se"    "ssq"           "vec"     "X"          "XpXinv"     "y"
```

To execute the function, we simply type it in with arguments at the command prompt or in another source file.

```
> myreg(X=X,y=y)
$b
           [,1]
[1,]   5.083871
[2,]  -1.485084
[3,]  -2.220859

$std_errors
[1] 1.0193862 0.8327965 1.3820287
```

`myreg` returns a list with b and the standard errors.

Objects are passed by copy in R rather than by reference. This means that if we give the command `myreg(Z,d)`, a copy of Z will be assigned to the "local" variable X in the function myreg and a copy of d to y. In addition, variables created in the function (e.g., XpXinv and res in myreg) are created only during the execution of the function and then erased when the function returns to the calling environment.

The arguments are passed and copied in the order supplied at the time of the call so that you must be careful. The statement `myreg(d,Z)` will not execute properly. However, if we explicitly name the arguments as in `myreg(X=Z,y=d)` then the arguments can be given in any order.

Many functions have default arguments and R has what is called "lazy" function evaluation, which means that if an argument is not needed it is not checked. See Introduction to R for a more discussion on default and other types of arguments. If a local variable cannot be found while executing a function, R will look in the environment or workspace that the function was called from. This can be convenient but it can also be dangerous!

Many functions are dependent on other functions. If a function called within a function is only used by that calling function and has no other use, it can be useful to define these utility functions in the calling function definition. This means that they will not be visible to the user of the function.

Example:

```
Myfun= function(X,y) {
#
# define utilty function needed
#
Util=function(X) { … }
#
# main body of myfun
#
…
}
```

A.2.10 Debugging Functions

It is a good practice to define your functions in a file and "source" them into R. This will allow you to recreate your set of function objects for a given project without having to save the workspace.

To debug a function, you can use the brute force method of placing print statements in the function. cat() can be useful here. For example, we can define a "debugging" version of myreg that prints out the value of se in the function. The cat command prints out a statement reminding us of where the "print" output comes from (note the use of fill=TRUE which insures that a new line will be generated on the console).

```
myreg=function(y,X){
#
# purpose: compute lsq regression
#
# arguments:
#     y -- vector of dep var
#     X -- array of indep vars
#
# output:
#     list containing lsq coef and std errors
#
XpXinv=chol2inv(chol(crossprod(X)))
bhat=XpXinv%*%crossprod(X,y)
res=as.vector(y-X%*%bhat)
ssq=as.numeric(res%*%res/(nrow(X)-ncol(X)))
se=sqrt(diag(ssq*XpXinv))
cat("in myreg, se = ",fill=TRUE)
print(se)
list(b=bhat,std_errors=se)
}
```

When run, this new function will produce the output:

```
> myregout=myreg(y,X)
in myreg, se =
[1] 1.0193862 0.8327965 1.3820287
```

R also features a simple debugger. If you "debug" a function, you can step through the function and inspect the contents of local variables. One can also modify their contents.

```
> debug(myreg)
> myreg(X,y)
debugging in: myreg(X, y)
debug: {
    XpXinv = chol2inv(chol(crossprod(X)))
    bhat = XpXinv %*% crossprod(X, y)
    res = as.vector(y - X %*% bhat)
    ssq = as.numeric(res %*% res/(nrow(X) - ncol(X)))
    se = sqrt(diag(ssq * XpXinv))
    cat("in myreg, se = ", fill = TRUE)
    print(se)
    list(b = bhat, std_errors = se)
}
Browse[1]>
debug: XpXinv = chol2inv(chol(crossprod(X)))
Browse[1]> X
[1]  1  2  3  4 10  9  8  7  6
Browse[1]> #OOPS!
debug: bhat = XpXinv %*% crossprod(X, y)
Browse[1]> XpXinv
            [,1]
[1,] 0.002777778
Browse[1]> Q
> undebug(myreg)
```

If there are loops in the function, the debugging command "c" can be used to allow the loop to finish. "Q" quits the debugger. You must turn off the debugger with the undebug command! If you want to debug other functions called by `myreg`, you must `debug()` 'em first!

A.2.11 Elementary Graphics

Graphics in R can be quite involved as the graphics capabilities are very extensive. For some examples of what is possible issue the commands `demo(graphics)`, `demo(image)` and `demo(persp)`. Let's return to our first example – a histogram of a distribution. The most downloaded R package is the ggplot2 by Hadley Wickham. Here we provide a short introduction to the standard built-in or "base" graphics capabilities.

```
hist(rnorm(1000),breaks=50,col="magenta")
```

This creates a histogram with 50 bars and with each bar filled in the color "magenta" (type `colors()` to see the list of available colors). This plot can be improved by inclusion of plot parameters to change the x and y axis labels and as well as the "title" of the plot.

```
hist(rnorm(1000),breaks=30,col="magenta",xlab="theta",ylab="",
    main="Non-parametric Estimate of Theta Distribution")
```

produces

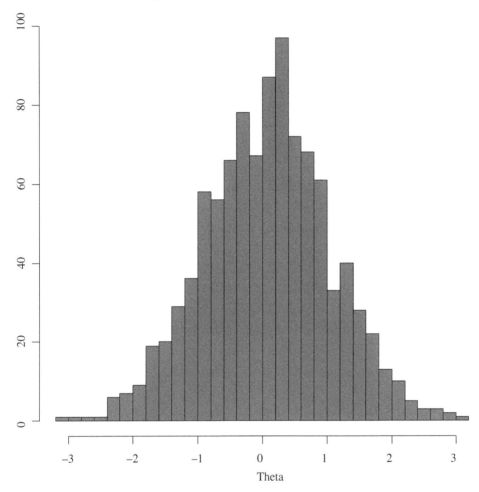

Three other basic plots are useful:

`plot(x,y)`	Scatterplot of x vs y
`plot(x)`	Sequence plot of x
`matplot(X)`	Sequence plots of columns of X
`acf(x)`	acf of time series in x

The col, xlab, ylab, and main parameters work on all of these plots.

In addition, the parameters

`type="l"`	Connects scatterplot points with a lines
`lwd=x`	Specifies the width of lines (1 is default, > 1 is thicker)
`lty=x`	Specifies type of line (e.g. solid vs dashed)
`xlim/ylim=c(z,w)`	Specifies x/y axis runs from z to w

are useful. `?par` displays all of the graphic parameters available.

It is often useful to display more than one plot per page. To do this, we must change the global graphic parameters with the command, `par(mfrow=c(x,y))`. This specifies an array of plots x by y plotted row by row.

```
par(mfrow=c(2,2))
X=matrix(rnorm(5000),ncol=5)
X=t(t(X)+c(1,4,6,8,10))

hist(X[,1],main="Histogram of 1st col",col="magenta",xlab="")
plot(X[,1],X[,2],xlab="col 1", ylab="col 2",pch=17,col="red",
    xlim=c(-4,4),ylim=c(0,8))
title("Scatterplot")
abline(c(0,1),lwd=2,lty=2)
matplot(X,type="l",ylab="",main="MATPLOT")
acf(X[,5],ylab="",main="ACF of 5th Col")
```

`title()` and `abline()` are examples of commands which modify the current "active" plot. Other useful functions are `points()` and `lines()` to add points and points connected by lines to the current plot.

The commands above will produce

A.2.12 System Information

The following commands provide information about the operating system:

`memory.limit()`	Current memory limit
`memory.size()`	Current memory size
`system.time(R expression)`	Times execution of R expression
`proc.time()[3]`	Current R session cpu usage in seconds
`getwd()`	Obtain current working directory
`setwd()`	Set current working directory
`Rprof(file="filename")`	Turns on profiling and writes to filename
`Rprof("")`	Turns off profiling
`summaryRprof(file="filename")`	Summarizes output in profile file

Histogram of 1st col

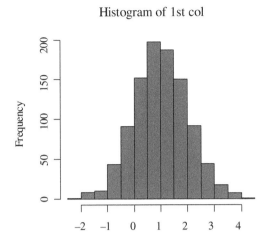

Scatterplot

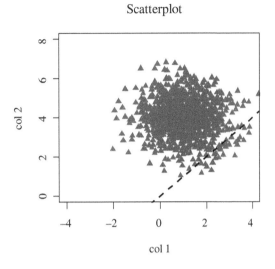

MATPLOT

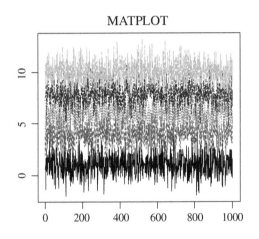

ACF of 5th Col

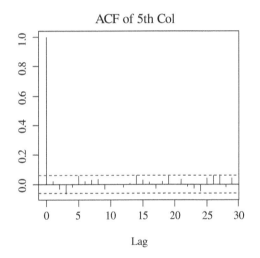

Examples of usage are given below.

```
> memory.size()
[1] 191135504
> getwd()
[1] "C:/userdata/per/class/37904"
> x=matrix(rnorm(1e07),ncol=1000)
> memory.size()
[1] 332070456
> memory.limit()
```

```
[1] 1992294400
> begin=proc.time()[3]
> z=crossprod(x)
> end=proc.time()[3]
> print(end-begin)
[1] 6.59
>test=function(n){x=matrix(rnorm(n),ncol=1000);z=crossprod(x);
                  cz=chol(z)}
> Rprof("test.out")
> test(1e07)
> Rprof()
> summaryRprof("test.out")
$by.self
           self.time self.pct total.time total.pct
rnorm           4.40     48.9       4.40      48.9
crossprod       4.16     46.2       4.16      46.2
matrix          0.22      2.4       4.72      52.4
.Call           0.12      1.3       0.12       1.3
as.vector       0.10      1.1       4.50      50.0
chol            0.00      0.0       0.12       1.3
test            0.00      0.0       9.00     100.0

$by.total
           total.time total.pct self.time self.pct
test             9.00     100.0      0.00      0.0
matrix           4.72      52.4      0.22      2.4
as.vector        4.50      50.0      0.10      1.1
rnorm            4.40      48.9      4.40     48.9
crossprod        4.16      46.2      4.16     46.2
.Call            0.12       1.3      0.12      1.3
chol             0.12       1.3      0.00      0.0

$sampling.time
[1] 9
```

The profile shows that virtually all of the time in the test function was in the generation of normal random numbers and in computing cross-products. The Cholesky root of a 1000×1000 matrix is essentially free! crossprod is undertaking 5 billion floating point multiplies ($1/2$ of $10,000 \times 1,000*1,000$).

A.2.13 More Lessons Learned from Timing

If you are going to fill up an array with results, preallocate space in the array. Do not append to an existing array.

```
> n=1e04
> x=NULL
> zero=c(rep(0,5))
> begin=proc.time()[3]
> for (i in 1:n) {x=rbind(x,zero) }
> end=proc.time()[3]
> print(end-begin)
```

```
[1] 6.62
> x=NULL
> begin=proc.time()[3]
> x=matrix(double(5*n),ncol=5)
> end=proc.time()[3]
> print(end-begin)
[1] 0.07
```

A.3 USING BAYESM

Once *bayesm* has been installed, you must "load" the package for each R session. This is done with the command library(bayesm).

Once loaded, you can use any of the more than 40 functions in *bayesm* just by inserting the function call in your source code. For all except the most trivial problems, it is best to edit a source file with your code and then "source" this file into R.

The following tips are useful in using *bayesm*:

1. If you don't know much about R, read the beginning of this appendix. *bayesm* has many functions defined in it. The "turn-key" or "end-user" functions start with the letter r, for example, rmnpGibbs is the Gibbs Sampler for the multinomial probit model.

2. Check the examples. Each function has an example file. To view the example for a function, use the R command ?, for example, ?rmnpGibbs. The example will be listed at the bottom of the displayed help text. You can also find the examples in the R program directory tree, for example, C:\Program Files\R\rwXXXX\library\bayesm\R-ex. You may have to unzip these files.

3. The best way is work with *bayesm* functions is to copy the examples into a .R file and then edit the file to read in your own data and run the function. At first, use as many defaults as possible (e.g., Priors) to make sure that the function is working properly on your example.

A.4 OBTAINING HELP WITH BAYESM

Once *bayesm* has been loaded, you can obtain help in various ways.

To see a list of all the functions in *bayesm*, use the command, library(help= bayesm). This command will produce only brief summaries of each function in *bayesm*.

To learn more, use the help() or ? commands as for any function in R. This will provide the standard information on usage, arguments and output from the function. The most useful aspect of these files is the examples section. For example, ?rmnlIndepMetrop produces:

```
rmnlIndepMetrop          package:bayesm          R Documentation

MCMC Algorithm for Multinomial Logit Model
Description:
```

'rmnIndepMetrop' implements Independence Metropolis for the MNL.

```
Usage:
rmnlIndepMetrop(Data, Prior, Mcmc)

Arguments:
   Data: list(p,X,y)
  Prior: list(A,betabar)   optional
   Mcmc: list(R,keep,nu)
```

Details:
 Model: $Pr(y=j) = exp(x\_j'beta)/sum\_k\{e^\{x\_k'beta\}\}$.

 Prior: $beta \sim N(betabar,A^\{-1\})$

 list arguments contain:
   'p' number of alternatives
   'X' nobs*m x nvar matrix
   'y' nobs vector of multinomial outcomes (1,..., m)
   'A' nvar x nvar pds prior prec matrix (def: .01I)
   'betabar' nvar x 1 prior mean (def: 0)
   'R' number of MCMC draws
   'keep' MCMC thinning parm: keep every keepth draw (def: 1)
   'nu' degrees of freedom parameter for independence t density
   (def: 6)

Value:
 a list containing:
betadraw: R/keep x nvar array of beta draws
 acceptr: acceptance rate of Metropolis draws

See Also:
 'rhierMnlRwMixture'

Examples:
 ##
 if(nchar(Sys.getenv("LONG_TEST")) != 0) # set env var LONG_TEST
    to run
 {

 set.seed(66)
 n=200; p=3; beta=c(1,-1,1.5,.5)
 simout=simmnl(m,n,beta)
 A=diag(c(rep(.01,length(beta)))); betabar=rep(0,length(beta))

 R=2000
 Data=list(y=simout$y,X=simout$X,p=p);
 Mcmc=list(R=R,keep=1) ; Prior=list(A=A,betabar=betabar)
 out=rmnlIndepMetrop(Data=Data,Prior=Prior,Mcmc=Mcmc)
 cat(" Betadraws ",fill=TRUE)
 mat=apply(out$betadraw,2,quantile,probs=c(.01,.05,.5,.95,.99))
 mat=rbind(beta,mat); rownames(mat)[1]="beta"; print(mat)
 }
```
```

The examples section could be clipped out and used as a template for running your own data.

The "manual" for *bayesm* is available in the help pane in RStudio.

A.5 TIPS ON USING MCMC METHODS

All of the important functions in *bayesm* are implementations of various MCMC methods. The following tips are useful:

1. If you are unfamiliar with MCMC methods, read Chapter 3. Try some of our test examples first, before trying your own data.

2. The "output" of an MCMC method is a set of draws of the parameters. You must decide how many draws to make and also how to analyze the draws produced. Unlike most classical methods, the MCMC methods provide an estimate of the entire posterior distribution, not just a few moments. Summarize the distribution by using histograms or quantiles. Resist the temptation to simply report the posterior mean and posterior standard deviation! For nonnormal distributions, these moments have little meaning!

3. Most of the MCMC methods implemented in *bayesm* run very fast so it is possible to make 10,000s of draws even for relatively large datasets in less than $1/2$ hour. Use this power where possible. Many use far too few draws to reduce simulation error to a minimum.

4. If you are having problems with using too much memory, set keep in the Mcmc parameter list to more than 1.

A.6 EXTENDING AND ADAPTING OUR CODE

We hope that many researchers will find our functions useful. Some will want to adapt and extend our code. To find the R and C source code, go to the CRAN website, click on the "packages" menu link. Zipped files with source for the package are available. When this file is "unzipped" it will create a directory structure. Look at the directories "R" for source for the functions and "src" for the C/C++ code.

References

A. Ainslie and P. Rossi. Similarities in choice behavior across categories. *Marketing Science*, 17:91–106, 1998.

J. Albert and S. Chib. Bayesian analysis of binary and polychotomous response data. *Journal of the American Statistical Association*, 88:669–679, 1993.

G. Allenby and J. Ginter. Using extremes to design products and segment markets. *Journal of Marketing Research*, 32:392–403, 1995.

G. Allenby and P. Lenk. Modeling household purchase behavior with logistic normal regression. *Journal of the American Statistical Association*, 89:1218–1231, 1994.

G. Allenby and P. Lenk. Reassessing brand loyalty, price sensitivity, and merchandizing effects on consumer brand choice. *Journal of Business and Economic Statistics*, 13:281–289, 1995.

G. M. Allenby and P. E. Rossi. Quality preceptions and asymmetric switching between brands. *Marketing Science*, 10(3):185–204, 1991.

G. M. Allenby and P. E. Rossi. A Bayesian approach to estimating household parameters. *Journal of Marketing Research*, 30(2):171–182, 1993.

G. M. Allenby and P. E. Rossi. Marketing models of consumer heterogeneity. *Journal of Econometrics*, 89:57–78, 1999.

G. Allenby, R. Leone, and L. Jen. A dynamic model of purchase timing with application to direct marketing. *Journal of the American Statistical Association*, 94:365–374, 1999.

G. M. Allenby, T. S. Shively, S. Yang, and M. J. Garratt. A choice model for packaged goods: Dealing with discrete quantities and quantity discounts. *Marketing Science*, 23:95–108, 2004.

G. M. Allenby, J. Brazell, J. Howell, and P. E. Rossi. Valuation of patented product features. *Journal of Law and Economics*, 57:629–663, 2014a.

G. M. Allenby, J. D. Brazell, J. P. Howell, and P. E. Rossi. Economic valuation of product features. *Quantitative Marketing and Economics*, 12(4):421–456, 2014b.

G. M. Allenby, N. Hardt, and P. E. Rossi. Economic foundations of conjoint analysis. In P. E. Rossi and J. P. Dube, editors, *Handbook of the Economics of Marketing*, Chapter 3, pages 151–192. Elsevier, 2019a.

G. M. Allenby, N. Hardt, and P. E. Rossi. Economic foundations of conjoint analysis. In *Handbook of the Economics of Marketing*, volume 1, pages 151–192. Elsevier, 2019b.

J. Barnard, X. Meng, and R. McCulloch. Modelling covariance matrices in terms of standard deviations and correlations. *Statistica Sinica*, 10:1281–1312, 2000.

J. Berger. *Statistical Decision Theory and Bayesian Analysis*. Springer-Verlag, 1985.

R. Berger and R Wolpert. *The Likelihood Principle*. Institute of Mathematical Statistics, 1984.

J. M. Bernardo and A. F. M. Smith. *Bayesian Theory*. John Wiley & Sons, 1994.

S. Berry, J. Levinsohn, and A. Pakes. Automobile prices in market equilibrium. *Econometrica*, 63(4):841–890, 1995.

P. G. Bissiri, C. C. Holmes, and S. G. Walker. A general framework for updating belief distributions. *Journal of the Royal Statistical Society*, 78(5):1103–1130, 2016.

R. Blattberg and E. George. Shrinkage estimation of price and promotional elasticities. *Journal of the American Statistical Association*, 86:304–315, 1991.

D. M. Blei, A. Y. Ng, and M. I. Jordan. Latent dirichlet allocation. *Journal of Machine Learning Research*, 3:993–1022, 2003.

D. M. Blei, A. Kucukelbir, and J. D. McAuliffe. Variational inference: A review for statisticians. *Journal of the American Statistical Association*, 112(518):859–877, 2017.

C. Blundell, J. Cornebise, K. Kavukcuoglu, and D. Wierstra. Weight uncertainty in neural network. In *International conference on machine learning*, pages 1613–1622. PMLR, 2015.

P. Boatwright R. McCulloch and P. Rossi. Account-level modeling for trade promotions. *Journal of the American Statistical Association*, 94:1063–1073, 1999.

G. E. P. Box and N. R. Draper. *Empirical Model-Building and Response Surfaces*. John Wiley & Sons, 1987.

J. Brazell, C. Diener, E. Karniouchina, W. Moore, V. Severin, and P.-F. Uldry. The no-choice option and dual response choice designs. *Marketing Letters*, 17(4):255–268, 2006.

B. J. Bronnenberg and V. Mahajan. Multimarket data: Joint spatial dependence in market shares and promotional variables. *Marketing Science*, 20(3):284–299, 2001.

J. Büschken and G. M. Allenby. Sentence-based text analysis for customer reviews. *Marketing Science*, 35(6):953–975, 2016.

J. Büschken and G. M. Allenby. Improving text analysis using sentence conjunctions and punctuation. *Marketing Science*, 39(4):727–742, 2020.

G. Casella and R. Berger. *Statistical Inference*. Duxbuy, 2002.

G. Chamberlain. Analysis of covariance with qualitative data. *Review of Economic Studies*, 47:225–238, 1980.

G. Chamberlain. Panel data. In Z. Griliches and M. Intrilligator, editors, *Handbook of Econometrics*, volume 2. North-Holland, 1984.

G. Chamberlain and G. W. Imbens. Nonparametric applications of Bayesian inference. *Journal of Business and Economic Statistics*, 21(1):12–18, 2003.

S. R Chandukala, J. P. Dotson, J. D. Brazell, and G. M. Allenby. Bayesian analysis of hierarchical effects. *Marketing Science*, 30(1):123–133, 2011.

K. Chang, S. Siddarth, and C. Weinberg. The impact of heterogeneity in purchase timing and price responsiveness on estimates of sticker shock. *Marketing Science*, 18:178–192, 1999.

S. Chib. Bayes inference in the Tobit censored regression model. *Journal of Econmetrics*, 51:79–99, 1992.

S. Chib. Marginal likelihood from Gibbs output. *Journal of the American Statistical Association*, 90:1313–1321, 1995.

S. Chib and E. Greenberg. Hierarchical analysis of sur models with extensions. *Journal of Econometrics*, 68:339–360, 1995a.

S. Chib and E. Greenberg. Understanding the metropolis-hastings algorithm. *The American Statistician*, 49:339–360, 1995b.

S. Chib and E. Greenberg. Understanding the metropolis-hastings algorithm. *The American Statistician*, 49(4):327–335, 1995c.

S. Chib and E. Greenberg. Analysis of multivariate probit models. *Biometrika*, 85(2):347–361, 1998.

M. Clyde and H. Lee. Bagging and the Bayesian bootstrap. In T. S. Richardson and T. S. Jaakkola, editors, *Proceedings of the 8th International Workshop on Artificial Intelligence and Statistics*, volume R3 of *Proceedings of Machine Learning Research*, pages 57–62. PMLR, 04–07 Jan 2001. URL https://proceedings.mlr.press/r3/clyde01a.html. Reissued by PMLR on 31 March 2021.

A. Deaton and J. Muellbauer. *Economics and Consumer Behavior*. Cambridge University Press, 1980.

T. DiCiccio, R. Kass, A. Raftery, and L. Wasserman. Computing Bayes factors by combining simulation and asymptotic approximations. *Journal of the American Statistical Association*, 92:903–915, 1997.

J. Diebolt and C. P. Robert. Estimation of finite mixture distributions through Bayesian sampling. *Journal of the Royal Statistical Society, Series B*, 56(2):363–375, 1994.

J.-P. Dubé and S. Misra. Personalized pricing and consumer welfare. *Journal of Political Economy*, 131(1):131–189, 2023.

Y. Edwards and G. M. Allenby. Multivariate analysis of multiple response data. *Journal of Marketing Research*, 40(3):321–334, 2003.

B. Efron and C. Morris. Data analysis using Stein's estimator and its generalizations. *Journal of the American Statistical Association*, 70:311–319, 1975.

M. H. Farrell, T. Liang, and S. Misra. Deep learning for individual heterogeneity: An automatic inference framework. *arXiv preprint arXiv:2010.14694*, 2020.

S. Fruhwirth-Schnatter. Markov chain Monte Carlo estimation of classical and dynamic switching and mixture models. *Journal of the American Statistical Association*, 96(453):194–209, 2001.

S. Fruhwirth-Schnatter. Estimating marginal likelihoods for mixture and Markov switching models using bridge sampling techniques. *Econometrics Journal*, 7:143–167, 2004.

Y. Gal and Z. Ghahramani. Dropout as a Bayesian approximation: Representing model uncertainty in deep learning. In *International conference on machine learning*, pages 1050–1059. PMLR, 2016.

A. Gelfand and D. Dey. Bayesian model choice: Asymptotics and exact calculations. *Journal of the Royal Statistical Society, Series B*, 56:501–514, 1994.

A. Gelman and D. Rubin. Inference from iterative simulation using multiple sequences. *Statistical Science*, 7:457–511, 1992.

A. Gelman, J. Carlin, H. Stern, and D. Rubin. Efficient metropolis jumping rules. In J. Bernardo, J. Berger, A. Dawid, and A. Smith, editors, *Bayesian Statistics*, volume 5, pages 599–608. Oxford University Press, 1996.

A. Gelman, J. B. Carlin, H. S. Stern, and D. B. Rubin. *Bayesian Data Analysis*. Chapman and Hall, 2004.

M. Gentzkow. Valuing new goods in a model with complementarity: Online newspapers. *American Economic Review*, 97(3):713–744, 2007.

E. I. George and R. E. McCulloch. Variable selection via Gibbs sampling. *Journal of the American Statistical Association*, 88(423):881–889, 1993.

J. Geweke. Bayesian inference in econometric models using Monte Carlo integration. *Econometrica*, 57:1317–1339, 1989.

J. Geweke. Bayesian reduced rank regression in econometrics. *Journal of Econometrics*, 75:121–146, 1996.

J. Geweke. Getting it right: Joint distribution tests of posterior simulators. *Journal of the American Statistical Association*, 99:799–804, 2004.

T. J. Gilbride and G. M. Allenby. A choice model with conjunctive, disjunctive, and compensatory screening rules. *Marketing Science*, 23(3):391–406, 2004. ISSN 07322399.

T. J. Gilbride, G. M. Allenby, and J. D. Brazell. Models for heterogeneous variable selection. *Journal of Marketing Research*, 43(3):420–430, 2006. ISSN 00222437. URL http://www.jstor.org/stable/30162416.

W. R. Gilks. Discussion of the paper by Richardson and green. *Journal of the Royal Statistical Society, Series B*, 59:731–792, 1997.

W. Gilks, N. Best, and K. Tan. Adaptive rejection metropolis sampling within Gibbs. *Applied Statistics*, 44:455–472, 1995.

G. Golub and C. Van Sloan. *Matrix Computations*. Johns Hopkins University, 1989.

I. Goodfellow, Y. Bengio, and A. Courville. *Deep Learning*. MIT Press, 2016.

T. L. Griffiths and M. Steyvers. Finding scientific topics. *Proceedings of the National Academy of Sciences of the United States of America*, 7(suppl 1):5228–5235, 2004.

V. Hajivassiliou, D. L. McFadden, and P. Ruud. Simulation of multivariate normal rectangle probabilities and their derivatives. *Journal of Econometrics*, 72:85–134, 1996.

W. K. Hastings. Monte Carlo sampling methods using Markov chains and their applications. *Biometrika*, 57:97–109, 1970.

J. Heckman and B. Singer. A method for minimizing the impact of distributional assumptions in econometric models. *Econometrica*, 52(2):271–320, 1984.

J. Hobert, C. Robert, and C. Goutis. Connectedness conditions for the convergence of the Gibbs sampler. *Statistics & Probability Letters*, 33:235–240, 1997.

K. Imai and D. A. van Dyk. A Bayesian analysis of the multinomial probit model using marginal data augmentation. *Journal of Econometrics*, 124:311–334, 2005.

S. Jarner and R. Tweedie. Necessary conditions for geometric and polynomial ergodicity of random walk-type Markov chains. *Bernoulli*, 9(4):559–578, 2001.

V. Johnson and J. Albert. *Ordinal Data Modeling*. Springer-Verlag, 1999.

M. Joo, M. L. Thompson, and G. M. Allenby. Optimal product design by sequential experiments in high dimensions. *Management Science*, 65(7):3235–3254, 2019.

M. P. Keane. A computationally practical simulation estimator for panel data. *Econometrica*, 62(1):95–116, 1994.

H. Kim and G. M. Allenby. Integrating textual information into models of choice and scaled response data. *Marketing Science*, 41(4):387–402, 2022.

J. Kim, G. M. Allenby, and P. E. Rossi. Modeling the consumer demand for variety. *Marketing Science*, 21(3):229–250, 2002. ISSN 07322399.

Y. Kim, N. Hardt, J. Kim, and G. M. Allenby. Conjunctive screening in models of multiple discreteness. *International Journal of Research in Marketing*, 39(4):1209–1234, 2022.

G. Koop. *Bayesian Econometrics*. John Wiley & Sons, 2003.

T. Lancaster. *An Introduction to Modern Bayesian Econometrics*. Blackwell, 2004.

P. Lenk and W. DeSarbo. Bayesian inference for finite mixtures of generalized linear models. *Psychometrika*, 65:93–119, 2000.

P. Lenk, W. DeSarbo, P. Green, and M. Young. Hierarchical Bayes conjoint analysis: Recovery of part-worth heterogeneity from reduced experimental design. *Marketing Science*, 15:173–191, 1996.

Q. Li and N. Lin. The Bayesian elastic net. *Bayesian Analysis*, 5(1):151–170, 2010. doi: https://doi.org/10.1214/10-BA506.

J. Liu. *Monte Carlo Strategies in Scientific Computing*. Springer-Verlag, 2001.

P. Manchanda, A. Ansari, and S. Gupta. The "shopping basket": A model for multicategory purchase incidence decisions. *Marketing Science*, 18(2):95–114, 1999.

P. Manchanda, P. E. Rossi, and P. K. Chintagunta. Response modeling with nonrandom marketing-mix variables. *Journal of Marketing Research*, 41:467–478, 2004.

G. Marsaglia. Random numbers of C: End at last? accessed electronically, 1999.

G. Marsaglia and W. Tsang. A simple method for generating gamma variables. *ACM Transactions on Mathematical Software*, 26:363–372, 2000.

M. Matsumoto and T. Nishimura. Mersenne twister: A 623-dimensionally equidistributed uniform pseudo-random number generator. *ACM Transactions on Modeling and Computer Simulation*, 8:3–30, 1998.

T. Matthew, C.-S. Chen, J. Yu, and M. Wyle. Bayesian and empirical Bayesian forests. In *International conference on machine learning*, pages 967–976. PMLR, 2015.

R. E. McCulloch and P. E. Rossi. An exact likelihood analysis of the multinomial probit model. *Journal of Econometrics*, 64:207–240, 1994.

R. E. McCulloch, N. G. Polson, and P. E. Rossi. A Bayesian analysis of the multinomial probit model with fully identified parameters. *Journal of Econometrics*, 99:173–193, 2000.

D. L. McFadden. Econometric models of probabilistic choice. In M. Intrilligator and Z. Griliches, editors, *Structural Analysis of Discrete Choice*, pages 1395–1457. North-Holland, 1981.

X. Meng and D. Van Dyk. Seeking efficient data augmentation schemes via conditional and marginal augmentation. *Biometrika*, 86:301–320, 1999.

X. Meng and W. Wong. Simulating ratios of normalizing constants via a simple identity. *Statistica Sinica*, 6:831–860, 1996.

A. Montgomery. Creating micro-marketing pricing strategies using supermarket scanner data. *Marketing News*, 16:315–337, 1997.

A. Montgomery and E. Bradlow. Why analyst overconfidence about the functional form of demand models can lead to overpricing. *Marketing Science*, 18:569–583, 1999.

A. L. Montgomery and P. E. Rossi. Estimating price elasticities with theory-based priors. *Journal of Marketing Research*, 36(4):413–423, 1999.

C. Morris. Parametric empirical Bayes inference. *Journal of the American Statistical Association*, 78:47–65, 1983.

R. Muirhead. *Aspects of Multivariate Statistical Theory*. John Wiley & Sons, 1982.

R. Neelameghan and P. K. Chintagunta. A Bayesian model to forecast new product performance. *Marketing Science*, 18:115–136, 1999.

M. A. Newton and A. E. Raftery. Approximate Bayesian inference with the weighted likelihood bootstrap. *Journal of the Royal Statistical Society: Series B (Methodological)*, 56(1):3–26, 1994.

M. A. Newton, N. G. Polson, and J. Xu. Weighted Bayesian bootstrap for scalable posterior distributions. *Canadian Journal of Statistics*, 49(2):421–437, 2021.

A. Nobile. A hybrid Markov chain for the Bayesian analysis of the multinomial probit model. *Statistics and Computing*, 8:229–242, 1998.

B. K. Orme. Assessing the monetary value of attribute levels with conjoint analysis. Technical report, Sawtooth Software, Inc., 2001.

B. K. Orme. *Getting Started with Conjoint Analysis*. Research Publishers, LLC, 2009.

B. K. Orme. *Getting Started with Conjoint Analysis: Strategies for Product Design and Pricing Research*. Research Publishers LLC, Manhattan Beach, CA, 2020.

B. Orme. Estimating WTP given competition in conjoint analysis. Technical report, Sawtooth Software, 2021.

M. J. Pachali, P. Kurz, and T. Otter. How to generalize from a hierarchical model. *Quantitative Marketing and Economics*, 18:343–380, 2020.

T. Park and G. Casella. The Bayesian Lasso. *Journal of the American Statistical Association*, 103(482):681–686, 2008.

N. G. Polson and V. Sokolov. Bayesian regularization: From Tikhonov to horseshoe. *WIREs Computational Statistics*, 11(4):e1463, 2019. doi: https://doi.org/10.1002/wics.1463. URL https://wires.onlinelibrary.wiley.com/doi/abs/10.1002/wics.1463.

S. Richardson and P. Green. On Bayesian analysis of mixtures with an unknown number of components. *Journal of the Royal Statistical Society, Series B*, 59:731–792, 1997.

C. Robert and G. Casella. *Monte Carlo Statistical Methods*. Springer-Verlag, 2004.

G. O. Roberts and J. S. Rosenthal. Optimal scaling for various Metropolis-Hastings algorithms. *Statistical Science*, 16(4):351–367, 2001.

V. Rockova and E. I. George. EMVS: The EM approach to Bayesian variable selection. *Journal of the American Statistical Association*, 109(506):828–846, 2014. doi: https://doi.org/10.1080/01621459.2013.869223.

P. E. Rossi, R. E. McCulloch, and G. M. Allenby. The value of purchase history data in target marketing. *Marketing Science*, 15(4):321–340, 1996.

P. E. Rossi, Z. Gilula, and G. M. Allenby. Overcoming scale usage heterogeneity: A Bayesian hierarchical approach. *Journal of the American Statistical Association*, 96(453):20–31, 2001.

D. B. Rubin. The Bayesian bootstrap. *The Annals of Statistics*, 9(1):130–134, 1981.

G. Schwarz. Estimating the dimension of a model. *Annals of Statistics*, 6:461–464, 1978.

P. B. Seetharaman, A. Ainslie, and P. K. Chintagunta. Investigating household state dependence effects across categories. *Marketing Science*, 36(4):488–500, 1999.

P. Sheatsley. Questionnaire construction and item writing. In P. H. Rossi, J. D. Wright, and A. B. Anderson, editors, *Handbook of Survey Research*, Chapter 6, pages 195–230. Academic Press, New York, 1983.

T. Steenburgh, A. Ainslie, and P. Engebretson. Massively categorical variables: Revealing the information in zipcodes. *Marketing Science*, 22:40–57, 2003.

M. Stephens. Dealing with label-switching in mixture models. *Journal of the Royal Statistical Society, Series B*, 62:795–809, 2000.

T. Tanner and W. Wong. The calculation of posterior probabilities by data augmentation. *Journal of the American Statistical Association*, 82:528–549, 1987.

F. Ter Hofstede, M. Wedel, and J. Steenkamp. Identifying spatial segments in international markets. *Marketing Science*, 21:160–178, 2002.

L. Tierney. Markov chains for exploring posterior distributions. *Annals of Statistics*, 22:1701–1728, 1994.

K. E. Train. *Discrete Choice Methods with Simulation*. Cambridge University Press, 2003.

M. Trajtenberg. The welfare analysis of product innovations, with an application to computed tomography scanners. *Journal of Political Economy*, 97(2):444–479, 1989.

D. VanDyk and X. Meng. The art of data augmentation. *Journal of Computational and Graphical Statistics*, 10:1–50, 2001.

J. M. Villas-Boas and R. S. Winer. Endogeneity in brand choice models. *Management Science*, 45(10):1324–1338, 1999.

W. K. Viscusi and J. Huber. Reference-dependent valuations of risk: Why willingness-to-accept exceeds willingness-to-pay. *Journal of Risk and Uncertainty*, 44:19–44, 2012.

M. Welling and Y. W. Teh. Bayesian learning via stochastic gradient Langevin dynamics. In *Proceedings of the 28th international conference on machine learning (ICML-11)*, pages 681–688, 2011.

S. Yang and G. Allenby. A model for observation, structural, and household heterogeneity. *Marketing Letters*, 11(2):137–149, 2000.

S. Yang and G. M. Allenby. Modeling interdependent consumer preferences. *Journal of Marketing Research*, 40:282–294, 2003.

S. Yang, G. Allenby, and G. Fennell. Modeling variation in brand preference: The roles of objective environment and motivating conditions. *Marketing Science*, 21:14–31, 2002.

S. Yang, Y. Chen, and G. M. Allenby. Bayesian analysis of simultaneous demand and supply. *Quantitative Marketing and Economics*, 1:251–275, 2003b.

D. S. Yeager, J. A. Krosnick, L. Chang, H. S. Javitz, M. S. Levendusky, A. Simpser, and R. Wang. Comparing the accuracy of RDD telephone surveys and internet samples conducted with probability and non-probability samples. *Public Opinion Quarterly*, 75(4):709–747, 2011.

A. Zellner. *An Introduction to Bayesian Inference in Econometrics*. John Wiley and Sons, Inc., New York, 1971.

A. Zellner. On assessing prior distributions and Bayesian regression. In A. Zellner and P. Goel, editors, *Bayesian Inference and Decision Techniques*. North-Holland, 1986.

A. Zellner and P. Rossi. Bayesian analysis of dichotomous quantal response models. *Journal of Econometrics*, 25:365–394, 1984.

Index